DEVELOPMENTS IN PETROLEUM SCIENCE 75

An Introduction to Multiphase, Multicomponent Reservoir Simulation

DEVELOPMENTS IN PETROLEUM
SCIENCE   75

An Introduction to Multiphase, Multicomponent Reservoir Simulation

Matthew Balhoff
Director, Center for Subsurface Energy and the Environment;
Professor, Hildebrand Department of Petroleum and Geosystems
Engineering, The University of Texas, Austin, TX, United States;
Bank of America Professorship in Petroleum Engineering

ELSEVIER

Elsevier
Radarweg 29, PO Box 211, 1000 AE Amsterdam, Netherlands
The Boulevard, Langford Lane, Kidlington, Oxford OX5 1GB, United Kingdom
50 Hampshire Street, 5th Floor, Cambridge, MA 02139, United States

ISBN: 978-0-323-99235-0
ISSN: 0376-7361

For information on all Elsevier publications visit our website at
https://www.elsevier.com/books-and-journals

Publisher: Candice Janco
Acquisitions Editor: Jennette McClain
Editorial Project Manager: Zsereena Rose Mampusti
Production Project Manager: Bharatwaj Varatharajan
Cover Designer: Christian J Bilbow

Typeset by TNQ Technologies

Working together
to grow libraries in
developing countries

www.elsevier.com • www.bookaid.org

Dedication

To my wife, Julie, and Thomas

Contents

Preface   xiii
Acknowledgments   xvii
Nomenclature   xix

1. Review of reservoir rock and fluid properties

1.1	Introduction	1
1.2	Overview of reservoir engineering principles	1
1.3	Definitions	2
	1.3.1 Phases and components in subsurface porous media	2
	1.3.2 Porosity, saturation, density, and concentrations	3
1.4	Phase behavior	4
1.5	Rock and Fluid Properties	6
	1.5.1 Formation properties	6
	1.5.2 Gaseous phase properties	7
	1.5.3 Oleic phase properties	9
	1.5.4 Aqueous phase properties	12
1.6	Petrophysical properties	14
	1.6.1 Darcy's law	14
	1.6.2 Relative permeability	16
	1.6.3 Capillary pressure	20
	1.6.4 Capillary pressure scanning curves	22
1.7	Reservoir initialization	23
1.8	Pseudocode	28
	1.8.1 Relative permeability	29
	1.8.2 Capillary pressure	30
	1.8.3 Initialization	30
	1.8.4 Preprocess	30
1.9	Exercises	32
	References	35

2. Phase mass balances and the diffusivity equation

2.1	Introduction	37
2.2	Phase mass balances	37
	2.2.1 Mass balance of a phase in Cartesian coordinates	38
2.3	The continuity equation	40
2.4	The diffusivity equation	41
	2.4.1 General multiphase flow	41
	2.4.2 Single-phase flow	42

	2.5	Analytical solutions	48
		2.5.1 1D heat equation in a finite medium	48
		2.5.2 1D heat equation in a semi-infinite medium	50
		2.5.3 Solution in cylindrical coordinates (around a wellbore)	51
	2.6	Exercises	54
	References		55

3. Finite difference solutions to PDEs

	3.1	Introduction	57
	3.2	Taylor series and finite differences	57
		3.2.1 First-order forward difference approximation	59
		3.2.2 First-order backward difference approximation	60
		3.2.3 Second-order, centered difference approximation	61
		3.2.4 Approximations to the second derivative	61
		3.2.5 Generalization to higher-order approximations	64
	3.3	Discretization of the parabolic diffusivity (heat) equation	68
	3.4	Boundary and initial conditions	70
		3.4.1 Dirichlet boundary condition	71
		3.4.2 Neumann boundary condition	71
		3.4.3 Robin boundary conditions	72
	3.5	Solution methods	72
		3.5.1 Explicit solution to the diffusivity equation	72
		3.5.2 Implicit solution to the diffusivity equation	76
		3.5.3 Mixed methods and Crank–Nicolson	77
		3.5.4 Linear systems of equations	83
	3.6	Stability and convergence	84
	3.7	Higher-order approximations	85
	3.8	Pseudocode for 1D, single-phase flow	88
	3.9	Exercises	89
	References		91

4. Multidimensional reservoir domains, the control volume approach, and heterogeneities

	4.1	Introduction	93
	4.2	Gridding and block numbering in multidimensions	93
		4.2.1 Grid block indexing in 2D and 3D	94
		4.2.2 Grid dimensions	95
		4.2.3 Irregular geometry and inactive grids	96
	4.3	Single-phase flow in multidimensions and the control volume approach	98
		4.3.1 Accumulation	99
		4.3.2 Flux terms	100
		4.3.3 Sources and sinks (wells)	101
		4.3.4 Single-phase flow	101

4.4	**Wells, boundary conditions, and initial conditions**		102
	4.4.1	Constant rate wells	102
	4.4.2	Neumann boundary conditions	102
	4.4.3	Dirichlet conditions	103
	4.4.4	Corner blocks	104
	4.4.5	Initial conditions	107
4.5	**Reservoir heterogeneities**		107
	4.5.1	Fluid properties	109
	4.5.2	Geometric properties	109
	4.5.3	Accumulation terms	113
4.6	**Matrix arrays**		113
	4.6.1	Accumulation and compressibility	113
	4.6.2	Transmissibility	114
	4.6.3	Source terms	114
	4.6.4	Gravity	115
4.7	**Pseudocode for single-phase flow in multidimensions**		120
	4.7.1	Preprocessing	120
	4.7.2	Interblock transmissibility	120
	4.7.3	Well Arrays	121
	4.7.4	Grid Arrays	121
	4.7.5	Main code	121
	4.7.6	Postprocessing	122
4.8	**Exercises**		124
	References		126

5. Radial flow, wells, and well models

5.1	**Introduction**		127
5.2	**Radial flow equations and analytical solutions**		127
5.3	**Numerical solutions to the radial diffusivity equation**		129
	5.3.1	Gridding	129
	5.3.2	Discretization	130
5.4	**Wells and well models in Cartesian grids**		135
	5.4.1	Well constraints	135
	5.4.2	Steady-state radial flow around a well	136
	5.4.3	Mass balance on the well-residing grid block	137
	5.4.4	Extension to horizontal wells and anisotropy	139
5.5	**Inclusion of the well model into the matrix equations**		142
5.6	**Practical considerations**		147
5.7	**Pseudocode for single-phase flow with constant BHP wells**		147
5.8	**Exercises**		148
	References		150

6.   Nonlinearities in single-phase flow through subsurface porous media

6.1	Introduction	151
6.2	Examples of nonlinearities in single-phase flow problems	151
	6.2.1 Gas flow	152
	6.2.2 Non-Newtonian flow	153
	6.2.3 Forchheimer flow	155
6.3	Numerical methods for nonlinear problems	156
	6.3.1 Explicit update of fluid and reservoir properties	157
	6.3.2 Picard iteration	157
	6.3.3 Newton's method	161
6.4	Pseudocode for Newton's method	169
6.5	Exercises	171
References		172

7.   Component transport in porous media

7.1	Introduction	175
7.2	Transport mechanisms	175
	7.2.1 Advection	175
	7.2.2 Hydrodynamic dispersion	176
	7.2.3 Reactive transport and other source terms	182
7.3	Component mass balance equations	183
	7.3.1 Single-phase flow	184
	7.3.2 Overall compositional equations	184
7.4	Analytical solutions	186
	7.4.1 1D Cartesian ADE in a semi-infinite domain	186
	7.4.2 Semianalytical solution to two-phase flow	189
7.5	Exercises	198
References		199

8.   Numerical solution to single-phase component transport

8.1	Introduction	201
8.2	Finite difference solution to the ADE in 1D for a single component	201
8.3	Discretization of advective terms	204
	8.3.1 Cell-centered	204
	8.3.2 Upwinding	205
	8.3.3 Matrices	205
8.4	Wells and boundary conditions	206
	8.4.1 Wells	206
	8.4.2 No flux boundary condition	207

	8.4.3	Constant concentration (Dirichlet)	211
8.5	**Solution methods**		212
	8.5.1	Implicit pressure, explicit concentration (IMPEC)	212
	8.5.2	Implicit pressure, implicit concentration	214
	8.5.3	Fully implicit	218
8.6	**Stability**		219
8.7	**Numerical dispersion**		221
8.8	**Channeling and viscous fingering**		224
8.9	**Multicomponents, multidimensions, and additional forms**		225
8.10	**Pseudocode for component transport**		226
8.11	**Exercises**		229
	References		230

9. Numerical solution to the black oil model

9.1	**Introduction**		231
9.2	**The black oil model**		231
9.3	**Finite difference equations for multiphase flow**		233
9.4	**Solution methods**		237
	9.4.1	Implicit pressure, explicit saturation	237
	9.4.2	Simultaneous solution method	243
	9.4.3	Fully implicit method	247
9.5	**Interblock transmissibilities and upwinding**		249
9.6	**Stability**		256
9.7	**Wells and well models**		256
	9.7.1	Constant rate injector wells	258
	9.7.2	Constant rate producer wells	259
	9.7.3	Constant BHP injector wells	261
	9.7.4	Constant BHP producer wells	261
	9.7.5	Time-dependent well constraints	261
9.8	**Pseudocode for multiphase flow**		272
	9.8.1	Preprocessing	272
	9.8.2	Block properties	273
	9.8.3	Interblock properties	273
	9.8.4	Well productivity index	273
	9.8.5	Well arrays	273
	9.8.6	Grid arrays	274
	9.8.7	Main code	274
	9.8.8	Postprocessing	274
9.9	**Exercises**		279
	References		282

10. Numerical solution to multiphase, multicomponent transport

10.1	**Introduction**		283
10.2	**Compositional equations for multiphase flow**		284
10.3	**Finite difference equations**		286

10.4	**Solution method**		287
	10.4.1	Flash calculations	287
	10.4.2	Equations of state	291
	10.4.3	Phase saturation	298
	10.4.4	Two-phase compressibility	299
	10.4.5	Phase viscosity	301
	10.4.6	Relative permeability and transmissibility	304
	10.4.7	Wells and source terms	306
	10.4.8	Pressure and composition solution	308
10.5	**Oleic–aqueous bipartitioning components**		311
10.6	**Pseudocode for multiphase, multicomponent transport**		313
10.7	**Exercises**		315
References			317
Index			319

Preface

The flow of fluids in subsurface porous media is important in many applications including the production of hydrocarbons, carbon storage, hydrogen storage, aquifer remediation, and production of geothermal energy. Accurate modeling of these processes is of critical importance for predictions and decision-making. For example, in hydrocarbon production, models can be used to make business decisions, such as: (1) Should a field be bought or sold? (2) Where, when, how many, and what type of wells should be drilled? (3) What should be the constraint (well rate or bottomhole pressure) of the wells? (4) If, when, and what type of secondary (and tertiary) recovery should be pursued? (5) When should a well be shut-in or converted to an injector and what fluids should be injected?

Modeling of subsurface phenomena is challenging for many reasons. The subsurface reservoir is thousands of feet below the surface and can be massive (thousands of acres in area, or in the case in the Ghawar oil field, over a million acres). Our understanding of a reservoir's size, lithology, permeability, porosity, fluid properties, etc., is an estimate. Reservoirs are generally very heterogeneous in their permeability, porosity, saturation, lithology, etc., and can change significantly over small or large length scales. Predictions may be required for years or decades into the future, or even millennia in the case of carbon storage.

Subsurface models vary in complexity and can be as simple as analytical or reduced-order models such as tank balances and the capacitance resistance model (Sayarpour et al., 2009). Such models are simplifications but often provide very valuable information and can even be predictive. The fundamental equations that describe flow and transport in subsurface media are multidimensional, multicomponent, multiphase, nonlinear, coupled partial differential equations (PDEs) with spatially heterogeneous and time-dependent properties. These equations, without major simplification and assumptions, are not amenable to analytical solution. Numerical reservoir simulators are the most advanced tools we have to solve these PDEs and predict flow and transport in subsurface porous media. These simulators are the closest thing we have to a *crystal ball*.

There are many types of reservoir simulators, with varying complexity and features, but generally they involve discretizing the reservoir into N grids, blocks, or elements. One can think of a reservoir simulator as a giant Rubik's

Cube, with each block in the cube being a grid in the model. The simulator can have thousands, millions, or even billions of grids and each grid has unique, constant (or simple function) properties, such as permeability, porosity, saturation, composition, pressure, etc. Balance (mass, energy, momentum) equations that are imposed are on each block which are dependent on adjacent block properties. As a result, the complicated PDEs reduce to a system of N algebraic equations and N unknowns.

Many commercial (e.g., CMG, ECLIPSE, INTERSECT, Nexus), academic, or open source (BOAST, MRST, UTCHEM, UTCOMP, IPARS, TOUGH) and proprietary, in-house simulators have been developed by teams of experts over decades. These simulators vary in their applicability but are based on the same basic fundamentals. These simulators are often relatively easy for the beginner to use, which can be as much of a problem as it is a feature. Failure to understand the principles and basic equations of numerical simulation (what is *under the hood*) can lead one to not recognize the model's limitations and lead to costly or even unsafe decisions. The mathematics are complicated and can be daunting for even PhD scientists and engineers. Many outstanding books have been written on the subject; Aziz and Settari (1979), Ertekin et al. (2001), Chen (2007), Lie (2019), and Abou-Kassem et al. (2020) are just a few of my favorites. Many of these books are best suited for advanced graduate students or professionals with some experience in simulation.

I have taught the fundamentals of reservoir simulation for 15 years to over a thousand undergraduates and first-year graduate students. Breaking down the complexities of simulation to students new to the subject is challenging, to put it mildly. In this book, I have attempted to organize my notes, teaching style, and "lessons learned" in a concise text for the beginner. Many advanced and modern topics are intentionally not included, but the interested reader should read the dozens of advanced books and thousands of publications that cover them.

This book includes two important features. The first is the inclusion of dozens of small (e.g., 4–9 block) example problems that are solved by hand and calculator, largely without the use of a computer. To quote Albert Einstein, "example isn't another way to teach; it is the only way to teach." I have found these examples essential for the beginner to understand the basics of reservoir simulation. In addition to example problems, each chapter includes additional exercises for the reader to attempt.

The second feature of the book is the emphasis on writing computer code with the end-goal of the reader developing their own multiphase, multidimensional, and multicomponent reservoir simulator. The final product will be a simulator that will produce identical (or nearly identical) results as the aforementioned commercial, academic, and in-house simulators. The user's code can be and should be validated against these simulators, analytical solutions, or the small example problems provided in the text. The book is organized in such a way that the code starts relatively simple (1D, single phase, homogeneous) and complexities (multidimensions, heterogeneities, multiphase, etc.) are added along the way. Pseudocode is provided in each chapter, with some explanation and discussion, to help the user develop their own code. The most computationally efficient, vectorized, or elegant pseudocodes are not always provided. In fact, this is often intentional, as sometimes the less elegant codes are better for understanding the logic and mathematics. The developer of the simulator is encouraged to optimize their code once they have a working code that they understand.

The simulator developer is encouraged to be patient and avoid frustration as best as possible. I have written hundreds of subroutines and codes for my reservoir simulation courses over the years and can say with confidence that every one of them had errors and bugs in the initial version. These errors have taken anywhere from minutes to days (or even weeks) to debug. However, every single time I have fixed an error, I have come away with a better understanding of reservoir simulation and reservoir engineering in general. When the developer obtains results that are nonphysical or disagree with analytical solutions, example problems, or commercial simulators, they should ask what physically or mathematically could cause such a discrepancy. In my experience, 99% of the coding errors are in the formation of the few matrices and vectors that are used to solve the problem. The error(s) can almost always be identified by comparison to the matrices/vectors created by hand in the examples with a small number of grids.

Your final reservoir simulator (albeit accurate and flexible) will probably not be as computationally efficient, scalable, user-friendly, or have nearly as many features as a commercial simulator. However, you will develop an excellent understanding of the details and limitations of these simulators. And, just maybe, you will join a team or have a career developing the next-generation commercial, in-house, or academic simulator.

References

Abou-Kassem, Hussein, J., Rafiqul Islam, M., Farouq-Ali, S.M., 2020. Petroleum Reservoir Simulation: The Engineering Approach. Elsevier.

Aziz, K., Settari, A., 1979. Petroleum Reservoir Simulation. 1979. Applied Science Publ. Ltd., London, UK.

Chen, Z., 2007. Reservoir simulation: mathematical techniques in oil recovery. Society for Industrial and Applied Mathematics.

Ertekin, T., Abou-Kassem, J.H., King, G.R., 2001. Basic Applied Reservoir Simulation, 7. Society of Petroleum Engineers, Richardson.

Lie, K.-A., 2019. An introduction to reservoir simulation using MATLAB/GNU Octave: User guide for the MATLAB Reservoir Simulation Toolbox (MRST). Cambridge University Press.

Sayarpour, M., Zuluaga, E., Shah Kabir, C., Lake, L.W., 2009. The use of capacitance–resistance models for rapid estimation of water flood performance and optimization. Journal of Petroleum Science and Engineering 69 (3–4), 227–238.

Acknowledgments

First and foremost, I would like to thank the many current and former, graduate and undergraduate, students who helped in the development of this book. Although impossible to list them all, I would like to specifically recognize Nkem Egboga, Yashar Mehmani, Hamza Salim Al Rawahi, Travis Salomaki, Moises Velasco, Jianping Xu, and Sarah Razmara. I would like to thank Mary Wheeler for introducing me to the subject matter of reservoir simulation and the many colleagues for which I have had discussions including Larry Lake, Kamy Sepehrnoori, Gary Pope, Russ Johns, David DiCarlo, and Cheng Chen. I also acknowledge Cooper Link, Joanna Castillo, and Jostine Ho for helping with the many illustrations. I would like to thank my father, who taught me to be an engineer and helped me numerically solve the Diffusivity equation for the first time, my mother, who taught me to persistent and dedicated, and my sisters. Finally, this book would not be possible without the endless support and love of my wife, Julie.

Nomenclature

a	cross-sectional area, ft^2; empirical coefficient for mechanical dispersion
A	accumulation term ($V_i\phi/\Delta t$), ft^3/day; parameter for cubic EOS
A_κ'	parameter for fugacity coefficient of component κ in cubic EOS
b	empirical exponent for mechanical dispersion
B	parameter for cubic EOS
B_α	formation volume factor for phase α, RB/STB or ft^3/scf
B_κ'	parameter for fugacity coefficient of component κ in cubic EOS
c_α	compressibility of phase α, psi^{-1}
c_f	formation compressibility, psi^{-1}
c_p	pore compressibility, psi^{-1}
c_r	rock matrix compressibility, psi^{-1}
c_B	bulk compressibility, psi^{-1}
c_t	total compressibility, psi^{-1}
C	constant for effective shear rate in porous media for non-Newtonian flow
C_κ	concentration of component κ, lb$_m$/ft^3
$C_{j,i}$	coefficient for block i in IMPES method (j=1,2,3)
d_p	grain diameter, ft
D	depth, ft; hydrodynamic dispersion coefficient, ft^2/day
D_1	capillary diffusion coefficient, ft^2/day
D_m	molecular diffusion coefficient, ft^2/day
$D_{m,eff}$	effective diffusion coefficient in porous medium, ft^2/day
D_L	longitudinal mechanical dispersion coefficient, ft^2/day
D_r	restricted diffusion coefficient ($D_{m,eff}/D_m$), ft^2/day
D_T	transverse mechanical dispersion coefficient, ft^2/day
f	weighting factor for capillary pressure scanning curve
f_α	fractional flow of phase α
$f_{\kappa,\alpha}$	fugacity of component κ of phase α, psia
F	residual of grid balance equation, ft^3/day; formation resistivity factor
g	gravitational constant (32 ft/s^2)
G	Gravity vector, ft^3/day
h	reservoir thickness, ft
i	grid block index
j	grid block index, x-direction
J_α	productivity index of phase α, ft^3/psi-day
J	total productivity index, ft^3/psi-day

k	permeability, mD; rate constant, 1/day; grid block index in y-direction
k_{app}	apparent permeability, mD
k_B	Boltzmann constant (1.38×10^{-23} J/K)
k_g	Klinkenberg apparent permeability for gas flow, mD
k_H	geometric mean of permeability, mD
$k_{r,\alpha}$	relative permeability of phase α
K_κ	K-values/equilibrium ratio of component κ
l	grid block index, z-direction
L	reservoir length, ft
m	mass, lb_m
$\dot{m}_x$	mass flux (in x-direction), lb_m/ft^2-day
M	mobility ratio ($k_{rw}\mu_o/k_{ro}\mu_w$)
M_κ	molecular weight of component κ
M_α	molecular weight of phase α
n	shear-thinning index for power-law or Carreau model
N	number of grid blocks in reservoir model
N_c	number of components in the reservoir model
$N_{\kappa,\alpha}$	flux (advective, diffusive, or dispersive) of component κ in phase α
N_p	cumulative amount of oil produced, bbl or ft^3
N_x	number of grid blocks in x-direction of reservoir model
N_y	number of grid blocks in y-direction of reservoir model
N_z	number of grid blocks in z-direction of reservoir model
O	scaling order
p	pressure (exact), psia
p_b	bubble point pressure, psia
p_c	capillary pressure, psi
$p_{c,\kappa}$	critical pressure of component κ, psia
p_e	capillary entry pressure, psia
p_κ^{sat}	vapor pressure of component κ, psia
P	pressure (numerical approximation), psia
P_{wf}	bottomhole pressure of well, psia
P_{lim}	limiting pressure of producing well, psia
$\tilde{q}$	source term (1/time)
q_{wf}	well flowing rate (ft^3/day)
Q	source vector, ft^3/day
r	radial direction in cylindrical coordinates, ft
r_e	drainage radius, ft
r_w	wellbore radius, ft
r_{eq}	equivalent radius for well models, ft
r_κ	effective radius of component κ, ft
R	ideal gas constant (10.73 psi-ft^3/lbmole-R)
R_s	solution gas-oil ratio, scf/STB)
s	skin factor
S_α	saturation of phase α
$S_{\alpha r}$	residual saturation of phase α
S_{wf}	water saturation at shock front
t	time, day

T	total transmissibility, ft^3/psi-day; reservoir temperature, °F or °R
$\mathbf{T_\alpha}$	phase transmissibility, scf/psi-day
$\mathbf{T_\kappa}$	component transmissibility, md-/cp-ft^2
$\mathbf{T_c}$	critical temperature, °R
$\mathbf{T_{r,\kappa}}$	reduced temperature of component κ ($T/T_{c,\kappa}$)
u	Darcy velocity, ft/day
U	Dispersive transmissibility in component balance matrix equations, ft^3/day
v	interstitial/frontal velocity, ft/day
$\mathbf{V_i}$	volume of grid block i, ft^3
$\mathbf{V_p}$	pore volume, ft^3
$\mathbf{V_m}$	molar volume, ft^3/lbmole
w	width, ft
$\mathbf{W_\kappa}$	bulk concentration of component κ, lb_m/ft^3
x	mole fraction
X	solution vector (pressures and saturations) for SS method
z	compressibility factor/z-factor; elevation, ft

Greek letters and symbols

α	phase (e.g., oleic, gaseous, aqueous)
α_L	longitudinal dispersivity, ft
α_T	transverse dispersivity, ft
β	non-Darcy coefficient (ft^{-1}); spatial differencing scheme for ADE
β_α	capillary pressure corrected formation volume factor
γ	shear rate (1/s); Euler's constant (0.5772)
γ_0	specific gravity of oil
γ_g	specific gravity of gas
γ_r	coefficient for geometric progression of grids in cylindrical coordinates
$\delta_{\kappa,\lambda}$	binary interaction parameter between components κ and λ
δP	change in the pressure vector between two iterations, psia
Δ	discriminant in cubic EOS
Δl	tortuous length of porous medium, ft
ε	small perturbation parameter
η	dimensionless diffusivity/Fourier number ($\alpha\Delta t/\Delta x^2$ or $D\Delta t/\Delta x^2$)
θ	dip angle, degrees
$\mathfrak{J}$	Jacobian for Newton–Raphson methods
κ	component or pseudocomponent (e.g., water, oil, gas)
λ	time constant for Carreau model, s
λ_α	mobility of phase α ($kk_{r\alpha}/\mu_\alpha$), mD/cp
μ	viscosity, cp
μ_0	zero-shear viscosity, cp
μ_∞	infinite-shear viscosity, cp
υ	moles gaseous phase/moles hydrocarbon
ρ	density, lb_m/ft^3; variable transformation for cylindrical coordinates (r^2)
$\rho_{r,\alpha}$	reduced density of phase α

σ_κ	parameter for component κ in cubic EOS
$\sigma_{\kappa,\alpha}$	volume fraction of component κ in phase α
τ	tortuosity in porous medium
ϕ	porosity
$\phi_{\kappa,\alpha}$	fugacity coefficient of component κ in phase α
ψ	pseudopressure, psi^2/cp
ω	temporal differencing scheme for ADE
ω_κ	weight fraction of component κ; acentric factor of component κ

Superscripts and subscripts

cen	centered
GOC	gas-oil contact line, ft
g	gaseous phase
o	oleic phase
rc	reservoir conditions
sc	standard conditions (14.7 psi, 60 F)
up	upwinding
WOC	water-oil contact line, ft
x	x-direction
y	y-direction
z	z-direction
υ	iteration number in nonlinear problems

Dimensionless variables

C_D	concentration
x_D	distance (x/L)
p_D	pressure $((C-C_{init})/(C_{inj}-C_{init}))$
$N_{Pe,d}$	Peclet number based on diffusion coefficient (ud_p/D_m)
N_{Pe}	Peclet number based on dispersion coefficient $(uL/D\phi)$
N_{Re}	Reynolds number $(\rho v d_p/\mu)$
N_g^0	gravity number $(kk_{ro}^0\Delta\rho g/u\mu_o)$
N_{pD}	oil recovery (N_p/V_p)
t_D	time/pore volumes injected $(ut/L\phi)$
η	Fourier number $(\alpha\Delta t/\Delta x^2$ or $D\Delta t/\Delta x^2)$
π	local Peclet number $(u\Delta x/D\phi)$
ζ	Courant number $(u\Delta t/\Delta x\phi)$
τ_D	time/pore volumes injected $(ut/L\phi)$

Chapter 1

Review of reservoir rock and fluid properties

1.1 Introduction

Reservoir simulation requires an accurate description of fluid and reservoir properties combined with the effect pressure, temperature, and composition has on them. Many texts (e.g. McCain, 1991; Pedersen et al., 2006; Dandekar, 2013) provide an excellent, thorough discussion of fluid and rock properties in the subsurface; here a concise review is provided. The reservoir fluids are composed of thousands of unique chemical molecules, are confined in a complex pore space, and partition between multiple fluid phases, which make modeling of subsurface reservoirs very challenging.

   In this chapter, an overview of basic reservoir engineering principles is first presented followed by a few basic definitions. Phase behavior principles are introduced, followed by definitions and equations for fluid, rock, and petrophysical properties. Finally, approaches for determining the initial pressure and saturation fields in the reservoir are discussed.

1.2 Overview of reservoir engineering principles

Petroleum reservoirs consist of porous rock thousands of feet below the earth's surface. The void space in the rock contains fluids, including hydrocarbons (oil and gas) and water that can be extracted through wells. Upon exploration and drilling, the pressure and temperature in the reservoir are relatively high (often thousands of psia and $>100°F$, respectively) and drilling a well with a lower pressure results in a driving force (*drawdown pressure*) for fluid production from the reservoir to the well. During *primary production* the fluids expand; the reservoir pressure, drawdown pressure, and production rate decrease with time until production of fluids is no longer economical. At the end of primary production, the reservoir is either abandoned or *secondary production* methods are employed. During secondary production, an aqueous phase containing salts and other dissolved solids (*brine*) or hydrocarbon gas is injected through

An Introduction to Multiphase, Multicomponent Reservoir Simulation.
https://doi.org/10.1016/B978-0-323-99235-0.00010-5
1

injector wells to repressurize the reservoir and drive mobile hydrocarbons toward the producer wells. When the injected fluid is an aqueous phase, it is often referred to as *waterflooding*. Secondary recovery can occur for decades and continues until the process becomes uneconomical, usually due to the decrease in production rate of hydrocarbons and increase in *water cut* (volume percent of produced fluids that is brine). Finally, *tertiary recovery* or *enhanced oil recovery* (EOR) involves injection of other fluids (steam, carbon dioxide, surfactants, polymers, microbes, etc.) not originally present in the reservoir. The purpose of EOR is to recover *unswept* (bypassed) and/or *residual* (capillary trapped) hydrocarbons usually by reducing the *interfacial tension* between phases, increasing the *viscosity* of the displacing fluid, or decreasing the viscosity of the hydrocarbons. Enhanced oil recovery methods are only pursued if economically viable.

1.3 Definitions

1.3.1 Phases and components in subsurface porous media

A *phase* (α) is a region of space with uniform physical properties. Subsurface reservoirs may have several fluid phases; in this text up to three are considered: liquid aqueous phase (w), liquid oleic phase (o), and gaseous phase (g). Other fluid phases such as *microemulsions* and *supercritical fluids* are not discussed in this text. A solid phase (s) consists of the rock matrix. Fig. 1.1 shows a cartoon of a reservoir with three fluid phases sealed by caprock.

A *component* (κ) is a unique chemical species. For example, water (H_2O), carbon dioxide (CO_2), methane (CH_4), decane ($C_{10}H_{22}$), sodium chloride (NaCl), and calcium carbonate ($CaCO_3$) are all examples of components that may exist in a reservoir. A phase may contain many components and the same component may be present in multiple phases. *Compositional reservoir simulators* (see Chapters 7, 8 and 10) are used to model flow and transport of individual components. However, subsurface reservoirs may contain thousands

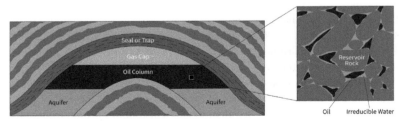

FIGURE 1.1 Three fluid phases, aqueous (*blue*), oleic (*brown*), and gaseous (*yellow*), and a solid phase (rock matrix) in a geological trap. Importantly the oleic and gaseous zones contain connate water due to capillary forces.

of unique components, modeling of which is neither practical nor computationally feasible. In practice, components of similar size and physical and/or chemical properties are lumped together to form *pseudocomponents*. Most compositional simulators model three to tens of pseudocomponents.

A special case of the compositional model is the *black oil* (or β) model. A black oil model contains three pseudocomponents (water, oil, and gas) in up to three fluid phases (aqueous, oleic, and gaseous). The oil pseudocomponent consists of all components present in a liquid hydrocarbon phase when brought to *standard conditions* (sc) of temperature ($T_{sc} = 60°F$) and pressure ($p_{sc} = 14.7$ psia). Likewise, the gas and water pseudocomponent consists of all components in a hydrocarbon gaseous phase and aqueous phase, respectively, at standard conditions. In the black oil model, the gas component may be dissolved in the oleic phase at reservoir conditions ($T > T_{sc}$; $p > p_{sc}$) and oil may or may not be volatilized in the gaseous phase, but the aqueous phase is assumed to contain no oil or gas and no water is in the gaseous/oleic phases. In reality, there is a very small solubility (< 100 ppm) of hydrocarbons in the aqueous phase and removal of the hydrocarbons in shallow aquifers is often the goal in *aquifer remediation* strategies.

1.3.2 Porosity, saturation, density, and concentrations

Porosity (ϕ) is defined as the pore volume divided by the bulk volume of the reservoir or porous domain, $\phi = V_p/V_b$, and is reported as a fraction or percentage. The *saturation* of a phase ($S_\alpha = V_\alpha/V_p$) is the volume fraction of the void space in the porous medium occupied by that phase. Phase saturations must sum to unity,

$$\sum_{\alpha=1}^{N_p} S_\alpha = S_w + S_o + S_g = 1, \tag{1.1}$$

where N_p is the total number of fluid phases and subscripts w, o, and g refer to the aqueous, oleic, and gaseous phases, respectively.

Density (ρ) is a mass (e.g., lb$_m$) per unit volume (e.g., ft^3) and usually refers to a phase (ρ_a) but may also be used to describe a component (ρ_κ). The density of a fluid is a function of pressure, temperature, and composition. The density of pure water (H$_2$O without dissolved solids) at standard conditions is 62.37 lb$_m$/ft^3, and the density of hydrocarbon liquids is usually (but not always) less than water.

The *pore concentration* of a component (C_κ) is defined as the amount of component κ in the pore volume (C_κ = mass of κ/pore volume). Thus, the units of C_κ are lb$_m$/ft^3. Concentrations can also be used to define the amount of component within a phase, $C_{\kappa,\alpha}$ = amount of κ/volume of phase α. Finally, the *bulk concentration*, W_κ, is defined as the mass of component in the total bulk volume,

$$W_\kappa = \underbrace{\left(C_{\kappa,w}S_w + C_{\kappa,o}S_o + C_{\kappa,g}S_g \right) \phi}_{C_\kappa} + C_{\kappa,s}(1 - \phi), \qquad (1.2)$$

where $C_{\kappa,s}$ is the concentration of component adsorbed onto the solid (rock) phase, s.

Mass ($\omega_{\kappa,\alpha}$), molar ($x_{\kappa,\alpha}$), and volume ($\sigma_{\kappa,\alpha}$) fractions are also used. A few important relationships between concentration and mass, molar, and volume fractions are provided in Eq. (1.3),

$$\omega_{\kappa,\alpha} = \frac{C_{\kappa,\alpha}}{\rho_\alpha} \qquad (1.3a)$$

$$x_{\kappa,\alpha} = \omega_{\kappa,\alpha}\frac{M_\alpha}{M_\kappa} \qquad (1.3b)$$

$$\sigma_{\kappa,\alpha} = \omega_{\kappa,\alpha}\frac{\rho_\alpha}{\rho_\kappa} = x_{\kappa,\alpha}\frac{\rho_{M,\alpha}}{\rho_{M,\kappa}} \qquad (1.3c)$$

where $\omega_{\kappa,\alpha}$ is the mass of component divided by the mass of phase, $x_{\kappa,\alpha}$ is the moles of component divided by the moles of phase, $\rho_{M,\alpha}$ is the molar density (e.g., lbmoles/ft^3) of the phase, and $\rho_{M,\kappa}$ is the molar density of the component. The molecular weight of the phase is the mole fraction weighted sum of components in the phase,

$$M_\alpha = \sum_{\kappa=1}^{N_c} x_{\kappa,\alpha}M_\kappa. \qquad (1.4)$$

Mass, molar, and volume fractions can also be defined for the entire pore volume. Mass fractions are often reported in wt% or ppm (parts per million). 1.0 wt% corresponds to 10,000 ppm.

1.4 Phase behavior

Hydrocarbons may reside in a gaseous phase, oleic phase, or both phases depending on the reservoir pressure, temperature, and overall composition. The reservoir is *undersaturated* if only one hydrocarbon phase is present (e.g., oleic or gaseous) but *saturated* if both oleic and gaseous phases are present at reservoir temperature and pressure. A phase, or P-T, diagram (Fig. 1.2) for a hydrocarbon mixture of specified composition can be used to determine which phases (oleic, gaseous, or both) are present at a given temperature and pressure.

The *bubble point* is the pressure and temperature at which the first bubble of gas is formed from a liquid phase. In Fig. 1.2, the reservoir fluid is an

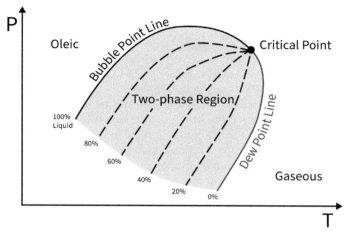

FIGURE 1.2 Generic phase, P-T, diagram for a multicomponent hydrocarbon mixture.

undersaturated oil (only oleic phase present) for pressures above the *bubble point line*. The *dew point* is the pressure and temperature at which the first drop of liquid forms from a gaseous phase. The reservoir fluid is an undersaturated gas (only gaseous phase present) if the pressure and temperature are below the lower *dew point line. The critical point* is the temperature and pressure at which the bubble point and dew point lines meet. Two hydrocarbon (oleic and gaseous) phases exist if the pressure and temperature are within the *two-phase envelope*.

Reservoir pressures decline during primary recovery while reservoir temperature often remains approximately constant (*isothermal*).[1] As a result, the reservoir may begin and remain as an undersaturated single phase or transition from a single phase to two phases during recovery. If the initial reservoir pressure is to the left of the critical point and above the bubble point line, the initial single phase is oleic and a decline in pressure results in crossing the bubble point line and a gaseous phase is also formed. If the initial pressure is to the right of the critical point and above the *upper dew point line*, a gaseous phase is present and a decline in pressure results in crossing the upper dew point line and the formation of a second (oleic) phase, a process referred to as *retrograde condensation*. Reservoirs have many different classification systems, one being the type of fluid: dry gas, wet gas, condensate, volatile oil, and black oil.

---

1. Not all reservoirs are isothermal. For example, in thermal enhanced oil recovery steam or other heating mechanisms are used to heat the reservoir fluids, usually for the purpose of decreasing viscosity.

1.5 Rock and Fluid Properties

1.5.1 Formation properties

The *formation compressibility* (c_f) is a reservoir property defined as the fractional change in pore volume with a change in reservoir pressure. Geertsma (1957) and Dandekar (2013) state that formation compressibility is a result of: (1) rock-matrix compressibility (c_r), a fractional change in volume of solid rock material (grains) with unit change in pressure; (2) bulk compressibility (c_B), a fractional change in bulk volume of rock with unit change in pressure; (3) pore compressibility (c_p), a fractional change in pore volume with a unit change in pressure. These compressibilities are illustrated in Fig. 1.3. Of these, the pore compressibility is almost always much greater than the rock and bulk compressibility,

$$c_f \approx c_p = \frac{1}{V_p}\left(\frac{\partial V_p}{\partial p}\right)_T = \frac{1}{\phi}\left(\frac{\partial \phi}{\partial p}\right)_T \tag{1.5}$$

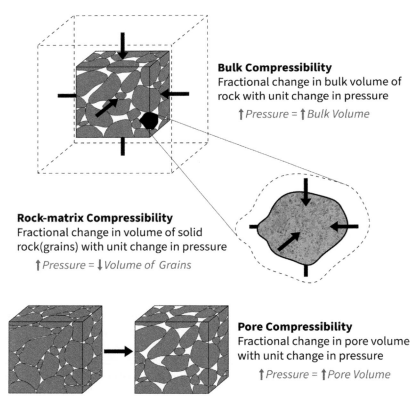

Bulk Compressibility
Fractional change in bulk volume of rock with unit change in pressure
↑ Pressure = ↑ Bulk Volume

Rock-matrix Compressibility
Fractional change in volume of solid rock(grains) with unit change in pressure
↑ Pressure = ↓ Volume of Grains

Pore Compressibility
Fractional change in pore volume with unit change in pressure
↑ Pressure = ↑ Pore Volume

FIGURE 1.3 Illustration of three types of formation compressibility and the effect of pore pressure.

Typical values of the formation compressibility are $c_f = 3 \times 10^{-6}$ to 30×10^{-6} psi^{-1} and assuming a constant value (independent of pressure) is often an acceptable approximation. Other important geomechanical properties are the Poisson's ratio and Young's, bulk, and shear modulus. However, geomechanics is not discussed in detail in this text despite its importance in many applications.

1.5.2 Gaseous phase properties

The *PVT* (pressure—volume—temperature) relationship for the gaseous phase is governed by a real gas law, e.g., $pV_m = zRT$, where V_m is the molar volume and the inverse of the molar density. The compressibility or z-factor, z, is a positive, dimensionless correction factor for the nonideality of reservoir gases at high pressure and temperature. The z-factor is unity at standard conditions ($z^{sc} = 1.0$) where the gas nearly acts ideally. For real gases, the z-factor can be measured experimentally or estimated from correlations (Standing and Katz, 1942; Hall and Yarborough, 1973; Dranchuk and Abou-Kassem, 1975) given the composition, temperature, and pressure. The z-factor is 1.0 at very low pressures such as standard pressure (14.7 psia), is less than 1.0 at moderate pressures, and is higher than 1.0 at high pressures.

The density of a gaseous phase can be determined using a real gas law,

$$\rho_g = \frac{M_g p}{zRT},$$

(1.6)

where M_g is the molecular weight of the gaseous phase, R is the gas constant (10.73 psia-ft^3/lbmole-°R), T is the absolute temperature (°R), and p is the absolute pressure (psia). A consequence of Eq. (1.6) is that 1.0 lbmole of any gas occupies 379.4 ft^3 at standard conditions. The *specific gravity* of a gas (γ_g) is the ratio of the gas density at *sc* to air density at *sc*, which is equivalent to the ratio of their respective molecular weights,

$$\gamma_g = \left(\frac{\rho_g}{\rho_{air}} \right)_{sc} = \frac{M_g}{M_{air}} = \frac{M_g}{29},$$

(1.7)

As reservoir fluids are brought to the surface, the respective phases (aqueous, oleic, and gaseous) undergo a change in volume. This volume change is a result of (a) expansion as the pressure drops from reservoir pressure to standard conditions, (b) thermal compression as the temperature decreases from reservoir temperature to standard temperature, (c) expulsion of a second phase as the pressure/temperature change (Fig. 1.4), and (d) change in composition or partitioning between phases.

The *formation volume factor* (B_α) is defined as the volume of the phase at reservoir conditions (*rc*) divided by the volume of the phase at standard

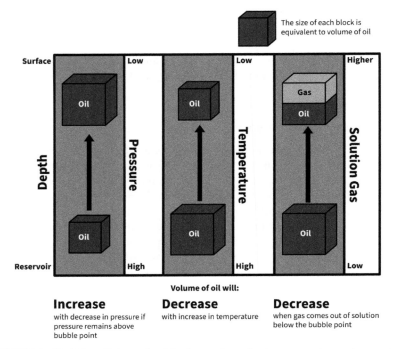

FIGURE 1.4 Depiction of change in oleic phase volume due to pressure, temperature, and solution gas changes.

conditions. The formation volume factor (B_g) of the gaseous phase is always much less than 1.0 because gases are very compressible and expand significantly as the pressure is reduced from the reservoir to the surface (for gases this greatly overwhelms any effect of temperature change, expulsion of volatile liquid, or composition change). Using the gas law,

$$B_g = \frac{V_{rc}}{V_{sc}} = \frac{zTp_{sc}}{z_{sc}T_{sc}p} = 0.0282 \frac{zT}{p} \frac{\text{ft}^3}{\text{scf}} = 0.005035 \frac{zT}{p} \frac{\text{res bbl}}{\text{scf}}, \qquad (1.8)$$

where the temperature, T, is defined in degrees Rankine and pressure, p, is in psia. The units *scf* are *standard cubic feet*, i.e. the amount of fluid that occupies one cubic foot at standard conditions. Fig. 1.5A shows the pressure dependency on B_g and the z-factor.

The *gas compressibility* (c_g) is a reservoir property defined as the fractional change in gas volume with a change in reservoir pressure.

$$c_g = -\frac{1}{V_g} \frac{\partial V_g}{\partial p}\bigg|_T = \frac{1}{B_g} \frac{\partial B_g}{\partial p}\bigg|_T = \frac{1}{p} - \frac{1}{z} \frac{\partial z}{\partial p}\bigg|_T \qquad (1.9)$$

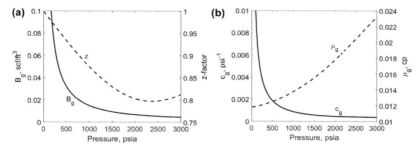

FIGURE 1.5   Pressure-dependent reservoir properties for gaseous phase: (A) formation volume factor and z-factor and (B) compressibility and viscosity. The gas molecular weight is 22 lbm/lbmole and the temperature is 150°F.

Gas compressibility at reservoir conditions is relatively high ($10^{-2}-10^{-4}$ psi^{-1}) compared to the compressibility of the formation and aqueous, oleic phases. Gas compressibility decreases with increasing pressure as shown in Eq. (1.9) and Fig. 1.5B.

The *gaseous phase viscosity* (μ_g) is relatively low (typical value $\sim 10^{-2}$ cp) compared to the aqueous and oleic phases. It is a function of composition, temperature, and pressure and decreases with decreasing reservoir pressure (as molecules spread apart at lower pressures, the gas molecules have less resistance and thus flow more easily). The decrease in pressure during production increases the mobility of gas compared to other fluid phases. Common correlations of gas viscosity are given by Carr et al. (1954) and Lee et al. (1966). Fig. 1.5B shows the pressure dependency on gas compressibility and viscosity.

1.5.3 Oleic phase properties

The *specific gravity* of an oil (γ_o) is defined as the ratio of the oil density and water density at standard conditions, $\gamma_o = \rho_{o,sc}/\rho_{w,sc}$. Specific gravity is often reported as API *gravity,*

$$\text{API}^o = \frac{141.5}{\gamma_o} - 131.5 \tag{1.10}$$

The specific and API gravity of water are 1.0 and 10°, respectively. The American Petroleum Institute (API) classification for petroleum fluids states that light crude oils have an API gravity of greater than 31.1°, medium oils have an API gravity between 22.3 and 31.1°, heavy crude oil between 10 and 22.3°, and extra heavy oils have an API gravity of less than 10° and are therefore heavier than water.

The oil formation volume factor, B_o, is defined as the volume of the oleic phase at reservoir conditions (e.g., RB, reservoir barrels) divided by the amount of oil at standard conditions (e.g., STB, stock tank barrels). The volume of the oleic phase at reservoir conditions includes any dissolved gas that comes out of solution at the surface,

$$B_o = \frac{V_{oleic,rc}}{V_{o,sc}} = \frac{V_{o,rc} + V_{dg,rc}}{V_{o,sc}} [\equiv] \frac{\text{bbl}}{\text{STB}}, \tag{1.11}$$

where V_{dg} is the volume of solution gas dissolved in the oleic phase. The formation volume factor of oleic phase is always greater than 1.0 because at standard conditions the volume is less than at reservoir conditions due to dissolved gas (at reservoir conditions) expelled from solution at the low surface pressure. The removal of gas greatly overwhelms any expansion of the oleic phase due to the pressure reduction.

The *solution gas-oil ratio*, R_s, is defined as the standard volume (e.g., scf, standard cubic feet) of gas component divided by the standard volume (e.g., STB) of oil component in the oleic phase at a given P, T,

$$R_s [=] \frac{\text{std.vol.gas}}{\text{std.vol.oil}} = \frac{V_{dg,sc}}{V_{o,sc}} [\equiv] \frac{\text{scf}}{\text{STB}}. \tag{1.12}$$

Above the bubble point, R_s is constant because no gas is expelled from solution with a decrease in pressure (oleic phase is undersaturated). Below the bubble point, the oleic phase is saturated, and as the pressure is reduced, gas is expelled from the oleic phase; therefore, R_s decreases with decreasing pressure. Empirical correlations for R_s include the Standing (1947), Lasater (1958), and Vasquez and Beggs (1980) correlations. Fig. 1.6A shows the pressure dependency on formation volume factor and solution gas ratio.

An equation for the density of the oleic phase at reservoir conditions can be derived by recognizing the mass of the oleic phase at reservoir conditions equals the mass of oil remaining at standard conditions plus the mass of gas at standard conditions that was expelled from the oleic phase,

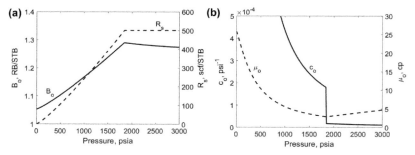

FIGURE 1.6 Pressure-dependent reservoir properties for oleic phase: (A) formation volume factor and solution gas ratio and (B) oil compressibility and viscosity. The bubble point is 1666 psia. The oil specific gravity is 0.822, gas molecular weight is 22 lbm/lbmole, and the temperature is 150°F.

$$\rho_{oleic,rc} = \frac{\rho_{o,sc} + R_s \rho_{g,sc}}{B_o}.$$ (1.13)

The *oil compressibility* (c_o) is a reservoir property defined as the fractional change in volume of the oleic phase with a change in reservoir pressure as given by,

$$c_o = -\frac{1}{V_o} \frac{\partial V_o}{\partial p}\bigg|_T = \frac{-1}{B_o}\left[\left(\frac{\partial B_o}{\partial p}\right)_T - B_g\left(\frac{\partial R_s}{\partial p}\right)_T\right],$$ (1.14)

Below the bubble point, the oil compressibility increases with decreasing pressure. Correlations, such as those by Vasquez and Beggs (1980), are often used. For an undersaturated oil (reservoir pressures above the bubble point), the second term in Eq. (1.14) vanishes because R_s is constant with pressure. The oil compressibility is a weak function of pressure above the bubble point, but assuming a constant value is usually an acceptable approximation. A typical value of the undersaturated oil compressibility is on the order of 1.0×10^{-5} psi^{-1}.

If the compressibility is assumed constant above the bubble point, Eq. (1.14) can be integrated to give,

$$B_o = B_{o,b}e^{-c_o(p-p_b)} \approx B_{o,b}[1 - c_o(p - p_b)],$$ (1.15)

where $B_{o,b}$ is the formation volume factor at the bubble point, p_b. The linear approximation can be found from a Taylor series and is valid when the oil compressibility is small in addition to being constant. Above the bubble point, the formation volume factor decreases with increasing pressure due to the fluid compressibility. A constant value can often be used as an approximation because B_o is a relatively weak function of pressure above the bubble point. Typical values of B_o for an undersaturated oil are 1.1−1.3 RB/STB.

The *viscosity of the oleic phase* (μ_o) increases with increasing pressure above the bubble point, mostly as a result of molecules being closer together at high pressures. Below the bubble point, however, viscosity increases with decreasing pressure because low-molecular-weight gas molecules, with low viscosity, are expelled from solution as the pressure decreases leaving only large molecules in solution. The viscosity of the oleic phase can range several orders of magnitude, from less than 1 cp to thousands (or millions) of cp. Viscosity is also a strong function of temperature. While many reservoirs are produced isothermally, injection of heat (e.g., steam) is often used to reduce the viscosity in reservoirs of very viscous oils. Correlations, such as those by Beggs and Robinson (1975), are useful for predicting viscosity. Fig. 1.6B shows the pressure dependency on oil compressibility and viscosity.

PVT properties can be measured in the laboratory or estimated from correlations. Generally, the results are collected in a PVT table at pressures ranging from atmospheric (14.7 psia) to the initial reservoir pressure at the reservoir temperature.

1.5.4 Aqueous phase properties

The aqueous phase is assumed to have negligible solubility of hydrocarbon components but usually contains dissolved solids including salts and minerals and may also contain carbon dioxide in solution. Reduction in reservoir pressure results in very little, if any, hydrocarbon gas evolution from the aqueous phase. Furthermore, the compressibility of the aqueous phase is small so the density, ρ_w, and formation volume factor, B_w, are weak functions of pressure. Correlations for B_w as a function of P, T and dissolved solids concentration have been developed (McCain, 1991), but values of B_w are usually near 1.0 at reservoir pressure.

The aqueous phase compressibility, c_w, is defined as the fractional change in phase volume with a change in pressure,

$$c_w = -\frac{1}{V_w}\left(\frac{\partial V_w}{\partial p}\right)_T = -\left(\frac{\partial \ln V_w}{\partial p}\right)_T = \frac{1}{\rho_w}\frac{\partial \rho_w}{\partial p}\bigg|_T = B_w\frac{\partial}{\partial p}\left(\frac{1}{B_w}\right) = \frac{-1}{B_w}\frac{\partial B_w}{\partial p}.$$

(1.16)

Each of the above definitions for compressibility can be shown to be equivalent. A typical value is $c_w \sim 1.0 \times 10^{-6}$ psi^{-1} but depends on reservoir temperature, pressure, and composition. The aqueous phase compressibility is approximately an order of magnitude less than a typical undersaturated oil compressibility. Integration of Eq. (1.16) gives the formation volume factor of water,

$$B_w = B_w^0 e^{-c_w(p-p^0)} \approx B_w^0\left[1 - c_w(p - p^0)\right],$$

(1.17)

where B_w^0 is the formation volume factor of the aqueous phase at a reference pressure, p^0, e.g., the pressure at the original water−oil contact line (OWOC). The linear approximation in Eq. (1.17) is based on a Taylor series and is almost always valid for water due to the low compressibility. For a water compressibility of 1.0×10^{-6} psi^{-1}, a change in reservoir pressure by 1000 psi from its reference value would result in less than 0.1% change in B_w, thus approximating B_w as a constant is generally valid.

The viscosity of the aqueous phase (μ_w) is a function of reservoir pressure and composition (e.g., concentration of dissolved solids) and a strong function of temperature. The viscosity of fresh water at standard temperature and pressure is ~ 1.0 cp. At reservoir conditions, the viscosity of the aqueous phase (brine) is less than unity, often near 0.5 cp. Correlations have been developed to predict the aqueous phase viscosity (brine) as a function of temperature and composition (Kestin et al., 1981; McCain, 1991; Sharqawy et al., 2010). Example 1.1 demonstrates some calculations using PVT properties.

Example 1.1 Properties of Mixtures.
A light, undersaturated oleic phase with three components, methane (C1), n-pentane (n-C5), and n-decane (n-C10), is initially at 150°F and 1500 psia and has a molar composition of $x_{C1} = 0.3$, $x_{C5} = 0.3$, and $x_{C10} = 0.4$. The bubble point pressure, p_b, is 1208 psia, the formation volume factor at bubble point, $B_{o,b}$, is 1.291 RB/STB, and the solution gas ratio at bubble point, R_{sb}, is 435 scf/STB. The undersaturated oil compressibility is 2.5×10^{-5} psi^{-1}.

The fluid is brought to standard conditions (60°F and 14.7 psia) where the stock tank oil has a composition of $x_{C1,o} = 0.005$, $x_{C5,o} = 0.37$, and $x_{C10,o} = 0.625$ and an API gravity of 69.4°. The stock tank gas has a composition of $x_{C1,g} = 0.822$, $x_{C5,g} = 0.177$, and $x_{C10,g} = 0.001$.

Calculate:

a. Specific gravity and density (lb$_m$/ft^3) of both the stock tank oil and gas
b. Mass fractions of each component in the oleic phase at reservoir conditions, 150°F and 1500 psia
c. Formation volume factor of the oleic phase at reservoir conditions
d. Density (lb$_m$/ft^3) of oleic phase at reservoir conditions

Solution:

(a) The API gravity can be used to compute the oil specific gravity and density using Eq. (1.10),

$$\gamma_o = \frac{141.5}{API^0 + 131.5} = \frac{141.5}{69.4^0 + 131.5} = 0.704$$

$$\rho_{o,sc} = \rho_{w,sc}\gamma_o = 62.4\frac{lb_m}{ft^3}\cdot 0.704 = 43.94\frac{lb_m}{ft^3}$$

The molecular weight of the gas (Eq. (1.4)) can be used to compute the specific gravity and density using Eqs. (1.7) and (1.6), respectively,

$$M_g = \sum_\kappa M_\kappa x_{\kappa,g} = 16\frac{lb_m}{lbmole}\cdot 0.822\frac{lbmole\ CH_4}{lbmole} + 72\frac{lb_m}{lbmole}\cdot 0.177\frac{lbmole\ C_5H_{12}}{lbmole}$$

$$+142\frac{lb_m}{lbmole}\cdot 0.001\frac{lbmole\ C_{10}H_{22}}{lbmole} = 26.04\frac{lb_m}{lbmole}$$

$$\gamma_g = \frac{M_g}{29} = \frac{26.04\frac{lb_m}{lbmole}}{29\frac{lb_m}{lbmole}} = 0.898$$

$$\rho_{g,sc} = \frac{M_g}{379.4\frac{ft^3}{lbmole}} = \frac{26.04\frac{lb_m}{lbmole}}{379.4\frac{ft^3}{lbmole}} = 0.069\frac{lb_m}{ft^3}$$

(b) The molecular weight of the oleic phase is given by,

Continued

Example 1.1 Properties of Mixtures.—cont'd

$$M_g = \sum_\kappa M_\kappa x_\kappa = 16\frac{lb_m}{lbmole} \cdot 0.3\frac{lbmole\ CH_4}{lbmole} + 72\frac{lb_m}{lbmole} \cdot 0.3\frac{lbmole\ C_5H_{12}}{lbmole}$$

$$+142\frac{lb_m}{lbmole} \cdot 0.4\frac{lbmole\ C_{10}H_{22}}{lbmole} = 83.2\frac{lb_m}{lbmole}$$

which is then used to compute mass fractions of each component

$$\omega_\kappa = x_\kappa \frac{M_\kappa}{M_\alpha}$$

$$\omega_{C1} = 0.3\frac{16}{83.2} = 0.058\frac{lbm\ C1}{lb_m\ oleic}; \omega_{C5} = 0.3\frac{72}{83.2} = 0.260\frac{lbm\ C5}{lb_m\ oleic};$$

$$\omega_{C10} = 0.4\frac{142}{83.2} = 0.683\frac{lbm\ C10}{lb_m\ oleic}$$

(c) Formation volume factor of an undersaturated oil can be computed using Eq. (1.11)

$$B_o = B_{ob}\exp[c_o(p_b - p)] = 1.291\frac{bbl}{STB}\exp[2.5 \times 10^{-5}psi^{-1}(1208psi - 1500psi)]$$

$$= 1.282\frac{bbl}{STB}$$

(d) The density of the oleic phase can be computed using Eq. (1.13)

$$\rho_o = \frac{\rho_{o,sc} + \rho_{g,sc}Rs}{B_o} = \frac{43.94\frac{lbm}{scf} + 0.069\frac{lbm}{scf} \cdot 435\frac{scf}{STB} \cdot \frac{1}{5.615}\frac{STB}{scf}}{1.282\frac{bbl}{STB}} = 38.4\frac{lbm}{ft^3}$$

A total compressibility is a saturated-weighted sum of phase and rock compressibilities,

$$c_t = c_f + \sum_{\alpha=1}^{N_p} S_\alpha c_\alpha = c_f + S_w c_w + S_o c_o + S_g c_g \tag{1.18}$$

This definition of total compressibility is very convenient for modeling multiphase flow.

1.6 Petrophysical properties

1.6.1 Darcy's law

In 1856, Henry Darcy performed experiments with water in a porous bed of sand grains (Darcy, 1856). He recognized the flow rate was proportional to the hydrostatic head (height of a column of water at a given depth) of water. The initial experiments were conducted without varying the bed permeability, fluid viscosity, length, nor medium; however, it has since been shown that all these factors impact flow rate (or velocity). Darcy's law in 1D, neglecting gravitational effects, can be written as,

$$u_x = \frac{q}{a} = -\frac{k}{\mu}\frac{\partial p}{\partial x}, \tag{1.19a}$$

where, u_x is the Darcy velocity in the x-direction, q is the flow rate, a is the cross-sectional area, and k is the permeability (scalar). More generally (3D with gravity), Darcy's law for flow through porous media is given by,

$$\vec{u} = -\frac{\mathbf{k}}{\mu}(\nabla p - \rho g \nabla D) = -\frac{\mathbf{k}}{\mu}\nabla\Phi, \tag{1.19b}$$

where $\mathbf{k}$ is the permeability tensor, p is the fluid pressure, g is the gravitational constant ($g = 9.80$ m/s^2 = 32 ft/s^2), D is the depth, and Φ is the potential. The depth, D, increases with distance below the surface so D is always positive in subsurface applications (elevation increases above the surface so elevation is always negative in the subsurface). The terms in parenthesis of Eq. (1.19b) (pressure gradient minus hydrostatic head) are referred to as the potential gradient, $\nabla\Phi$.

In the most general form of Darcy's law, the permeability is taken as a second-order 3 × 3 tensor. For anisotropic media, permeability is dependent on the direction of flow (i.e., x, y, or z). In Cartesian coordinates, the 3×3 permeability tensor can be written,

$$\mathbf{k} = \begin{pmatrix} k_{xx} & k_{xy} & k_{xz} \\ k_{yx} & k_{yy} & k_{yz} \\ k_{zx} & k_{zy} & k_{zz} \end{pmatrix} \Rightarrow \mathbf{k} = \begin{pmatrix} k'_x & 0 & 0 \\ 0 & k'_y & 0 \\ 0 & 0 & k'_z \end{pmatrix}, \tag{1.20}$$

where the coordinate system can always be rotated, at some angle onto its *principal axis* such that the permeability tensor can be transformed into a diagonal one (all off-diagonal terms are zero). Usually, it is assumed in cell-centered finite difference reservoir simulation studies that the diagonal tensor in Eq. 1.20 can be used. In this text, diagonal tensors are assumed and the apostrophe (') is dropped for convenience. Furthermore, in many subsurface applications, $k_x \approx k_y$, and k_x is 1−10 times larger than the vertical permeability, k_z. For isotropic media, permeability is not dependent on the flow direction, thus the permeability tensor reduces to a scalar, k.

Darcy's law is semiempirical, originating from experiments. The linear relationship between velocity and pressure (or potential) gradient is only valid under certain assumptions such as single-phase, creeping flow (Reynolds number, $N_{Re} \ll 1$), Newtonian fluid (viscosity independent of shear rate and time), and no fluid slip at the particle walls. Under these assumptions and more, Darcy's law can be derived directly from the Navier−Stokes equations using techniques such as volume averaging (Whitaker, 1999) and homogenization (Hornung, 1997). However, when these assumptions are not valid, Darcy's law is not appropriate and other nonlinear models for velocity have been proposed (Forchheimer, 1901; Bird et al., 1960; Klinkenberg, 1941). Some of these models are discussed more in Chapter 6.

1.6.2 Relative permeability

1.6.2.1 Definition and relationship with phase saturation

When multiple fluid phases flow in a porous medium the ability of each phase to flow is reduced from the single-phase permeability, k, of the medium. The effective permeability of the phase, k_α, is the single-phase permeability multiplied by a dimensionless *relative permeability*, $k_{r,\alpha}$. Relative permeability of a phase is primarily a function of the phase saturation, although there are other potential dependencies (Blunt, 2017). Darcy's law can be modified for multiphase flow by including relative permeability,

$$\vec{u}_\alpha = -\frac{\mathbf{k}k_{r,\alpha}}{\mu_\alpha}(\nabla p_\alpha - \rho_\alpha g \nabla D) = -\frac{\mathbf{k}k_{r,\alpha}}{\mu_\alpha}\nabla \Phi_\alpha, \tag{1.21}$$

where u_α and μ_α are the velocity and viscosity, respectively, of phase α.

Relative permeability curves for a typical oleic/aqueous and liquid (oleic plus aqueous)/gaseous system in a water-wet medium are depicted in Fig. 1.7A and B, respectively.

1.6.2.2 Empirical models

The relative permeability curves are nonlinear with saturation, and the curves do not pass through the origin. Below a minimum, *residual saturation* (e.g., S_{wr}, S_{or}, S_{gr}), the phase is immobile and does not flow (relative permeability of

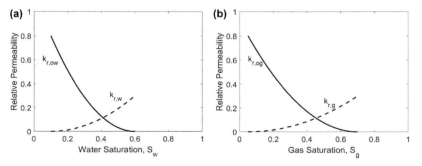

FIGURE 1.7 Typical relative permeability curves in a water-wet medium for (A) oleic-aqueous and (B) liquid-gaseous.

zero). The term *irreducible water saturation* is also used for S_{wr}. The relative permeability of the complementary phase at residual saturation is the *endpoint permeability* (e.g., k^0_{rw}, k^0_{ro}, k^0_{rg}). The endpoint relative permeability for the nonwetting phase is often close to 1.0 at residual wetting-phase saturation (Fig. 1.7A) because the nonwetting phase occupies the larger, more conductive pores in the porous medium. The endpoint relative permeability of the wetting phase is usually much less than 1.0 because wetting phase occupies the smaller, less conductive pores.

In reservoir simulation, relative permeability values can be interpolated from tabulated experimental data or computed from a simple theoretical model obtained via best fit to the experimental data. Several models have been proposed (Chierici, 1984; Honarpour et al., 1986; Alpak et al., 1999). The Brooks–Corey model (1964) is one of the simplest and most widely used relative permeability models for two-phase flow. For example, for the oleic-aqueous curve in Fig. 1.7A,

$$k_{rw} = k^0_{rw} S^{n_w}; \quad k_{ro} = k^0_{ro}(1 - S)^{n_o}, \tag{1.22}$$

where the normalized saturation, S, is defined by,

$$S = \frac{S_w - S_{wr}}{1 - S_{wr} - S_{or}}. \tag{1.23}$$

Usually, the residual saturations of the phases along with endpoint relative permeabilities are determined directly by inspection of the relative permeability versus saturation data. The empirical power-law coefficients (n_o, n_w) are then found through curve fitting. Typical values for n_w and n_o are between 1.0 and 4.0.

Computation of three-phase relative permeability is more complicated than two-phase relative permeability and existing correlations have limited predictive success. A common approach is to obtain experimental data in two-phase systems (e.g., oil-water and gas-oil, with residual water present). The pair of two-phase data is then used to fit empirical three-phase models. The Stone I model (Stone, 1970) is a popular model which successfully fits data at low gas and water saturation but often fails to match over a wide range of saturations (Kianinejad and DiCarlo, 2016). In the Stone I model, the aqueous and gaseous relative permeability equations are similar to the Corey model because it assumed those phases do not interact with each other, but both interact with the oleic phase,

$$k_{rw} = k^0_{rw} S^{N_w}_{wD}; \quad k_{rg} = k^0_{rg} S^{N_g}_{gD}, \tag{1.24}$$

where the normalized saturations are defined as,

$$S_{wD} = \frac{S_w - S_{wr}}{1 - S_{wr} - S_{orw}}; \quad S_{gD} = \frac{S_g - S_{gr}}{1 - S_{gr} - S_{org} - S_{wr}} \tag{1.25}$$

and S_{orw} and S_{org} are the residual oil saturations in the two-phase experiments with water and gas, respectively. The relative permeability curve of the oleic phase is more complicated because the phase interacts with both the aqueous and gaseous phases,

$$
k_{ro}\left(S_w, S_o, S_g\right) = \begin{cases} 0 & \text{if} \quad S_{oD} < 0 \\[2mm] \dfrac{S_o^* k_{row} k_{rog}}{k_{row}^0 \left(1 - S_w^*\right)\left(1 - S_g^*\right)} & \text{if} \quad 0 < S_{oD} < 1 \\[2mm] k_{row}^0 & \text{if} \quad S_{oD} > 1 \end{cases}
$$

(1.26)

where the normalized oil saturation is defined by,

$$
S_{oD} = \frac{S_o - S_{orw}}{1 - S_{wr} - S_{orw} - S_{gr}}.
$$

(1.27)

The oil-water and oil-gas relative permeabilities are defined as,

$$
k_{row} = k_{row}^0 (1 - S_{wD})^{N_{ow}}; \quad k_{rog} = k_{rog}^0 (1 - S_{gD})^{N_{og}},
$$

(1.28)

where the normalized saturations are

$$
S_o^* = \frac{S_o - S_{om}}{1 - S_{wr} - S_{om}}; \quad S_w^* = \frac{S_w - S_{wr}}{1 - S_{wr} - S_{om}}; \quad S_g^* = \frac{S_g}{1 - S_{wr} - S_{om}},
$$

(1.29)

and the residual oil saturation, S_{or}, was originally proposed to be the minimum value of S_{orw} and S_{org}. However, Fayers and Matthews (1982) have proposed to use a weighted average of the two residual oil saturations.

$$
S_{om} = (1 - a)S_{orw} + aS_{org}; \quad a = \frac{S_g}{1 - S_{wr} - S_{org}}
$$

(1.30)

There are several other three-phase relative permeability models including those by Stone II (1973), Jerauld (1997), Hustad and Hansen (1996), Blunt (2000), and Beygi et al. (2015).

Example 1.2 Three-phase relative permeability models.
A three-phase water-oil-gas system is described by the following empirical parameters obtained from oil-water and oil-gas two-phase experiments, $S_{wr} = 0.1$, $S_{orw} = 0.4$, $S_{org} = 0.2$, $S_{gr} = 0.05$, $k_{rw}^0 = 0.3$, $k_{row}^0 = 0.8$, $k_{rog}^0 = 0.8$, $k_{rg}^0 = 0.3$, $N_w = 2.0$, $N_g = 2.0$, $N_{ow} = 2.0$, $N_{og} = 2.0$. The two-phase relative permeability curves for oil-water and oil-gas are shown in Fig. 1.8.
Compute three-phase relative permeabilities (k_{rw}, k_{ro}, k_{rg}) at $S_w = 0.3$, $S_o = 0.5$, $S_g = 0.2$.

Example 1.2 Three-phase relative permeability models.—cont'd

Solution

(a) First, compute the normalized saturations for relative permeability

$$S_{wD} = \frac{S_w - S_{wr}}{1 - S_{wr} - S_{orw}} = \frac{0.3 - 0.1}{1 - 0.1 - 0.4} = 0.4$$

$$S_{gD} = \frac{S_g - S_{gr}}{1 - S_{gr} - S_{org} - S_{wr}} = \frac{0.2 - 0.05}{1 - 0.05 - 0.2 - 0.1} = 0.231$$

$$S_{oD} = \frac{S_o - S_{orw}}{1 - S_{wr} - S_{orw}} = \frac{0.5 - 0.4}{1 - 0.1 - 0.4} = 0.2$$

Calculate relative permeabilities of water and gas using Brooks–Corey relationships,

$$k_{rw} = k_{rw}^0 S_{wD}^{N_w} = 0.3 \cdot (0.4)^{2.0} = 0.048$$

$$k_{rg} = k_{rg}^0 S_{gD}^{N_g} = 0.3 \cdot (0.231)^{2.0} = 0.016$$

Compute the oil relative permeability. First compute relative permeability of oil to water and gas, respectively,

$$k_{row} = k_{row}^0 (1 - S_{wD})^{N_{ow}} = 0.8 \cdot (0.6)^{2.0} = 0.288$$

$$k_{rog} = k_{rog}^0 (1 - S_{gD})^{N_{og}} = 0.8 \cdot (1 - 0.231)^{2.0} = 0.473$$

Compute "a" and "S_{om}"

$$a = \frac{S_g}{1 - S_{wr} - S_{org}} = \frac{0.2}{1 - 0.1 - 0.2} = 0.2857$$

$$S_{om} = (1 - a)S_{orw} + aS_{org} = (1 - 0.2857) \cdot 0.4 + 0.2857 \cdot 0.2 = 0.3429$$

Compute normalized saturations,

$$S_o^* = \frac{S_o - S_{om}}{1 - S_{wr} - S_{om}} = \frac{0.5 - 0.3429}{1 - 0.1 - 0.3429} = 0.282$$

$$S_w^* = \frac{S_w - S_{wr}}{1 - S_{wr} - S_{om}} = \frac{0.3 - 0.1}{1 - 0.1 - 0.3429} = 0.359$$

$$S_g^* = \frac{S_g}{1 - S_{wr} - S_{om}} = \frac{0.2}{1 - 0.1 - 0.3429} = 0.359$$

Finally, compute the relative permeability of oil,

$$k_{ro} = \frac{S_o^* k_{row} k_{rog}}{k_{row}^0 (1 - S_w^*)(1 - S_g^*)} = \frac{0.282 \cdot 0.288 \cdot 0.473}{0.8(1 - 0.359)(1 - 0.359)} = 0.117$$

The relative permeability curves can be generated using the above equations at different saturations and are shown in Fig. 1.8.

1.6.3 Capillary pressure

1.6.3.1 Definition and relationship with phase saturation

Capillary pressure is the difference between the pressure of two phases. In this text we define the oil-water capillary pressure as the oleic phase pressure minus the aqueous phase pressure, regardless of the rock wettability,

$$p_{c,ow}(S_w) = p_o - p_w. \tag{1.31}$$

Like relative permeability, it is generally assumed to be a function of saturation. Fig. 1.8 depicts a typical capillary pressure versus saturation curve for a mixed-wet medium.

Two curves are depicted in Fig. 1.8, one each for drainage and imbibition. A drainage process is defined here as aqueous phase saturation decreasing. It is usually assumed for initializing the reservoir because the reservoir was initially saturated with aqueous phase and hydrocarbons migrated from a source rock to the reservoir. An imbibition process, defined here as aqueous phase saturation increasing, is often assumed during secondary production because water from injector wells displaces the oleic phase. The imbibition curve is positive for all water saturations for a water-wet medium and negative for all saturations for an oil-wet medium. The imbibition curve crosses from positive to negative for mixed-wet media as shown in Fig. 1.8.

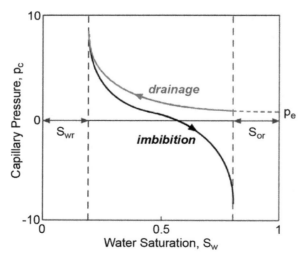

FIGURE 1.8 Drainage and imbibition capillary curves for a mixed-wet medium, where p_e is the entry pressure, S_{wr} is the irreducible water saturation, and S_{or} is the residual oil saturation.

For many reservoirs the capillary pressure is small (a few psi) compared to the reservoir pressure (thousands of psia). In these cases, capillary pressure can often be ignored for numerical solution of flow and the pressure of all phases assumed equal. Some reservoirs, such as gas-bearing shales, have large capillary pressures (order of magnitude of the reservoir pressure) and therefore must be included in the solution. Capillary pressure can also have the effect of smearing a sharp front between two immiscible phases as will be shown in Chapter 9. Capillary pressure is, however, always important for initializing the reservoir because it determines phase saturations at reservoir depths.

1.6.3.2 Capillary pressure models

Like relative permeability, capillary pressure values can be interpolated from tabulated experimental data or calculated directly from a curve-fitted model. Brooks and Corey (1964) also developed such models for drainage Eq. (1.32a) and imbibition Eq. (1.32b),

$$p_c = p_e S_{w*}^{-1/\lambda}, \tag{1.32a}$$

$$p_c = p_e \left(S_e^{-1/\lambda} - 1 \right), \tag{1.32b}$$

where p_e is the capillary entry pressure and λ is an empirical constant. The imbibition equation is for water-wet media but can be adapted for oil-wet or mixed-wet media. The normalized saturations are defined as,

$$S_{w*} = \frac{S_w - S_{wr}}{1 - S_{wr}}; \quad S_e = \frac{S_w - S_{wr}}{1 - S_{wr} - S_{or}} \tag{1.33}$$

The Van Genuchten (1980) model is another popular model for describing imbibition capillary pressure versus saturation.

$$p_c = \frac{1}{\gamma} \left(S_e^{-1/m} - 1 \right)^{1/n}, \tag{1.34}$$

where n, m, and γ are empirical constants. Example 1.2 demonstrates the use of empirical models to generate capillary pressure and relative permeability curves.

The Leverett J-function (Leverett, 1941) is a way to nondimensionalize capillary pressure curves. The J-function can then be applied to determine capillary pressure in a similar rock type but different permeability, porosity, or interfacial tension of the fluids,

$$J(S_w) = \frac{p_c(S_w)\sqrt{k/\phi}}{\sigma \cos \theta}, \tag{1.35}$$

where σ is the interfacial tension between the fluids and θ is the contact angle.

1.6.4 Capillary pressure scanning curves

Initialization of a reservoir is assumed to occur by primary drainage (oleic displaces aqueous phase) by migration from a source rock. During water-flooding, the capillary pressure transitions from a drainage curve to an imbibition curve. Capillary scanning curves are utilized in simulators to model this transition. A capillary scanning curve is a weighted average of the imbibition and drainage capillary pressure curves (Killough, 1976),

$$p_c(S_w) = \begin{cases} p_c(Dr, S_w) & \text{if } S_w \leq S_w(\text{hyst}) \\ f \cdot p_c(\text{Im}, S_w) + (1 - f) \cdot p_c(Dr, S_w) & \text{if } S_w > S_w(\text{hyst}) \end{cases}, \quad (1.36)$$

where the weighting factor, f, is determined by,

$$f = \left(\frac{S_w(\text{max}) - S_w(\text{hyst}) + \text{epspc}}{S_w(\text{max}) - S_w(\text{hyst})}\right)\left(\frac{S_w - S_w(\text{hyst})}{S_w - S_w(\text{hyst}) + \text{epspc}}\right) \quad (1.37)$$

and $S_w(\text{hyst})$ is the water saturation when a reversal occurred from decreasing water saturation in the block to increasing water saturation. This may be the initial water saturation, but water saturation could first decrease before increasing even during a waterflood. $S_w(\text{max})$ is the maximum obtainable water saturation, i.e., $1 - S_{or}$, and epspc is a small dimensionless number that determines the transition between imbibition and drainage. The closer the value to zero, the more quickly the curve transitions from drainage to imbibition.

Example 1.3 Capillary pressure curves.
An oil-water system has a capillary entry pressure, $p_{c,e} = 3.5$ psi and $\lambda = 2.0$. The residual saturations are $S_{wr} = 0.1$, $S_{or} = 0.2$. The capillary pressure data were measured for a 100 mD rock with 20% porosity and $\sigma_{ow} = 20$ dyn/cm.
    Estimate the capillary pressure at $S_w = 0.3$ for a similar rock type but 50 mD, 15% porosity, and $\sigma_{ow} = 25$ dyn/cm.

Solution
First, compute the normalized saturations at $S_w = 0.3$.

$$S_w^* = \frac{S_w - S_{wr}}{1 - S_{wr}} = \frac{0.3 - 0.1}{1 - 0.1} = 0.222$$

$$S_e = \frac{S_w - S_{wr}}{1 - S_{wr} - S_{or}} = \frac{0.3 - 0.1}{1 - 0.1 - 0.4} = 0.4$$

Compute the capillary pressures for drainage and imbibition and the original rock, IFT,

$$p_{cD} = p_e(S_w^*)^{-1/\lambda} = 3.5(0.222)^{-1/2} = 7.42 \text{ psia}$$

$$p_{cI} = p_e\left[(S_e)^{-1/\lambda} - 1\right] = 3.5\left[(0.4)^{-1/2} - 1\right] = 2.03 \text{ psia}$$

Example 1.3 Capillary pressure curves.—cont'd

Finally, estimate the capillary pressures at 50 mD, 15% porosity, and $\sigma_{ow} = 25$ dyn/cm. From Eq. (1.35), Leverett J-function,

$$p_{c,2} = p_{c,1} \frac{\sigma_2}{\sigma_1} \sqrt{\frac{k_1 \phi_2}{k_2 \phi_1}} = p_{c,1} \frac{25}{20} \sqrt{\frac{100 \cdot 0.15}{50 \cdot 0.2}} = 1.53 p_{c,1}$$

$$p_{cD} = 1.53 \cdot 7.42 = 11.35 \text{ psia}$$
$$p_{cI} = 1.53 \cdot 2.03 = 3.11 \text{ psia}$$

1.7 Reservoir initialization

Initial phase pressures and saturations in the reservoir are determined from phase densities, reservoir depth, and capillary pressure data. Fig. 1.9A is a cartoon that illustrates a water column from the surface to the aquifer. The aquifer is connected to the reservoir with a less dense oleic phase at the *original water-oil contact line* (OWOC) and at the *original gas-oil contact line* (OGOC) a gaseous phase (cap) meets the oleic phase.

Fig. 1.9B illustrates each of the phase pressures with depth. The aqueous phase pressure at a given depth is given by,

$$p_w(D) = p_{w,WOC} + \rho_w g(D - D_{WOC}), \tag{1.38}$$

where D is the depth and has the convention that $D = 0$ at the surface and is positive below the surface.[2] The hydrostatic gradient, $\rho_w g$, is 0.433 psi/ft for fresh water and slightly higher (e.g., 0.465 psi/ft) for brine.

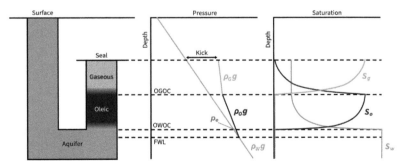

FIGURE 1.9 Schematic of (A) aqueous, oleic, and gaseous phases in subsurface and (B) phase pressures as a function of depth.

---

2. Elevation, z, is sometimes used instead of depth and has the opposite sign, i.e., it is negative below the surface.

The oleic and aqueous phase pressures are equal at the free water level (FWL). At the original water-oil contact line (OWOC), a depth that can often be estimated from well logs, the difference in phase pressures $(p_o - p_w)$ is the capillary entry pressure $(p_{e,ow})$ between oleic and aqueous phases. Using the aqueous phase pressure at the OWOC and the oleic phase density, oleic phase pressures at every depth can be computed,

$$p_o(D) = \left(p_{e,ow} + p_{w,WOC} \right) + \rho_o g(D - D_{WOC}). \tag{1.39}$$

Likewise, gaseous phase pressures can be computed from the original gas-oil contact (OGOC) line and gaseous phase density,

$$p_g(D) = \left(p_{o,GOC} + p_{e,og} \right) + \rho_g g(D - D_{GOC}) \tag{1.40}$$

Importantly, the densities of all phases in Eqs. (1.38), (1.39), and (1.40) are functions of the unknown pressure. A first approximation assumes a constant density for the phase at all depths. An improved, iterative method estimates the pressure at a depth which is then used to estimate density. That density is used to calculate pressure in an incrementally shallower depth, D-ΔD, and so forth.

Fig. 1.9C illustrates each of the phase saturations with depth. Phase saturations are not, in general, 100% in their respective zones; all three phases can exist at the same depth. Recall, however, that saturations of all phases must sum to unity. The aqueous phase saturation is 100% below the OWOC; above the OWOC line a transition zone exists where the water saturation decreases with decreasing depth, approaching the residual saturation, S_{wr}. The oil saturation is given by $S_o = 1 - S_w$ between the OWOC and OGOC. Above the OGOC, water is present with $S_w \geq S_{wr}$. Gas saturation is zero at, and below, the OGOC but increases with shallower depths in a gas-oil transition zone. In the transition zone, oil saturation approaches 0 at shallower depths. Above the oil-gas transition zone, $S_g = 1 - S_w$.

Mathematically, capillary pressure $(p_{c,ow}$ and $p_{c,og})$ at all depths is determined using the difference in phase pressures computed using Eqs. (1.38)−(1.40). The inverse of drainage capillary pressure curves (Eq. 1.32a), then, give the phase saturation at the respective depth. Example 1.3 demonstrates the initialization of phase pressures and saturations.

Example 1.4 Reservoir initialization.
A reservoir is isothermal at $T = 175°F$, has a depth of 7500 ft, and is 200 ft thick (reservoir is 7300 ft at shallowest location). The water-oil contact (WOC) line is at 7475 ft and the gas-oil contact (GOC) line is at 7450 ft. The density of the aqueous (brine) phase is 66.96 lb_m/ft^3, and the density of the oleic phase is 53 lb_m/ft^3. The gaseous phase has a molecular weight of 19 $lb_m/lbmole$ and a constant z-factor of 1.2 can be assumed. The reservoir is normally pressured. The interfacial tensions between the phases are $\sigma_{ow} = 20$ dyn/cm, $\sigma_{og} = 30$ dyn/cm, and $\sigma_{gw} = 50$ dyn/cm. Use a Brooks-Corey model for the drainage capillary pressure curve.

Example 1.4 Reservoir initialization.—cont'd

The capillary entry pressure, $p_{e,ow}$, is 3.5 psia for oil/water, $\lambda = 2$, and $S_{wr} = 0.2$. The drainage curve can be scaled using a Leverett J-function for oil/gas or gas/water.

(a) Determine initial pressure of the oleic and aqueous phases at the WOC line and at the GOC line and all three phases at 7449 ft (1.0 ft above the GOC line).

(b) Calculate the Free Water Level (FWL), the depth at which oleic and aqueous phase pressures are the same.

(c) Compute the phase saturations at the WOC, GOC, 7475 and 7449 ft.

(d) Compute the phase saturations at 7425 ft where only water and gas are present (oil saturation is zero).

(e) Make plots of depth (y-axis) versus pressure (x-axis) and depth (y-axis) versus water saturation (x-axis).

Solution

(a) Phase pressures are given by Eqs. (1.38), (1.39), and (1.40).

At the WOC line, aqueous phase pressure is given by hydrostatic gradient assuming the reservoir is normally pressured from the surface and oil pressure is offset by the capillary entry pressure,

$$p_{w,WOC} = p_{atm} + \rho_w g D = 14.7 \text{ psia} + \underbrace{\left(66.96 \frac{lb_m}{ft^3}\right)\left(\frac{1 \ ft^2}{144 \ in^2}\right)\left(32 \frac{ft}{s^2}\right)\left(\frac{lb_f}{32 \frac{ft - lb_m}{s^2}}\right)}_{0.465 \frac{psi}{ft}}$$

$(7475 \text{ ft}) = 3490.6 \text{psia } p_{o,WOC} = p_{w,WOC} + p_{e,ow} = 3490.6\text{psi} + 3.5 = 3494.1 \text{ psia}$

At the GOC line,

$$p_{w,GOC} = p_{w,WOC} + \rho_w g(D - D_{WOC}) = 3490.6 \ psia$$

$$+ \left(\frac{66.96 lb/ft^3}{144}\right)(7450 - 7475)\text{ft} = 3479.0 psia$$

$$p_{o,GOC} = \left(p_{w,WOC} + p_{e,ow}\right) + \rho_o g(D - D_{WOC}) = (3490.6 + 3.5)psia$$

$$+ \left(\frac{53 lb/ft^3}{144}\right)(7450 - 7475)\text{ft} = 3484.88 psia$$

Calculation of the gaseous phase pressure requires the density. The density changes with pressure/depth but assume a representative pressure of 3500 psi to compute gas density,

Continued

Example 1.4 Reservoir initialization.—cont'd

$$\rho_g = \frac{pMg}{zRT} = \frac{(3500 \ psi) \cdot \left(19 \ lb/lbmole\right)}{(1.2) \cdot \left(10.73 \ psi - ft^3/lbmole - R\right) \cdot (175 + 460R)} = 8.1 \frac{lb_m}{ft^3}$$

The gaseous phase pressure at the GOC is determined using the capillary entry pressure, which must be scaled (Leverett J-function) using the interfacial tension of oil/gas.

$$p_{g,GOC} = \left(p_{o,GOC} + p_{e,ow}\frac{\sigma_{og}}{\sigma_{ow}}\right) + \rho_g g(D - D_{GOC})$$

$$= \left(3484.88 + 3.5\frac{20 \ dynes/cm}{30 \ dynes/cm}\right) psia$$

$$+ \left(\frac{8.1 \ lb_m/ft^3}{144}\right)(7450 - 7450)ft = 3490.1 \ psi$$

At 7449 ft (1 ft above GOC line), all three phases are present.

$$p_{w,7449} = p_{w,WOC} + \rho_w g(D - D_{WOC}) = 3490.6 \ psia$$

$$+ \left(\frac{66.96 \ lb/ft^3}{144}\right)(7449 - 7475)ft = 3478.5 \ psia$$

$$p_{o,7449} = \left(p_{w,WOC} + p_{e,ow}\right) + \rho_o g(D - D_{WOC}) = (3490.6 + 3.5)psia$$

$$+ \left(\frac{53 \ lb/ft^3}{144}\right)(7449 - 7475)ft = 3484.5 \ psia$$

$$p_{g,7449} = \left(p_{o,GOC} + p_{e,ow}\frac{\sigma_{og}}{\sigma_{ow}}\right) + \rho_g g(D - D_{GOC})$$

$$= \left(3484.9 + 3.5\frac{20 \ dynes/cm}{30 \ dynes/cm}\right) psi + \left(\frac{8.1 \ lb_m/ft^3}{144}\right)(7449 - 7450)ft$$

$$= 3490.1 \ psi$$

(b) At the FWL, the aqueous and oleic phase pressures are equal,

$$p_{w,WOC} + \rho_w g(D_{FWL} - D_{WOC}) = \left(p_{w,WOC} + 3.5\right) + \rho_o g(D_{FWL} - D_{WOC})$$

Example 1.4 Reservoir initialization.—cont'd

$$D_{FWL} = D_{WOC} + \frac{3.5}{(\rho_w - \rho_o)g} = 7475 + \frac{3.5}{(66.96 - 53)}144 = 7511.10 \text{ft}$$

(c) Saturations can be computed using the inverse of the capillary pressure drainage curve,

$$p_{cD} = p_e(S_w^*)^{-1/\lambda} \Rightarrow S_w^* = \left(\frac{p_o - p_w}{p_e}\right)^{-\lambda} \Rightarrow S_w = S_{wr} + (1 - S_{wr})\left(\frac{p_o - p_w}{p_e}\right)^{-\lambda}$$

At the WOC line

$$S_{w,WOC} = 0.2 + (1 - 0.2)\left(\frac{3.5 \text{ psi}}{3.5 \text{ psi}}\right)^{-2.0} = 1.0; \quad S_{o,WOC} = 1 - S_{w,WOC} = 0$$

At the GOC line

$$S_{w,GOC} = 0.2 + (1 - 0.2)\left(\frac{3484.9 - 3478.95 \text{ psi}}{3.5 \text{ psi}}\right)^{-2.0} = 0.479; \quad S_{o,GOC}$$
$$= 1 - S_{w,GOC} = 0.521; S_{g,GOC} = 0$$

At 7449 (i.e. 1 ft above the GOC line)

A transition zone between oil and gas exists above the GOC line and all three phases are present. The capillary pressure model uses a liquid (aqueous plus oleic) phase saturation and the capillary entry pressure between oil and gas:

$$p_c = p_{e,og}(S_l^*)^{-1/\lambda} \Rightarrow S_l^* = \left(\frac{p_g - p_o}{p_{e,og}}\right)^{-\lambda} \Rightarrow S_l = S_{wr} + (1 - S_{wr})\left(\frac{p_g - p_o}{p_e}\right)^{-\lambda}$$

$$S_{l,7449\text{ft}} = 0.2 + (1 - 0.2)\left(\frac{3490.1 - 3484.5 \text{ psi}}{3.5\left(\frac{20}{30}\right) \text{ psi}}\right)^{-2.0}$$
$$= 0.913; \quad S_g = 1 - S_l = 0.087$$

Calculate the water saturation using the capillary pressure between oil and water and then back-calculate the oil saturation:

$$S_{w,7449\text{ft}} = 0.2 + (1 - 0.2)\left(\frac{3485.5 - 3478.5 \text{ psi}}{3.5 \text{ psi}}\right)^{-2.0} = 0.470;$$

$$S_o = S_l - S_w = 0.913 - 0.470 = 0.442$$

(d) Repeating this calculation at a depth of 7428.1 feet results in an oil saturation $S_o = 0$, $S_w = 0.352$, $S_g = 0.648$.

Above the oil/gas transition zone a different calculation must be performed because oil saturation would be negative (which is nonphysical). Only aqueous and gaseous phases are present, and therefore, a water-gas capillary pressure must be employed. Once again, the capillary entry pressure must be scaled by the appropriate interfacial tension (50 dyn/cm),

Continued

Example 1.4 Reservoir initialization.—cont'd

$$p_c = p_{e,wg}\left(S_w^*\right)^{-1/\lambda} \Rightarrow S_w^* = \left(\frac{p_g - p_w}{p_{e,wg}}\right)^{-\lambda} \Rightarrow S_w = S_{wr} + (1 - S_{wr})\left(\frac{p_g - p_w}{p_{e,wg}}\right)^{-\lambda}$$

$$p_{g,7425} = p_{g,GOC} + \rho_g g(D - D_{GOC}) = 3490.1 + \frac{8.1}{144}(7425 - 7450) = 3488.7$$

$$p_{w,7425} = p_{w,WOC} + \rho_w g(D - D_{WOC}) = 3490.6 + \frac{66.96}{144}(7425 - 7475) = 3467.3$$

$$S_{w,7445ft} = 0.2 + (1 - 0.2)\left(\frac{3488.7 - 3467.3 \text{ psi}}{3.5\frac{50}{20} \text{ psi}}\right)^{-2.0}$$

$$= 0.334; \quad S_g = 1 - S_w = 0.666$$

(e) A plot of phase pressures and saturations is given below. Notice there is a discontinuity in the derivative of the water saturation plot at a depth of 7446.432 ft. This is at the top of the oil/gas transition zone.

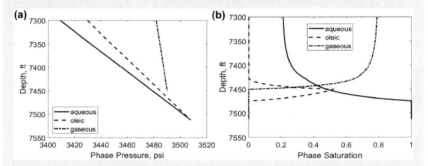

Figure Ex.1.4. Solution for three-phase static equilibrium for (A) phase pressures versus depth and (B) phase saturations versus depth.

1.8 Pseudocode

An important feature of this book is that it walks the reader through the development of their own multidimensional, multiphase, multicomponent reservoir simulator. Development of a simulator all at once can be daunting even for the expert reservoir engineer and programmer, so we develop our simulator in small steps. In each chapter, pseudocode is provided for subroutines or functions to complete individual tasks. The provided pseudocode is in no particular programming language and is more of an outline of the code.

In the exercises the reader is asked to write and validate codes in a programming language of their choice, using pseudocode as a guide. In some cases, the developed code/subroutine is complete and will be called as part of the final simulator. In other cases, the subroutines will be enhanced with additional features in later chapters.

Data for reservoir, fluid, petrophysical, well, numerical, etc., properties must be gathered into an input file and used in the code. An input file with all properties is provided for the reader for a synthetic reservoir, the Thomas oilfield at https://github.com/mbalhof/Reservoir-Simulation. The reader may use (and adapt) this input file or reformat into one the reader prefers. Regardless of the format, the same one should be used for different reservoirs and it should be easy to change the input data for more or less complex reservoirs. The user will need to read in all the data for use in the reservoir simulator program. It is recommended, but not required, that the reader uses *classes* and *dictionaries* to define variables that belong to a similar category, such as reservoir (e.g., permeability, porosity, OWOC, domain size), fluid (e.g., viscosity, formation volume factor, bubble point), petrophysical (e.g., relative permeability and capillary pressure curve parameters), well (e.g., locations, constraints, radius), numerical (e.g., number of grids, timestep size), and output variables (e.g., pressure, saturation, concentration). In many of the provided pseudocodes, only the class will be listed as an input or output. This will indicate that some or all the variables in that class may be used in the subroutine.

It should be emphasized that there is no single *correct* way to code a reservoir simulator and a straightforward approach is presented in this text. The proposed algorithms are not always the most computationally efficient or elegant; in fact, this is often intentional since the focus is on obtaining the correct answer and learning the fundamentals. One can always rewrite the code to make it faster or more elegant. It is recommended in this text to code "bottom up"; that is, create the most basic subroutines first and then create the subroutines that call them. The order in which subroutines are presented throughout this text is the recommended order that they should be written, but not a requirement. It is highly recommended that each subroutine is tested and validated upon completion.

1.8.1 Relative permeability

The purpose of this subroutine is to compute the relative permeability of all phases when provided the saturation of $N_p - 1$ phases and the petrophysical properties (e.g., residual saturations, endpoint relative permeabilities, Corey exponents). For two-phase flow the normalized saturation can be computed first using Eq. (1.23) and then the relative permeabilities using Eq. (1.22). The code can be generalized to three-phase flow by using Eqs. (1.24)–(1.30). Alternatively, if relatively permeability data are provided in a table(s), a subroutine could be developed to calculate relative permeability via interpolation. It is assumed here that scalar saturations are sent as inputs and scalar relative permeabilities as outputs, but the code could be generalized to allow for vector inputs and outputs.

1.8.2 Capillary pressure

Capillary pressure (both drainage and imbibition) between two phases is computed using empirical models such as Eqs. 1.32a and 1.32b when the saturation of one of the phases and petrophysical properties (e.g., residual saturations, capillary entry pressure, λ) are provided as inputs. Derivatives of capillary pressure may also be useful which can be computed analytically or approximately using finite differences. Both the drainage and imbibition capillary pressure curves, as well as the derivative of the imbibition curve, tend to infinity (or negative infinity) near the residual saturations which can cause numerical issues in the simulator. Some additional logic can be added to the code to limit the outputs to some maximum value or create a function that allows for a smooth transition from the capillary pressure curves to the allowed maximum. For added generality, the subroutine can compute a capillary pressure from a scanning curve.

1.8.3 Initialization

The reservoir pressures and phase saturations are initialized given reservoir, petrophysical, and fluid properties as well as the depth as inputs. The aqueous and oleic phase pressures at depth are first computed using Eqs. (1.38) and (1.39). Capillary pressure is computed as the difference of those pressures. Finally, water saturation is computed from the inverse of Eq. (1.32a).

1.8.4 Preprocess

Preprocessing includes all the initial steps that should be completed before the time-dependent numerical simulation. The first step is to read in all the data to the program. It is important that the data files are in a uniform format so that they can be changed/tailored for a specific reservoir and still read in properly. There may be multiple data files; for example, individual files for the permeability, porosity, depth of all grids in the reservoir in addition to a file that includes PVT data. It is also important that the data file (and code to read it in) is flexible to allow for different types of inputs (e.g., scalar permeability for homogenous reservoir or a data file of heterogeneities permeabilities).

Grid locations/centers (x, y, z), along with the distances between centers and grid sizes, are needed $(\Delta x, \Delta y, \Delta z)$ and should be stored in arrays or vectors. These might be provided in a data file or the programmer will have to code an algorithm to place the grids based on the domain size and number of grids. The simplest algorithm is to space the grids uniformly. Well data will include the location of the wells including the start and end of the perforations. This, along with the arrays for grid locations, can be used to determine which grids are perforated by each well. If PVT and petrophysical data are provided as data in a table as a function of pressure and saturation, respectively, they can be curve fit or an interpolating function used.

A loop through all N grids is used to calculate phase pressures and saturations by calling the initialization subroutine. Figures, such as contour or 3D surface plots of the initial pressure and saturation field, can be created. Plots of relative permeability and capillary pressure versus saturation can be generated by calling those subroutines over the appropriate range of saturations. PVT data can be plotted as a function of pressure if applicable.

```
SUBROUTINE PREPROCESS
Inputs: data file or spreadsheet of all input data
Outputs: reservoir, fluid, petrophysical, numerical, well properties

READ datafiles
INIT vectors of x,y,z locations of grid centers
INIT arrays of perforated grid blocks for each well
DEF curvefit for pressure-dependent PVT variables (if desired over
interpolation)

#INIT reservoir pressure and saturations
FOR i = 1 to N (# grid blocks)
     CALL INITIALIZE(i)
ENDFOR

PLOT PVT/Petrophysical data and initial pressure/saturation fields
with well locations
```

```
SUBROUTINE INITIALIZE
Inputs: grid number (i), numerical, fluid, petrophysical properties
Outputs: Pw(i), Po(i), Pc(i), Sw(i)

CALCULATE Pw(i), Po(i) using eqns 1.38-1.39
CALCULATE Pc(i) = Po(i)-Pw(i), eqn 1.31
CALCULATE Sw(i), inverse of eqn 1.32a
```

```
SUBROUTINE RELPERM
Inputs: Sw, So, petrophysical properties
Outputs: krw, kro, krg

IF two-phase
     CALCULATE S using eqn 1.23
     CALCULATE krw & kro using eqn 1.22
ELIF three-phase
     CALCULATE SwD, SgD,and SoD using eqns 1.25 and 1.27
     CALCULATE Som using eqn 1.30
     CALCULATE So*, Sw*, Sg* using 1.29
     CALCULATE krow & krog using 1.28
     CALCULATE krw, kro, krg using eqn 1.24 and 1.26
ENDIF
```

```
SUBROUTINE CAPPRESS
Inputs: Sw, petrophysical properties
Outputs: Pcimb, Pcdrain, Pc'imb, Pc'drain, Pcscan, Pc'scan

CALCULATE normalized saturations, Sw* and Se using eqn 1.33
CALCULATE Pcdrain using eqn 1.32a
CALCULATE Pc'drain analytically or numerically
CALCULATE Pcimb using eqn 1.32b
CALCULATE Pc'imb analytically or numerically
CALCULATE f using eqn 1.37
CALCULATE Pcscan using eqn. 1.36
CALCULATE Pc'scan analytically or numerically
```

1.9 Exercises

Exercise 1.1. Pressure-dependent porosity and formation volume factor.
Using the definitions of rock and fluid compressibility derive the following expressions for porosity and formation volume factor, respectively. Then use a Taylor series approximation to obtain linear approximations to the expressions. Assume constant and small compressibilities.

a. $\phi = \phi^0 \exp[c_f(p - p^0)]$
b. $B_\alpha = B_\alpha^0 \exp[c_\alpha(p^0 - p)]$

where ϕ^0 and B_a^0 are the porosity and formation volume factor, respectively, at a reference pressure, p^0.

Exercise 1.2. Darcy's law in multidimensions with anisotropic permeability. Calculate the velocity, u_x, u_y, and u_z (ft/day) of a single-phase fluid ($\mu_w = 1$ cp) flowing in three dimensions with pressure gradient given by $dp/dx = 1.0$ psi/ft, $dp/dy = 0.2$ psi/ft, $dp/dz = 0.3$ psi/ft. Gravity is negligible and permeability is given as a second-order tensor,

$$\mathbf{k} = \begin{pmatrix} 100 & 10 & 5 \\ 10 & 50 & 16 \\ 5 & 16 & 70 \end{pmatrix} \text{mD}$$

Exercise 1.3. Multiphase Darcy's law with gravity. A 1D reservoir with permeability of 500 mD and a dip angle of 30 degrees is saturated with oleic ($\mu_o = 5$ cp, $\rho_o = 50$ lb$_m$/ft^3) and aqueous ($\mu_w = 1$ cp, $\rho_w = 62$ lb$_m$/ft^3) phase. The relative permeabilities of aqueous and oleic phases are 0.15 and 0.10, respectively. The length of the reservoir is 1000 ft. The pressure at the bottom (deepest part) of the reservoir is 3000 psia and at the top (shallowest part) is 2800 psia. Assume capillary pressure is negligible. Calculate the Darcy velocity (in ft/day) of aqueous and oleic phases. Hint: be very careful with your

± signs and recognize that do due to density differences and gravity that flow of aqueous and oleic phases can be in opposite directions.

Exercise 1.4. PVT properties. PVT data are provided in the table below and the file 'PVT.xlsx' on https://github.com/mbalhof/Reservoir-Simulation. The densities of the oil and gas at standard conditions are 53 and 0.0458 lbm/ft^3, respectively. Using the data, identify, calculate, or estimate:

a. Bubble point
b. Oil compressibility at 3000 psia
c. Oil compressibility at 2000 psia
d. z-factor at 2000 psia
e. Gas compressibility at 2000 psia
f. Read in the data to a computer code and make plots of B_o, B_g, R_s, c_o, c_g, z versus pressure

PVT table for Exercise 1.4			
P	Bo	Bg	Rs
(psia)	(RB/STB)	(ft^3/scf)	(scf/STB)
1000	1.1414	0.0161	168.81
1200	1.1607	0.0132	209.60
1400	1.1807	0.0112	251.69
1600	1.2012	0.0097	294.92
1800	1.2222	0.0085	339.17
2000	1.2436	0.0076	384.36
2200	1.2655	0.0069	430.39
2400	1.2877	0.0063	477.22
2600	1.3103	0.0058	524.79
2800	1.3208	0.0054	550.57
3000	1.3175	0.0051	550.57
3200	1.3147	0.0048	550.57
3400	1.3121	0.0046	550.57
3600	1.3099	0.0044	550.57
3800	1.3079	0.0042	550.57
4000	1.3061	0.004	550.57

Exercise 1.5. Relative permeability subroutine. Write a code that computes the relative permeability of a two-phase or three-phase system when sent petrophysical properties and saturations. Validate your code against Example 1.2. Make plots of water/oil relative permeability versus water saturation (for two or three phases) and oil/gas relative permeability (three phases). For an additional challenge, make a ternary plot of oil relative permeability for a three-phase system with 100% water, oil, and gas at three vertices, respectively, of the diagram.

Exercise 1.6. Capillary pressure subroutine. Write a code to compute the capillary pressure between two phases given petrophysical properties and saturation using the Brooks-Corey models. The code should compute both imbibition and drainage capillary pressure as well as capillary pressure obtained from a scanning curve (when provided $S_{w,\text{hyst}}$ and espc). The code should also compute the derivative of all three capillary pressures at the provided saturation. Use the code to make plots of capillary pressure (drainage, imbibition, scanning) versus water saturation for $p_e = 3.5$ psia, $\lambda = 2.0$, $S_w(\text{hyst}) = 0.3$, and espc $= 0.1$. For added flexibility, allow your code to compute capillary pressure at a different permeability, porosity, and interfacial tension than the parameters were measured by using the Leverett J function.

Exercise 1.7. Initialization subroutine. Write a code that computes the aqueous and oleic phase pressures as well as water saturation at a given depth for a two-phase system. Other required input parameters include petrophysical (e.g., capillary pressure model parameters) and fluid (e.g., phase densities) properties as well as the OWOC line and aqueous phase pressure at the PWOC line. Validate your code against Example 1.4 for depths below the OGOC line and reproduce the plots of phase pressures and saturations versus depth below the OGOC line. For an additional challenge, extend your code to allow for three phases.

Exercise 1.8. Reading in data and preprocessing. An input file for the synthetic Thomas oilfield, Thomas.yml, is available for download at https://github.com/mbalhof/Reservoir-Simulation. The input file includes reservoir (e.g., domain size, permeability, porosity, WOC line), fluid (e.g., PVT properties), petrophysical (e.g., residual saturations, relative permeability, and capillary pressure constants), well (e.g., locations, constraints, boundary conditions), and numerical properties (e.g., number of grids, timestep size). Some of the properties (e.g. permeability) are included in a separate data file, referenced in Thomas.yml. If you prefer that the input file be in a different format, you may convert it. However, the input file should allow for easy edits and more or less complexities in the problem. Throughout this text, several input files will be used or adapted in various exercises.

Write a code, based on the pseudocode in Section 1.8, called *preprocess* to

a. Read in all the data from the input file, Thomas.yml, and all the associated data files.
b. Create vectors (x, y, and z) of grid center locations using the domain size and number of grids. Assume uniform grids.
c. Initialize the reservoir oleic/aqueous phase pressures in all blocks using grid depths.
d. Plot permeability, porosity, depth, oleic phase pressure, and aqueous phase saturation in all blocks.

The code should be flexible enough so that it can read in different files with different data. For example, permeability, porosity, thickness, and depth can be input as scalar constants or be spatially dependent and provided in a separate data file.

References

Alpak, F.O., Lake, L.W., Embid, S.M., 1999. Validation of a modified Carman-Kozeny equation to model two-phase relative permeabilities. In: Presented at the SPE Annual Technical Conference and Exhibition, Houston, Texas, 3–6 October.

Beggs, H.D., Robinson, J.R., 1975. Estimating the viscosity of crude oil systems. Journal of Petroleum Technology 27 (9), 1140–1141.

Beygi, M.R., Delshad, M., Pudugramam, V.S., Pope, G.A., Wheeler, M.F., 2015. Novel three-phase compositional relative permeability and three-phase hysteresis models. SPE Journal v20 (1), 21–34. https://doi.org/10.2118/165324-PA.

Bird, R.B., Stewart, W.E., Lightfoot, E.N., 1960. Transport Phenomena. John Wiley and Sons Inc., New York.

Blunt, M.J., 2000. An empirical model for three-phase relative permeability. SPE Journal 5 (4), 435–445. https://doi.org/10.2118/67950-PA. SPE-67950-PA.

Blunt, M.J., 2017. Multiphase Flow in Permeable Media: A Pore-Scale Perspective. Cambridge University Press.

Brooks, R.H., Corey, A.T., 1964. Hydraulic Properties of Porous Media. Hydrology Papers 3. Colorado State U., Fort Collins, Colorado.

Carr, N.L., Kobayashi, R., Burrows, D.B., 1954. Viscosity of hydrocarbon gases under pressure. Journal of Petroleum Technology 6 (10), 47–55.

Chierici, G.L., 1984. Novel relations for drainage and imbibition relative permeabilities. SPE Journal 24 (3), 275–276.

Dandekar, A.Y., 2013. Petroleum Reservoir Rock and Fluid Properties. CRC Press.

Darcy, H., 1856. Les Fontaines Publiques de la Ville de Dijon. Dalmont, Paris, p. 647 (and atlas).

Dranchuk, P.M., Abou-Kassem, H., 1975. Calculation of Z factors for natural gases using equations of state. Journal of Canadian Petroleum Technology 14 (3), 34–36.

Fayers, F.J., Matthews, J.P., 1982. Evaluation of normalized Stone's methods for estimating three-phase relative permeabilities. SPE Journal 24, 224–232.

Forchheimer, P., 1901. Wasserbewegung durch Boden. Zeitz Ver Deutsch Ing 45, 2145.

Geertsma, J., 1957. The effect of fluid pressure decline on volumetric changes of porous rocks. Transactions of the AIME 210.01, 331–340.

Hall, K.R., Yarborough, L., 1973. A new equation of state for Z-factor calculations. Oil & Gas Journal 71 (25), 82.

Honarpour, M., Koederitz, L., Harvey, A.H., 1986. Relative Permeability of Petroleum Reservoirs. CRC Press, Boca Raton, Florida.

Hornung, U., 1997. Homogenization and porous media. Interdisciplinary Applied Mathematics 6 (Springer).

Hustad, O., Hansen, A.-G., 1996. A consistent formulation for three-phaserelative permeability and phase pressure based on three sets of two-phase data. In: InRUTH—A Norwegian Research Program on Improved OilRecovery, Program Summary, Norwegian Petroleum Directorate. Norwegian Petroleum Directorate, Stavanger, Norway, pp. 183—194.

Jerauld, G., 1997. General three-phase relative permeability model for Prudhoe Bay. SPE Reservoir Engineering 12 (4), 255—263. https://doi.org/10.2118/36178-PA. SPE-36178-PA.

Kestin, J., Khalifa, H.E., Correia, R.J., 1981. Tables of the dynamic and kinematic viscosity of aqueous NaCl solutions in the temperature range 20—150 C and the pressure range 0.1—35 MPa. Journal of Physical and Chemical Reference Data 10 (1), 71—88.

Kianinejad, A., DiCarlo, D.A., 2016. Three-phase oil relative permeability in water-wet media: a comprehensive study. Transport in Porous Media 112, 665—687. https://doi.org/10.1007/s11242-016-0669-z.

Killough, J.E., 1976. Reservoir simulation with history-dependent saturation functions. SPE Journal 16, 37—48. https://doi.org/10.2118/5106-PA.

Klinkenberg, L.J., 1941. The permeability of porous media to liquids and gases. Drilling and Production Practice.

Lasater, J.A., 1958. Bubble point pressure correlations. Journal of Petroleum Technology 10 (5), 65—67.

Lee, A.L., Gonzalez, M.H., Eakin, B.E., 1966. The viscosity of natural gases. Journal of Petroleum Technology 18 (8), 997—1000.

Leverett, M.C., 1941. Capillary behavior in solids. Transactions of the AIME 142, 159—172.

McCain Jr., W.D., 1991. Reservoir-Fluid Property Correlations-State of the Art (includes associated papers 23583 and 23594).

Pedersen, K.S., et al., 2006. Phase Behavior of Petroleum Reservoir Fluids. CRC Press.

Sharqawy, M.H., Lienhard, J.H., Zubair, S.M., 2010. Thermophysical properties of seawater: a review of existing correlations and data. Desalination and Water Treatment 16 (1—3), 354—380.

Standing, M.B., Katz, D.L., 1942. Density of natural gases. Transaction AIME 146, 140.

Standing, M.B., 1947. A pressure-volume-temperature correlation for mixtures of California oils and gases. API Drilling and Production Practice 275—287.

Stone, H.L., 1970. Probability model for estimating three-phase relative permeability. Journal of Petroleum Technology 22 (2), 214—218. SPE-2116-PA.

Stone, H.L., 1973. Estimation of three-phase relative permeability and residual data. Journal of Canadian Petroleum Technology 12, 53—61.

van Genuchten, M.T., 1980. Closed-form equation for predicting the hydraulic conductivity of unsaturated soils. Soil Science Society of America Journal 44, 892—898.

Vazquez, M., Beggs, H.D., 1980. Correlations for fluid physical property prediction. Journal of Petroleum Technology 32 (6), 968—970.

Whitaker, M., 1999. Microscopy research and technique. Special Issue: Calcium Identification 46 (6), 342—347.

Chapter 2

Phase mass balances and the diffusivity equation

2.1 Introduction

In this chapter, the partial differential equations (PDEs) that describe multi-phase flow in porous media are derived using basic mass balance principles. Mass balances on each phase are first derived. These phase balances are then summed over all fluid phases to obtain an overall mass balance, known as the continuity equation. Phase mass balances can also be weighted by the inverse of phase density and then summed, which gives the diffusivity equation. Darcy's law can be substituted for phase velocity and the accumulation term of the diffusivity equation is often expanded to introduce a time derivative in pressure. These forms of the diffusivity equation are often referred to as the pressure equation. The derived PDEs, when combined with appropriate boundary conditions and input parameters, are the equations that are solved in reservoir simulators. A few analytical solutions to the diffusivity equation under certain simplifying assumptions and common boundary conditions are also presented. The analytical solutions are useful for verifying the accuracy of reservoir simulators in subsequent chapters.

2.2 Phase mass balances

A *control volume* is a three-dimensional reference volume that is either fixed in space or moves with constant velocity. A control volume, Fig. 2.1, can be of any shape or size.

Balance equations are often imposed on a control volume, where mass, energy, and/or momentum are conserved. The shape and size of the control volume are usually chosen in a way to make the derivation of these balance equations most useful. The balance is usually performed over a specified time interval.

An Introduction to Multiphase, Multicomponent Reservoir Simulation.
https://doi.org/10.1016/B978-0-323-99235-0.00008-7
37

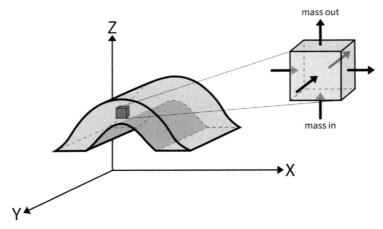

FIGURE 2.1 Generic reservoir geometry with cubic control volume.

2.2.1 Mass balance of a phase in Cartesian coordinates

Consider compressible flow of multiple fluid phases in a one-dimensional (1D) porous medium depicted by Fig. 2.2.

The control volume is taken to be a small slice of the porous domain, $\Delta V = \Delta x \Delta y \Delta z$, where $a = \Delta y \Delta z$ is the cross-sectional area of the domain and Δx is the length of the control volume. The *Law of Conservation of Mass* states that over any arbitrary control volume, the total mass that enters the control volume minus the mass that exits, plus any mass generated or consumed from a source, must be equal to the mass accumulated in the control volume.

Using a conservation equation, one can write a mass balance on each phase in the control volume. The mass of the phase that enters the control volume (at some point "x") is the mass flux (mass per unit area per time) times the cross-sectional area times the time (Δt) over which the mass balance is taken. In the

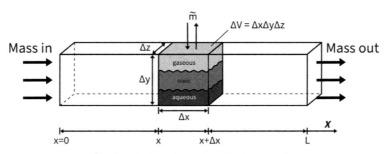

FIGURE 2.2 Flow through a 1D porous medium.

figure, mass exits downstream at a position $x + \Delta x$ (we note that flow can go in the opposite direction, left to right, or flow in to, or out of, the control volume from both sides as is common with wells). Fluid can be generated or consumed in the control volume from a source (e.g., injection well) or sink (e.g., producer well).

Collecting all terms results in a mass balance in 1D for phase α in the control volume,

$$\underbrace{m_\alpha|_{t+\Delta t} - m_\alpha|_t}_{\text{accumulation}} = \underbrace{\dot{m}_{\alpha,x}\Big|_x a\Delta t - \dot{m}_{\alpha,x}\Big|_{x+\Delta x} a\Delta t}_{\text{mass in-out}} + \underbrace{\widetilde{m}_\alpha \Delta V \Delta t}_{\text{generation}}, \qquad (2.1)$$

where $\dot{m}_{\alpha,x}$ is the mass flux (mass/area-time) of phase α, Δt is the time interval over which the mass balance is performed, $\widetilde{m}_\alpha$ is a source or sink (mass/volume-time) of mass of phase α, and m_α is the mass of fluids in the control volume at a given time. Since Eq. (2.1) is a "mass" balance, each term in the equation has units of mass, e.g., kg or lb$_\text{m}$. The variables in Eq. (2.1) are then defined as follows,

$$\begin{aligned}
\dot{m}_{\alpha,x} &= \rho_\alpha u_{x,\alpha} \\
\widetilde{m}_\alpha &= \rho_\alpha \widetilde{q}_\alpha \\
m_\alpha &= \rho_\alpha S_\alpha \Delta V \varphi
\end{aligned} \qquad (2.2)$$

where ΔV is the volume of the control volume, the source terms, $\widetilde{q}_\alpha$, have units of time^{-1}, S_α is the phase saturation, and $u_{x,\alpha}$, is the velocity in x-direction of the flowing phase.

The expressions in Eq. (2.2) are substituted into Eq. (2.1) and the equation is divided through by the control volume ($\Delta V = a\Delta x$) and the time interval (Δt) to give,

$$\frac{(\varphi \rho_\alpha S_\alpha)|_{t+\Delta t} - (\varphi \rho_\alpha S_\alpha)|_t}{\Delta t} = \frac{(\rho_\alpha u_{x,\alpha})|_x - (\rho_\alpha u_{x,\alpha})|_{x+\Delta x}}{\Delta x} + \rho_\alpha \widetilde{q}_\alpha. \qquad (2.3)$$

The control volume can be chosen to be infinitesimally small and the mass balance performed over an infinitesimally small timestep; the limit as both Δx and Δt go to zero results in partial derivatives. The one-dimensional mass balance on a phase in Cartesian coordinates is then obtained,

$$\frac{\partial}{\partial t}(\varphi \rho_\alpha S_\alpha) = -\frac{\partial}{\partial x}(\rho_\alpha u_{x,\alpha}) + \rho_\alpha \widetilde{q}_\alpha. \qquad (2.4a)$$

The general form of the phase mass balance in any coordinate system and dimensions is written,

$$\frac{\partial}{\partial t}(\varphi \rho_\alpha S_\alpha) = -\nabla \cdot \left(\rho_\alpha \overrightarrow{u}_\alpha\right) + \rho_\alpha \widetilde{q}_\alpha, \qquad (2.4b)$$

where ∇ is the del, or vector differential, operator for multidimensions and any coordinate system. Eq. (2.4) is the starting point for deriving the continuity and diffusivity equations.

2.3 The continuity equation

Eq. (2.4b) can be summed over all phases, N_p, to obtain an overall mass balance, which is known as the continuity equation,

$$\sum_{\alpha=1}^{N_p} \frac{\partial}{\partial t}(\varphi \rho_\alpha S_\alpha) = -\sum_{\alpha=1}^{N_p} \nabla \cdot \left(\rho_\alpha \overrightarrow{u}_\alpha\right) + \sum_{\alpha=1}^{N_p} \rho_\alpha \widetilde{q}_\alpha. \tag{2.5}$$

The continuity equation simply states that the total accumulation of mass from all phases is equal to the net flow of mass into the control volume plus the net amount of mass that enters the control volume from a source, such as a well. Example 2.1 demonstrates the expansion of the time (accumulation) derivative terms using the product and chain rules.

Example 2.1 Expansion of Accumulation Terms.
Expand the time derivative terms in the continuity equation using a product rule and show that the saturation terms do not vanish.

Solution
Use the product rule to expand the time derivatives,

$$\sum_{\alpha=1}^{N_p} \frac{\partial}{\partial t}(\varphi \rho_\alpha S_\alpha) = \sum_{\alpha=1}^{N_p} \left[\varphi \rho_\alpha \frac{\partial S_\alpha}{\partial t} + \rho_\alpha S_\alpha \frac{\partial \varphi}{\partial t} + \varphi S_\alpha \frac{\partial \rho_\alpha}{\partial t}\right].$$

Use the chain rule on the porosity and density derivatives,

$$= \sum_{\alpha=1}^{N_p} \left[\varphi \rho_\alpha \frac{\partial S_\alpha}{\partial t} + \rho_\alpha \varphi S_\alpha \left(\frac{1}{\varphi}\frac{\partial \varphi}{\partial p} + \frac{1}{\rho_\alpha}\frac{\partial \rho_\alpha}{\partial p}\right)\frac{\partial p}{\partial t}\right].$$

Substitute rock and fluid compressibilities using their definitions from Chapter 1,

$$= \sum_{\alpha=1}^{N_p} \left[\varphi \rho_\alpha \frac{\partial S_\alpha}{\partial t} + \rho_\alpha \varphi S_\alpha (c_f + c_\alpha)\frac{\partial p}{\partial t}\right].$$

Finally, expanding the summation for up to three phases gives,

$$\sum_{\alpha=1}^{N_p} \frac{\partial}{\partial t}(\varphi \rho_\alpha S_\alpha) = \varphi \rho_w \frac{\partial S_w}{\partial t} + \varphi \rho_o \frac{\partial S_o}{\partial t} + \varphi \rho_g \frac{\partial S_g}{\partial t} + \varphi \left[\rho_w S_w (c_f + c_w)\right.$$

$$\left. + \rho_o S_o (c_f + c_o) + \rho_g S_g (c_f + c_g)\right]\frac{\partial p}{\partial t}.$$

Example 2.1 Expansion of Accumulation Terms.—cont'd

Unfortunately, the terms with time derivatives of saturation are weighted by phase densities. However, when the time derivatives are weighted by inverse densities before summation (as is the case in the diffusivity equation),

$$\sum_{\alpha=1}^{N_p} \frac{1}{\rho_\alpha} \frac{\partial}{\partial t} (\varphi \rho_\alpha S_\alpha) = \varphi c_t \frac{\partial p}{\partial t}$$

Although saturation does not appear explicitly, the total compressibility is a saturation-weighted compressibility.

2.4 The diffusivity equation

2.4.1 General multiphase flow

The continuity Eq. (2.5) includes phase saturations explicitly in the time derivative, which limits its direct use in multiphase reservoir simulation (Chapters 9 and 10). However, the phase balances (Eq. 2.4) can be weighted by the inverse of phase density, ρ_α, and *then* summed over all phases,

$$\sum_{\alpha=1}^{N_p} \frac{1}{\rho_\alpha} \frac{\partial}{\partial t} (\varphi \rho_\alpha S_\alpha) = - \sum_{\alpha=1}^{N_p} \frac{1}{\rho_\alpha} \nabla \cdot \left(\rho_\alpha \overrightarrow{u}_\alpha \right) + \sum_{\alpha=1}^{N_p} \tilde{q}_\alpha. \tag{2.6}$$

The dimensions of Eq. (2.6) are time^{-1}. Note the important differences between Eqs. (2.5) and (2.6). Expansion of the time derivatives in Eq. (2.6) using a product rule gives,

$$\underbrace{\left[\varphi \frac{\partial}{\partial t} \sum_{\alpha=1}^{N_p} S_\alpha \right]}_{=0} + \left[\sum_{\alpha=1}^{N_p} \frac{S_\alpha}{\rho_\alpha} \frac{\partial}{\partial t} (\varphi \rho_\alpha) \right] = - \sum_{\alpha=1}^{N_p} \frac{1}{\rho_\alpha} \nabla \cdot \left(\rho_\alpha \overrightarrow{u}_\alpha \right) + \sum_{\alpha=1}^{N_p} \tilde{q}_\alpha, \tag{2.7}$$

where the first term in Eq. (2.7) vanishes, unlike in Eq. 2.5, because the sum of all phase saturations is constant (1.0) and the derivative of a constant is zero.

The multiphase form of Darcy's law (Eq. 1.21) can be substituted for phase velocities in Eq. (2.7),

$$\sum_{\alpha=1}^{N_p} \frac{S_\alpha}{\rho_\alpha} \frac{\partial}{\partial t} (\varphi \rho_\alpha) = \sum_{\alpha=1}^{N_p} \frac{1}{\rho_\alpha} \nabla \cdot \left(\rho_\alpha \frac{k_{r\alpha} \mathbf{k}}{\mu_\alpha} \nabla \Phi_\alpha \right) + \sum_{\alpha=1}^{N_p} \tilde{q}_\alpha, \tag{2.8}$$

where Φ_a is the potential of the phase. Eq. (2.8) is the most general form of the diffusivity equation, although more specific forms, including for single-phase flow, are often used.

The time derivatives in Eq. (2.8) can be further expanded using a product and chain rule to give,

$$\sum_{\alpha=1}^{N_p} \frac{S_\alpha}{\rho_\alpha} \left[\varphi \frac{\partial \rho_\alpha}{\partial p_\alpha} + \rho_\alpha \frac{\partial \varphi}{\partial p_\alpha} \right] \frac{\partial p_\alpha}{\partial t} = \sum_{\alpha=1}^{N_p} \frac{1}{\rho_\alpha} \nabla \cdot \left(\rho_\alpha \frac{k_{r\alpha} \mathbf{k}}{\mu_\alpha} \nabla \Phi_\alpha \right) + \sum_{\alpha=1}^{N_p} \widetilde{q}_\alpha, \quad (2.9)$$

where the pressure, p_α, may be unique to each phase, but one phase (e.g., nonwetting phase) is used as a reference for pressure. Using the definition of formation and phase compressibilities (Chapter 1), Eq. (2.9) can be written as,

$$\varphi c_t \frac{\partial p}{\partial t} = \sum_{\alpha=1}^{N_p} \frac{1}{\rho_\alpha} \nabla \cdot \left(\rho_\alpha \frac{k_{r\alpha} \mathbf{k}}{\mu_\alpha} (\nabla p_\alpha - \rho_\alpha g \nabla D) \right) + \sum_{\alpha=1}^{N_p} \widetilde{q}_\alpha \quad (2.10a)$$

where the definition of phase potential was substituted and c_t is the saturation-weighted total fluid/rock compressibility (Eq. 1.18). If gravity and capillary pressure are negligible,

$$\varphi c_t \frac{\partial p}{\partial t} = \sum_{\alpha=1}^{N_p} \frac{1}{\rho_\alpha} \nabla \cdot \left(\rho_\alpha \frac{k_{r\alpha} \mathbf{k}}{\mu_\alpha} \nabla p \right) + \sum_{\alpha=1}^{N_p} \widetilde{q}_\alpha. \quad (2.10b)$$

Eq. (2.10) is often referred to as the *pressure equation* because both temporal and spatial derivatives are written explicitly in terms of pressure. Saturation does not appear directly in Eq. (2.10) but is indirectly included in the total compressibility, c_t, relative permeability, and capillary pressure. Solution of the PDE results in pressure in the reservoir as a function of space and time.

2.4.2 Single-phase flow

A special case of the diffusivity equation is single-phase flow. By single-phase flow, it could mean only one fluid phase is present, such as brine flowing in an aquifer. However, it could also mean that only one phase is flowing but other fluid phases are present and immobile (i.e., capillary-trapped fluid). Such is the case during primary production and no aquifers present when an oleic phase is flowing and produced from wells but the aqueous phase is at irreducible/residual saturation and not flowing, or flows at negligible rates. Likewise, near the end of a waterflood, the aqueous phase is flowing but the oleic phase is at or near residual oil saturation and is immobile.

For the case of single-phase flow without gravity, Eq. (2.10) can be simplified,

$$\varphi c_t \frac{\partial p}{\partial t} = \frac{1}{\rho_\alpha} \nabla \cdot \left(\rho_\alpha \frac{k_{r\alpha}^0 \mathbf{k}}{\mu_\alpha} \nabla p \right) + \widetilde{q}_\alpha \quad (2.11)$$

where ρ_α now refers to the density of the single, flowing phase and the endpoint relative permeability, $k_{r\alpha}^0$, is now used because if other phases are present, they are assumed to be at residual saturation. Velocities and production rates of all other phases that may be present are assumed to be negligible.

2.4.2.1 Slightly compressible liquids

Slightly compressible fluids, such as an aqueous phase or an undersaturated oleic phase, are defined here as a compressibility $\leq \sim 5 \times 10^{-5}$ psi^{-1}. For these fluids, the diffusivity equation can be further simplified by employing the product rule on the spatial derivatives in Eq. (2.11),

$$\varphi c_t \frac{\partial p}{\partial t} = \frac{1}{\rho_\alpha} \left[\rho_\alpha \nabla \cdot \left(\frac{k_{r\alpha}^0 \mathbf{k}}{\mu_\alpha} \nabla p \right) + \left(\frac{k_{r\alpha}^0 \mathbf{k}}{\mu_\alpha} \nabla p \right) \cdot \nabla \rho_\alpha \right] + \widetilde{q}_\alpha, \tag{2.12}$$

and then the chain rule for the density gradient gives,

$$\varphi c_t \frac{\partial p}{\partial t} = \nabla \cdot \left(\frac{k_{r\alpha}^0 \mathbf{k}}{\mu_\alpha} \nabla p \right) + \left(\frac{k_{r\alpha}^0 \mathbf{k}}{\mu_\alpha} \right) \underbrace{\left(\frac{1}{\rho_\alpha} \frac{\partial \rho_\alpha}{\partial p} \right)}_{c_\alpha} (\nabla p)^2 + \widetilde{q}_\alpha. \tag{2.13}$$

The fluid compressibility, c_α, is small as is the square of the modulus of pressure gradient. Therefore, the second term on the right-hand side (RHS) of Eq. (2.13) is negligible which simplifies to,

$$\varphi c_t \frac{\partial p}{\partial t} = \nabla \cdot \left(\frac{k_{r\alpha}^0 \mathbf{k}}{\mu_\alpha} \nabla p \right) + \widetilde{q}_\alpha, \tag{2.14}$$

where gravity is neglected but can be included by substituting potential for pressure in the spatial derivatives.

Eq. (2.14) can be expanded in Cartesian coordinates,

$$\varphi c_t \frac{\partial p}{\partial t} = \frac{\partial}{\partial x} \left(\frac{k_{r\alpha}^0 k_x}{\mu_\alpha} \frac{\partial p}{\partial x} \right) + \frac{\partial}{\partial y} \left(\frac{k_{r\alpha}^0 k_y}{\mu_\alpha} \frac{\partial p}{\partial y} \right) + \frac{\partial}{\partial z} \left(\frac{k_{r\alpha}^0 k_z}{\mu_\alpha} \frac{\partial p}{\partial z} \right) + \widetilde{q}_\alpha \tag{2.15}$$

where a diagonal permeability tensor was assumed applicable and gravity is negligible. For flow in a 1D, homogenous medium and viscosity and no sources or sinks (wells) the equation further simplifies,

$$\frac{\partial p}{\partial t} = \alpha \frac{\partial^2 p}{\partial x^2} \tag{2.16}$$

where the diffusivity constant, α, is defined[1] as,

$$\alpha = \frac{k k_{r\alpha}^0}{\mu_\alpha \varphi c_t}. \tag{2.17}$$

Eq. (2.16) is a parabolic PDE; it is second-order in space and first-order in time. The PDE describes many phenomena in physics, mathematics, and engineering and is therefore commonly found in these fields. It is sometimes

---

1. The diffusivity constant is usually defined without endpoint relative permeability but it is included in this text for convenience of including modeling of one phase when another is immobile.

referred to as the *heat equation* because when pressure, p, is substituted with temperature, T, it describes heat flow in 1D. Eq. (2.16) can also be used to describe diffusion of component species (Chapters 7 and 8) when the concentration, C, is substituted in for pressure. The PDE described by Eq. (2.16) is the simplest form of the diffusivity equation in Cartesian coordinates. Although its derivation required many assumptions (1D, single-phase flow, homogeneous permeability, negligible gravity effects, small fluid compressibility, no sources/sinks, etc.), it is the starting point for numerical simulation of flow through porous media in Chapter 3.

Eq. (2.14) can also be expanded in cylindrical coordinates,

$$\varphi c_t \frac{\partial p}{\partial t} = \frac{1}{r}\frac{\partial}{\partial r}\left(r\frac{k_{r\alpha}^0 k_r}{\mu_\alpha}\frac{\partial p}{\partial r}\right) + \frac{1}{r^2}\frac{\partial}{\partial \theta}\left(\frac{k_{r\alpha}^0 k_\theta}{\mu_\alpha}\frac{\partial p}{\partial \theta}\right) + \frac{\partial}{\partial z}\left(\frac{k_{r\alpha}^0 k_z}{\mu_\alpha}\frac{\partial p}{\partial z}\right) + \tilde{q}_\alpha,$$

(2.18)

where again a diagonal permeability tensor is assumed applicable and gravity neglected. For flow in a 1D homogenous medium and no sources or sinks,

$$\frac{\partial p}{\partial t} = \frac{\alpha}{r}\frac{\partial}{\partial r}\left(r\frac{\partial p}{\partial r}\right),$$

(2.19)

where the diffusivity constant, α, was defined by Eq. (2.17).

Eq. (2.19) is commonly used to approximate flow around a wellbore as shown in Fig. 2.3. Analytical solutions are discussed in Section 2.5 and are the basis for important reservoir engineering techniques such as well testing. Analytical solutions are also used as benchmarks for numerical techniques.

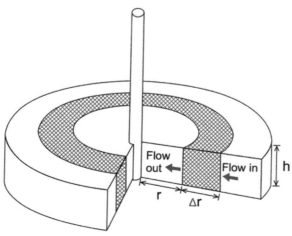

FIGURE 2.3 Schematic of near-wellbore flow in a reservoir. Cylindrical coordinates are appropriate because of the well geometry and radial flow occurs near the wellbore.

2.4.2.2 Compressible gases

Eq. (2.11) is valid for any single phase, including a gaseous phase. However, the approximation of small compressibility used to obtain the parabolic "heat equation" in Eq. (2.16) is not valid for gases because of their large compressibility. This is unfortunate because Eq. (2.16) has analytical solutions for certain boundary conditions, whereas Eq. (2.11) does not have an analytical solution in general and is very nonlinear because of the pressure-dependency on density.

However, a linearization procedure can be performed to transform Eq. (2.11) into one similar in form to Eq. (2.16) by introducing pseudopressure, ψ (Al-Hussainy et al., 1966),

$$\psi = \int_{p_{ref}}^{p} \frac{2p}{\mu_g z} dp, \tag{2.20}$$

where p_{ref} is a low reference pressure and z is the gas z-factor. The integral can be computed numerically and the relationship between pseudopressure and pressure is shown in Fig. 2.4A. The pseudopressure does not have much physical meaning and in fact is not a "pressure" at all. The units of ψ, psi^2/cp, are not even the same as pressure (at least as defined by Eq. 2.20). However, the mathematical transformation does result in a simpler, more linear PDE as shown by Al-Hussainy et al. (1966). In 1D, homogeneous, and in the absence of wells, the transformation results in,

$$\frac{\partial \psi}{\partial t} = \alpha_g \frac{\partial^2 \psi}{\partial x^2}, \tag{2.21a}$$

$$\frac{\partial \psi}{\partial t} = \frac{\alpha_g}{r} \frac{\partial}{\partial r} \left(r \frac{\partial \psi}{\partial r} \right), \tag{2.21b}$$

where Eqs. (2.21a) and (2.21b) are the transformed PDEs in Cartesian and cylindrical coordinates, respectively, and α_g is the diffusivity constant using only the gas (instead of total) compressibility because rock/liquid compressibilities are negligible in comparison to gas.

Eq. (2.21) *looks like* Eqs. (2.16) and (2.19) (the heat equation for slightly compressible liquids), but is written in terms of pseudopressure, ψ, instead of pressure, p. The procedure involves solving Eq. (2.21) after mapping boundary conditions to pseudopressure. Any analytical solutions (Section 2.5) for Eqs. (2.16) and (2.19) are applicable but in terms of ψ. The pseudopressure can then be transformed back to pressure using the definition, Eq. (2.20), or Fig. 2.4A. The analytical solutions do assume that the diffusivity constant is in fact *constant*, even though viscosity and compressibility are strong functions of pressure for gasses. However, gas viscosity increases while gas compressibility

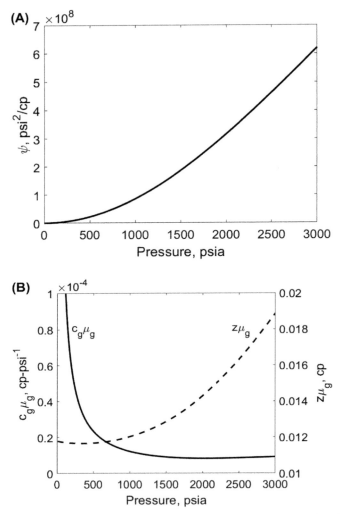

FIGURE 2.4 Pseudopressure relationships for gas (A) pseudopressure as a function of pressure, (B) gas diffusivity not a strong function of pseudopressure.

decreases with increasing pressure and the effect is partially canceled out. Fig. 2.4B shows the effect of pressure on $c_g\mu_g$, which is weak at relatively high pressures. At low pressure, $z\mu_g$ is nearly constant which makes the integral, Eq. (2.20), simple ($\psi = P^2/z\mu_g$) and is often referred to as the p-squared approach. Example 2.2 demonstrates the mapping between pressure and pseudopressure.

Example 2.2 Pseudopressure.
The expression, $1/\mu_g z$, for a particular gas can be approximated by the following expression,

$$\frac{1}{\mu_g z} \approx \begin{cases} 100 \text{ cp}^{-1} & \text{if } 0 < p < 1000 \text{ psia} \\ 2.0E - 6p^2 - 7.84E - 3p + 9.35E + 01 \text{ cp}^{-1} & \text{if } 1000 \text{ psia} > p > 4000 \text{ psia} \end{cases}$$

Using a base pressure of 0 psia, calculate the following
a. ψ at $p = 500$ psia
b. ψ at $p = 2000$ psia
c. p at $\psi = 5.0E07$ psia^2/cp
d. p at $\psi = 6.0E08$ psia^2/cp

Solution
First compute the integral definition of pseudopressure. The integral is usually computed numerically (e.g., trapezoidal rule), but the approximate expressions allow for analytical computation,

$$\psi = \int_0^p \frac{2p}{\mu_g z} dp = \frac{p^2}{\mu z} = 100p^2 \quad 0 < p < 1000 \text{psia}$$

$$\psi = \int_0^p 2p(2.0E - 6p^2 - 7.84E - 3p + 9.35E + 01) dp$$

$$= \int_0^p (4.0E - 6p^3 - 1.568E - 2p^2 + 1.87E + 02p) dp$$

$$= \frac{4.0E - 6}{4} p^4 - \frac{1.568E - 2}{3} p^3 + \frac{1.87E + 02}{2} p^2$$

$$= 1.0E - 6p^4 - 5.23E - 3p^3 + 9.35E + 01p^2 \quad 1000\text{psia} > p > 4000\text{psia}$$

Pseudopressures and pressures can then be computed
(a) $\psi = 100p^2 = 100 \cdot (500)^2 = 2.50E + 07 \frac{\text{psia}^2}{\text{cp}}$

(b) $\psi = 1.0E - 6(2000)^4 - 5.23E - 3(2000)^3 + 9.35E + 01(2000)^2$

$$= 3.48E08 \frac{\text{psia}^2}{\text{cp}}$$

(c) $p = \sqrt{\frac{\psi}{100}} = \sqrt{\frac{5.0E07}{100}} = 707$psia

(d) Find the positive, real root of the polynomial,

$$f(p) = 1.0E - 6p^4 - 5.23E - 3p^3 + 9.35E + 01p^2 - 6.0E08 = 0$$
$$p = 2631 \text{ psia}$$

2.5 Analytical solutions

2.5.1 1D heat equation in a finite medium

Eq. (2.16), and the pseudopressure equivalent Eq. 2.21a, is one of the simplest forms of the diffusivity equation and describes 1D Cartesian flow in a homogeneous reservoir without any sources or sinks (other than boundary conditions) in Cartesian coordinates. For a finite length, L, medium the PDE can be solved using Finite Fourier Transforms (FFT) for certain, simple boundary conditions. Fig. 2.5A illustrates a porous medium with a constant pressure imposed at $x = 0$, $p(0,t) = p_{B1}$ and at $x = L$, $p(L,t) = p_{B2}$. In Fig. 2.5B, the same boundary condition is imposed at $x = 0$, but a no-flow condition, $q(L,t) = dp/dx(L,t) = 0$, is imposed at $x = L$. Both have an initial condition of constant pressure, $p(x,0) = p_{init}$.

Boundary conditions of constant pressure are Dirichlet and those of constant flux, including no flux, are Neumann boundary conditions. The analytical solution to Eq. (2.16) for the boundary conditions described by Fig. 2.5A and B are given by Eqs. 2.22a and 2.22b, respectively.

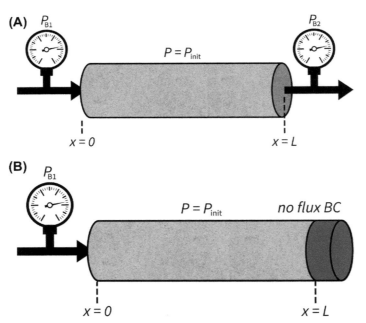

FIGURE 2.5 Schematics of 1D porous rocks core for (A) constant pressure boundary conditions on both boundaries and (B) constant pressure on the left boundary but no flux on the right. The pressure along the length of the cores at any time is governed by the 1D "heat equation."

$$p_D(x_D, t_D) = 1 - x_D - p_{Di} \frac{2}{\pi} \sum_{n=1}^{\infty} \left[\frac{(-1)^n}{n} sin(n\pi x_D) exp\left(-n^2\pi^2 t_D \right) \right]$$

$$-(1 - p_{Di}) \frac{2}{\pi} \sum_{n=1}^{\infty} \left[\frac{(-1)^n}{n} sin(n\pi(x_D - 1)) exp\left(-n^2\pi^2 t_D \right) \right],$$

(2.22a)

$$p_D(x_D, t_D) = 1 - \sum_{n=0}^{\infty} \frac{4}{(2n+1)\pi} exp\left(-\frac{(2n+1)^2}{4}\pi^2 t_D \right) sin\left(\frac{(2n+1)\pi}{2} x_D \right),$$

(2.22b)

where the equations are dimensionless and the dimensionless variables in Eq. (2.21a) are defined as $p_D = (p - p_{B2})/(p_{B1} - p_{B2})$ and $p_{Di} = (p_{init} - p_{B2})/(p_{B1} - p_{B2})$. In Eq. (2.21b), $p_D = (p - p_{init})/(p_{B1} - p_{init})$. In both equations, $x_D = x/L$, and $t_D = \alpha t/L^2$. The solutions are infinite series, but, in practice, only a finite number of terms are evaluated to obtain a solution with small error (higher order terms are sequentially smaller and less significant). The error is known as *truncation error* because the infinite series is truncated. Truncation error is discussed more in Chapter 3.

Fig. 2.6A and B show the dimensionless pressure versus distance at various dimensionless times, t_D, using Eqs. 2.22a and 2.22b, respectively. The solutions match the boundary and initial conditions and intuitively make sense. For example, in Fig. 2.7B, at all times, $t > 0$, the pressure at the left boundary ($x = 0$) is equal to the boundary condition, $p = p_{B1}$ ($p_D = 1$). At very early times, the pressure is close to the initial pressure, p_{init} ($p_D = 0$), far away from the boundary ($x_D >> 0$). At the right boundary, $x_D = 1$, the slope

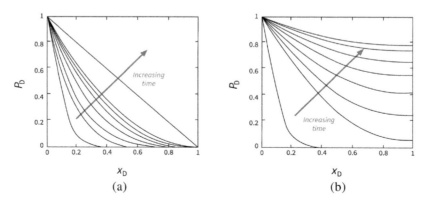

FIGURE 2.6 Analytical solution to the 1D heat Eq. (1.26) for (A) Dirichlet boundary conditions at both $x_D = 0$ ($p_D = 1$) and $x_D = 0$ ($p_D = 0$). and (B) Dirichlet boundary condition ($p_D = 1$) at $x_D = 0$ and Neumann (no flow) at $x_D = 1$.

of the curve (dp/dx) is zero, reflecting the no-flow boundary condition. At later times, the boundary conditions are still met on both ends but the pressure increases throughout the domain, approaching the p_{B1} boundary condition.

Fig. 2.6B might be better understood using the analogy of heat transfer in a 1D medium, which is governed by the same PDE. Consider a long iron rod initially ($t = 0$) at constant (e.g., 75°F, room) temperature. On the left end ($x = 0$), the rod is maintained at a constant temperature (e.g., 150°F) and the right end ($x = L$) the rod is sealed with insulation so that no heat can escape through that boundary. At early times, most of the rod would be at or near room temperature; only near the left end would the rod be hot and would be exactly 150°F at the boundary. At later times, the temperature of much of the rod would be hot but would decrease montotonically from $x = 0$ to $x = L$ and may still be near the initial temperature at $x = L$. Eventually, the entire rod would be hot (at or near 150°F), and the temperature of the rod would not change with time. At that time, the temperature would have reached *steady-state*, meaning there is no change in temperature anywhere with time. The steady-state solution of the diffusivity equation can be solved by neglecting the time derivative term in Eq. (2.16), since by definition there is no change with time at steady state. The PDE in 1D is reduced to a simple ODE which can be solved by separation of variables with the relevant boundary conditions.

2.5.2 1D heat equation in a semi-infinite medium

A domain is semi-infinite ($0 < x < \infty$) if there is no boundary on one end or the boundary is so far away that the pressure at the boundary is unaffected by a stimulus at the opposite boundary. Eq. (2.16) is still the governing PDE, but the boundary and initial conditions, $p_D = 1$ at $x = 0$; $p_D = p_{Di}$ at $x = \infty$, $p_D = p_{Di}$ at $t = 0$ are applied. The dimensionless solution can be found using Laplace transforms and is given by,

$$p_D = \text{erfc}(\xi_D), \tag{2.23}$$

where $\xi_D = x/2\sqrt{\alpha t}$, and p_D is defined as before. The complementary error function (*erfc*) is the solution to the integral,

$$erfc(\xi) = 1 - erf(\xi) = 1 - \int_0^\xi e^{-\chi^2} d\chi. \tag{2.24}$$

Unfortunately, the integral cannot be found analytically and must be evaluated numerically. The numerical solution is tabulated and readily available. The solution (Eq. 2.23) is plotted in Fig. 2.7; only one curve is shown because the solution is self-similar in space and time.

The semi-infinite boundary is applicable for many problems, including mass transfer. For example, the PDE and boundary conditions presented above

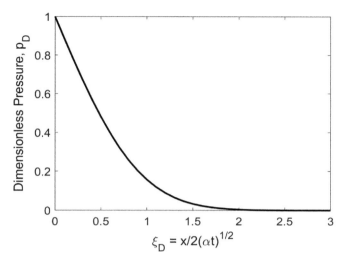

FIGURE 2.7 Analytical solution of pressure in a semi-infinite medium.

often describe the 1D diffusion of a chemical species through a 1D semi-infinite porous medium. Compositional simulation and mass transfer are discussed in Chapters 7, 8, and 10.

2.5.3 Solution in cylindrical coordinates (around a wellbore)

Eq. (2.19) describes flow in a cylindrical homogenous porous medium (Fig. 2.3). One application of the PDE is flow into or out of a wellbore with radius, r_w. If the well is operated at constant rate, q_w, there is no flow at the reservoir boundary ($r = r_e$), and the initial pressure, p_{init}, is uniform, the PDE has an analytical solution (Lee et al., 2003),

$$p_D(r_D, t_D) = \frac{2}{(r_{eD}^2 - 1)}\left(\frac{r_D^2}{4} + t_D\right) - \frac{r_{eD}^2 \ln r_D}{(r_{eD}^2 - 1)} - \frac{(3r_{eD}^4 - 4r_{eD}^4 \ln r_{eD} - 2r_{eD}^2 - 1)}{4(r_{eD}^2 - 1)^2}$$

$$+\pi \sum_{n=1}^{\infty}\left\{\frac{e^{-a_n^2 t_D} J_1^2(a_n r_{eD})[J_1(a_n)Y_0(a_n r_D) - Y_1(a_n)(J_0)(a_n r_D)]}{a_n\left[J_1^2(a_n r_{eD}) - J_1^2(a_n)\right]}\right\},$$

(2.25)

where J_0 and J_1 are the Bessel functions of the first kind, zero and first order, respectively, and Y_0 and Y_1 are the Bessel functions of the second kind, zero and first order, respectively, and the coefficients, a_n, are the roots of the equations,

$$J_1(a_n r_{eD})Y_1(a_n) - J_1(a_n)Y_1(a_n r_{eD}) = 0.$$

The dimensionless variables are defined as,

$$p_D = \frac{2\pi k k_{r\alpha}^0 h}{q_w \mu_\alpha}[p_{init} - p]; \quad t_D = \frac{\alpha t}{r_w^2}; \quad t_{DA} = \frac{t_D}{\pi r_{eD}^2}; \quad r_D = \frac{r}{r_w}; \quad r_{eD} = \frac{r_e}{r_w},$$

where the well rate, q_w, is at reservoir conditions.

Eq. (2.25) is a general solution and valid for all times and all radii for the aforementioned boundary conditions, but its usefulness is limited because the analytical solution is so complicated. Tables and figures, based on the solution, are often used to avoid calculation. However, the solution can be simplified for a few limiting conditions.

2.5.3.1 Pseudo-steady-state (pss) flow

For the special case of sufficiently long times ($t_{DA} > \sim 0.1$), production (or injection) rate of the wellbore impacts the drainage boundary. This flow regime is known as pseudo-steady-state flow and the infinite series in Eq. (2.25) can be neglected. If only the pressure at $r = r_w$ is of interest and the well radius is small compared to the drainage radius ($r_e \gg r_w$), the dimensionless pressure at the wellbore is given by,

$$p_{wD}(t_D) \approx \frac{2}{r_{eD}^2}t_D + \ln(r_{eD}) - \frac{3}{4} \tag{2.26}$$

Importantly, the equation shows that the wellbore pressure decreases linearly with time for a producer well.

2.5.3.2 Infinite-acting flow

At early times ($t_{DA} < \sim 0.1$), the reservoir is infinite-acting (i.e., the boundaries are so far from the well that there is effectively no boundary) and the appropriate boundary conditions are uniform initial pressure, $p(r,0) = p_{init}$, constant pressure far from the well, $p(\infty,t) = p_{init}$, and constant flow rate, q_w, in or out of the well at $r = 0$. The analytical solution to the PDE can be solved using a Boltzmann transformation and the derivation can be found in many texts (e.g., Matthews and Russel, 1967),

$$p_D = -\frac{1}{2}Ei\left(-\frac{r_D^2}{4t_D}\right) \approx -\frac{1}{2}\left[\ln\left(\frac{r_D^2}{4t_D}\right) + \gamma\right], \tag{2.27}$$

where γ ($= 0.5772$) is Euler's constant and the natural logarithm approximation on the right is valid if $t_D > 25r_D^2$. The exponential integral, Ei, is defined,

$$Ei(-x) = -\int\limits_x^\infty \frac{e^{-u}}{u}du \tag{2.28}$$

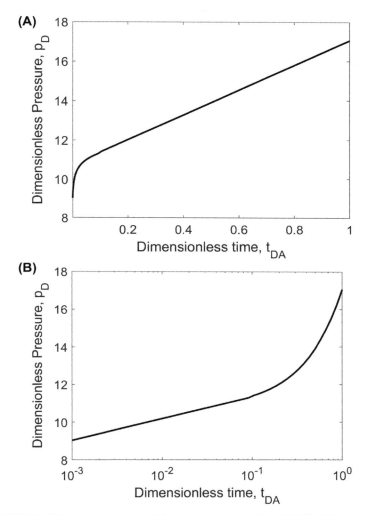

FIGURE 2.8 Dimensionless exponential integral solution to the radial diffusivity equation.

The integral can be computed numerically (results are tabulated). Fig. 2.8A and B show the dimensionless pressure at the wellbore versus dimensionless time, t_{DA}, for a cylindrical bounded reservoir on a linear and semi-log scale, respectively. Eqs. (2.27) and (2.26) were used for t_{DA} <0.1 and t_{DA} >0.1, respectively, in Fig. 2.8.

The pressure is linear with logarithmic time during infinite-acting flow and is linear with time during pseudo-steady-state flow. There is a small discontinuity in the solution at $t_{DA} = 0.1$ because a short transitional flow regime occurs between the infinite-acting and pseudo-steady-state flow regimes.

2.6 Exercises

Exercise 2.1. Expanded form of continuity equation. The general form of the continuity equation in vector notation is given by Eq. (2.5). Write the 3D expanded form of the equation for single-phase flow in the following coordinate systems: (a) Cartesian and (b) cylindrical.

Exercise 2.2. Derivation of continuity equation. Derive the 3D continuity equation in (a) Cartesian and (b) cylindrical coordinates by performance a mass balance on a differential element control volume.

Exercise 2.3. Derivation of diffusivity equation. Using the 3D form of the phase mass balances in cylindrical coordinates, substitute Darcy's law and assume a slightly compressible fluid to derive the Diffusivity equation. Simplify by assuming homogeneous rock and fluid properties.

Exercise 2.4. Plotting analytical solution. Write a computer code to solve for the dimensionless pressure (p_D), at a given dimensionless position (x_D) and time (t_D) for 1D, single-phase flow in Cartesian coordinates. The code should be flexible to allow for any combination of boundary conditions (Dirichlet–Dirichlet, Dirichlet–no flux, no flux–Dirichlet, and Dirichlet–infinite acting). Use your code to make plots of p_D versus x_D at various t_D analogous to Figs. 2.8 and 2.7.

Exercise 2.5. Analytical solution with gravity. The 1D diffusivity Eq. (2.16) can be written in terms of potential, Φ, instead of pressure and then the analytical solutions apply. A 1D reservoir (Cartesian coordinates) has a dip angle, θ, and is initially at static equilibrium, $\Phi = \Phi_{init}$ is constant. The dimensionless boundary and initial conditions are $\Phi_{D,B1} = 1$ at $x_D = 0$, no flux at $x_D = 1$, and $\Phi_{D,init} = 0$. The dimensionless gravity is $pg(\sin\theta)L/(\Phi_{B1}-\Phi_{init}) = 0.3$. First write the solution for $\Phi(x_D,t_D)$ and then plot the dimensionless *pressure* (not potential) versus distance at various dimensionless times.

Exercise 2.6. Plotting wellbore pressure. Write a computer code to compute the dimensionless pressure at the wellbore for a constant rate well in a cylindrical bounded reservoir. If the dimensionless time, t_{DA}, is less than 0.1 use the approximate form of the infinite-acting solution and if t_{DA} is greater than 0.1, then use the pseudo-steady-state solution. Make the plots of p_{wD} versus t_D shown in Fig. 2.9. For an additional challenge, code the full solution (Eq. 2.25) for flow around a well at all radii and times.

Exercise 2.7. Pseuodopressure. Write a computer code to that computes the pressure given the pseudopressure and vice versa using the integral transformation, Eq. (2.20), and numerical integration. Validate your code against Example 2.2. Then solve the p_D (r_D, t_D) of a gas during infinite-acting flow by first solving for pseudopressure and converting back to pressure. Using the fluid properties in Example 2.2 and $p_{init} = 3000$ psia, make a plot of dimensionless well pressure versus time for a gas. Assume that the well pressure remains above 1000 psia.

References

Al-Hussainy, R., Ramey Jr., H.J., Crawford, P.B., 1966. The flow of real gases through porous media. Journal of Petroleum Technology 18 (5), 624−636.

Lee, J., Rollins, J.B., Spivey, J.P., 2003. Pressure transient testing (eBook). SPE textbook series 9.

Matthews, C.S., Russell, D.G., 1967. Pressure Buildup and Flow Tests in Wells, Monograph Series 1. Society of Petroleum Engineers, Richardson, Texas.

Chapter 3

Finite difference solutions to PDEs

3.1 Introduction

Analytical solutions to the PDEs describing flow and transport in porous media, including subsurface reservoirs, are rare and only apply for simple boundary conditions, geometries, and limiting assumptions about the reservoir and fluids. Therefore, approximate numerical techniques must be used to solve these complicated PDEs. Many numerical options are available, including finite element and finite volume methods, but the finite difference method continues to be a popular approach utilized in reservoir simulation[1]. In the finite difference method, continuous partial derivatives are approximated by discrete finite differences and then substituted into the original PDE. In this chapter, finite difference approximation equations are derived and then substituted into the PDEs for single phase flow to obtain algebraic equations. Solution methods, including explicit, implicit, and Crank-Nicolson, are presented along with their stability criteria. Finally, pseudocode is presented for the reader to develop their first reservoir simulator.

3.2 Taylor series and finite differences

Any smooth function can be represented by a Taylor series expansion about a single point using an infinite sum of terms containing the function's derivatives,

$$f(x) = f(x_i) + f'(x_i)(x - x_i) + \frac{f''(x_i)(x - x_i)^2}{2!} + \frac{f'''(x_i)(x - x_i)^3}{3!} + \dots, \quad (3.1)$$

where the function, f, can be evaluated at any point x by expanding around another point x_i in the infinite series. It is of course impossible to utilize the infinite number of terms in the series to obtain an exact function evaluation, but approximate solutions can be found by truncating the series; that is, using a

---

1. It can be shown that the finite difference method is a subset of the finite element and finite volume methods.

An Introduction to Multiphase, Multicomponent Reservoir Simulation.
https://doi.org/10.1016/B978-0-323-99235-0.00007-5

finite number of terms. This approximation introduces *truncation error* in the evaluation of $f(x)$. Example 3.1 demonstrates the use of a Taylor series for estimating the formation volume factor of a slightly compressible fluid.

Example 3.1. Formation Volume Factor and Taylor Series Approximation

Calculate the formation volume factor for an undersaturated oleic phase at 3000 psia if the bubble point pressure is $p_b = 1208$ psi, formation volume factor at bubble point is, $B_{ob} = 1.291$ RB/STB, and oil compressibility (c_o) is 2.5×10^{-5} psi^{-1}. Use

a. the exact formula for the formation volume factor of a constant compressibility fluid.
b. Taylor series approximations using the first 1, 2, 3, 4, and 5 terms of the series, respectively. Create a table summarizing your results and compare to the true relative percent error in the approximations.

Solution
a. Exact formula

$$B_o = B_{ob} \exp\left[c_o(p_b - p)\right] = 1.291 \frac{\text{bbl}}{\text{STB}} \exp\left[2.5 \times 10^{-5}\text{psi}^{-1}(1208\text{psi} - 3000\text{psi})\right]$$

$$= 1.23444 \frac{\text{bbl}}{\text{STB}}$$

b. Approximation using a Taylor series,

$$B_o = B_{ob}\left[1 + c_o(p_b - p) + \frac{1}{2}c_o^2(p_b - p)^2 + \frac{1}{6}c_o^3(p_b - p)^3 + \frac{1}{24}c_o^4(p_b - p)^4 + \cdots\right]$$

Plugging in numbers,

$$B_o = 1.291 \left[\underbrace{1}_{1.0} + \underbrace{(2.5 \times 10^{-5})(1208 - 3000)}_{-0.0448} + \underbrace{\frac{1}{2}(2.5 \times 10^{-5})^2(1208 - 3000)^2}_{1.00352 \times 10^{-3}} \right. $$
$$\left. + \underbrace{\frac{1}{6}(2.5 \times 10^{-5})^3(1208 - 3000)^3}_{-1.49859 \times 10^{-5}} + \underbrace{\frac{1}{24}(2.5 \times 10^{-5})^4(1208 - 3000)^4}_{1.67824 \times 10^{-7}} + \cdots \right]$$

The absolute value of each term in the infinite series is smaller than the one that precedes it. Terms alternate from positive to negative.

Absolute error and absolute relative error are calculated as follows when only using the first term in the Taylor series,

$$E_t = \left|B_{o,true} - B_{o,approximate}\right| = |1.23444 - 1.291| = 0.056560388 \frac{\text{RB}}{\text{STB}}$$

$$e_t = \frac{\left|B_{o,true} - B_{o,approximate}\right|}{\left|B_{o,true}\right|} = 100 \times \frac{|1.23444 - 1.291|}{|1.23444|} = 4.5818675\%$$

The table below summarizes the values and errors using various numbers of terms in the series.

Terms	B_o (RB/STB)	E_t (RB/STB)	e_t (%)
1	1.291	0.056560388	4.5818675
2	1.2331632	0.001276412	0.1034001
3	1.234458744	1.9132E−05	0.0015499
4	1.234439398	2.14757E−07	0.0000174
5	1.234439614	1.92709E−09	0.0000002

Using the linear approximation (2 terms in the Taylor series) results in $\sim$0.1% error for this slightly compressible fluid.

3.2.1 First-order forward difference approximation

The Taylor series (Eq. 3.1) can be rearranged to solve for certain derivatives. Consider, for example, that the reservoir pressure, p, is a function of position, x. Using a Taylor series, the pressure at position x_{i+1} using a reference position of x_i can be evaluated,

$$p(x_{i+1}) = p(x_i + \Delta x) = p(x_i) + p'(x_i)(x_{i+1} - x_i) + \frac{p''(x_i)(x_{i+1} - x_i)^2}{2!} + \cdots,$$

$$(3.2)$$

where $\Delta x = x_{i+1} - x_i$. Rearranging Eq. (3.2) and using some algebra to solve for the first derivative of pressure,

$$p'(x_i) = \frac{\partial p}{\partial x}\bigg|_{x_i} = \frac{p(x_{i+1}) - p(x_i)}{\Delta x} + R_f \approx \frac{p(x_{i+1}) - p(x_i)}{\Delta x}, \qquad (3.3)$$

where the remainder term includes the truncated terms,

$$R_f = -\frac{p''(x_i)}{2}\Delta x - \frac{p'''(x_i)}{3!}(\Delta x)^2 - \frac{p''''(x_i)}{4!}(\Delta x)^3 + \cdots \qquad (3.4)$$

If Δx is taken to be sufficiently small, each term in the remainder is smaller than the preceding term. Neglecting all terms in the remainder, the derivative (Eq. 3.3) is approximated as a first-order finite forward difference. Although the direction of flow is assumed x here, it could be y or z (or r, t, z in cylindrical coordinates). Many texts use the general nomenclature h in place of Δx to indicate the spacing between points.

Since the first (and largest) term in the neglected remainder (Eq. 3.4) is proportional to Δx, the forward difference approximation (Eq. 3.3 without the remainder) is of first order accuracy, written as $O(\Delta x)$. This means that if Δx is decreased by a factor of two, the error in the derivative would decrease by a factor two, in the limit that Δx approaches 0.

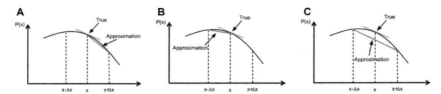

FIGURE 3.1 Approximation to a first derivative using (A) forward difference, (B) backward difference, and (C) centered difference.

Fig. 3.1A illustrates an intuitive schematic of a forward difference approximation to a first-order derivative. Recall that by definition, a function's derivative is the *slope of the line tangent to the curve* at a specific point, x_i. The forward difference approximation (Eq. 3.3) is the slope of a "secant" line that connects two points (x_i and x_{i+1}). In the limit that Δx approaches zero, the secant line becomes the tangent and, mathematically, all remainder terms become zero, thus resulting in no truncation error.

3.2.2 First-order backward difference approximation

The Taylor series can be written "backward" in space, i.e., $x_{i-1} = x_i - \Delta x$,

$$p(x_{i-1}) = p(x_i - \Delta x) = p(x_i) - p'(x_i)(\Delta x) + \frac{p''(x_i)(\Delta x)^2}{2!} - \dots, \qquad (3.5)$$

where the terms in the series now alternate between positive and negative as a result of the subtraction of Δx from x_i (negative terms that are raised to an even power are positive). Neglecting all terms higher than the first-order derivative and then rearranging to solve for the derivative gives the first-order backward finite difference approximation,

$$\left.\frac{\partial p}{\partial x}\right|_{x_i} = \frac{p(x_i) - p(x_{i-1})}{\Delta x} + R_b \approx \frac{p(x_i) - p(x_{i-1})}{\Delta x}, \qquad (3.6)$$

where the remainder term, R_b, is different than R_f in Eq. (3.3). The backward difference, like the forward difference, approximation contains $O(\Delta x)$ accuracy meaning that its associated truncation error is proportional to Δx. Therefore, the forward and backward differences have the same *order of accuracy*. This does not mean that the forward and backward approximations will give the same answer, nor will they have the same error. It does mean that both will *converge* to the true derivative at the same *rate*. Fig. 3.1B illustrates the backward difference approximation, which is the slope of the secant line between x_i and x_{i-1}.

3.2.3 Second-order, centered difference approximation

Approximations to the first derivative can also be determined using a second-order, centered finite difference, which can be derived by subtracting a backward Taylor series (Eq. 3.5) from a forward Taylor series (Eq. 3.2),

$$
p(x_{i+1}) - p(x_{i-1}) = \left[p(x_i) + p'(x_i)\Delta x + \frac{p''(x_i)}{2!}\Delta x^2 + \frac{p'''(x_i)}{3!}\Delta x^3 + \dots \right] -
$$
$$
\left[p(x_i) - p'(x_i)\Delta x + \frac{p''(x_i)}{2!}\Delta x^2 - \frac{p'''(x_i)}{3!}\Delta x^3 + \dots \right]
$$

(3.7)

where the odd numbered terms (zero-order derivative, second-order derivative, fourth-order derivative, etc.) cancel out upon subtraction,

$$
p(x_{i+1}) - p(x_{i-1}) = 2\left[p'(\Delta x_i) + \frac{p'''(\Delta x_i)^3}{3!} + \dots \right].
$$

(3.8)

Rearranging Eq. (3.8) for the first-order derivative gives the centered difference approximation,

$$
\left.\frac{\partial p}{\partial x}\right|_{x_i} = \frac{p(x_{i+1}) - p(x_{i-1})}{2\Delta x} + R_c \approx \frac{p(x_{i+1}) - p(x_{i-1})}{2\Delta x},
$$

(3.9)

where the leading term in the remainder, R_c, is proportional to Δx^2 because the second derivative term is canceled during the subtraction of the Taylor series. Thus, the centered difference approximation is $O(\Delta x^2)$ accurate, which means it converges faster to the true solution than either a forward or backward difference approximation. Recall that if Δx were decreased by a factor of two, the error in the derivative approximation would also decrease by a factor of two for either a forward or backward difference; however, the error would decrease by a factor of four for a centered difference.

Fig. 3.1C illustrates a centered difference approximation, which is a secant line between the points x_{i-1} and x_{i+1}. Note that a centered difference does require evaluation of the function at two points other than x_i (x_{i-1} and x_{i+1}) which may require additional computations. In practice, the forward, backward, and centered difference equations are all useful in different situations for approximating derivatives.

3.2.4 Approximations to the second derivative

A Taylor series can also be used to derive higher-order derivative approximations, including second-order derivatives. Like the first-order derivative, the second-order approximations can be a forward, backward, or centered difference. In this text, the centered difference for the second-order derivative will usually be used. There are a few different approaches to derive the centered difference approximation; here the Taylor series for $p(x_{i+1})$ and $p(x_{i-1})$ are summed,

$$p(x_{i+1}) + p(x_{i-1}) = \left[p(x_i) + p'(x_i)\Delta x + \frac{p''(x_i)}{2!}\Delta x^2 + \frac{p'''(x_i)}{3!}\Delta x^3 + O(\Delta x^4) \right] +$$
$$\left[p(x_i) - p'(x_i)\Delta x + \frac{p''(x_i)}{2!}\Delta x^2 - \frac{p'''(x_i)}{3!}\Delta x^3 + O(\Delta x^4) \right].$$

(3.10)

Several terms cancel in Eq. (3.10); rearranging for the second derivative gives

$$\left. \frac{\partial^2 p}{\partial x^2} \right|_{x_i} = \frac{p(x_{i+1}) - 2p(x_i) + p(x_{i-1})}{\Delta x^2} + R_{c2} \approx \frac{p(x_{i+1}) - 2p(x_i) + p(x_{i-1})}{\Delta x^2}$$

(3.11)

The leading term in the remainder in the approximation Eq. (3.14) is proportional to Δx^2, so the centered difference approximation for the second-order derivative has error of $O(\Delta x^2)$.

Eq. (3.11) can alternatively be derived by recognizing that the second-order derivative is simply the derivative of the first-order derivative. Substitution of the centered difference for the first-order derivatives gives,

$$p''(x_i) = \frac{d}{dx}(p'(x_i)) \approx \frac{p'(x_{i+\frac{1}{2}}) - p'(x_{i-\frac{1}{2}})}{\Delta x} \approx \frac{\frac{p(x_{i+1})-p(x_i)}{\Delta x} - \frac{p(x_i)-p(x_{i-1})}{\Delta x}}{\Delta x}, \quad (3.12)$$

where $x_{i+1/2}$ refers to the midpoint of x_i and x_{i+1}, and $x_{i-1/2}$ refers to the midpoint of x_i and x_{i-1}. Eq. (3.12) reduces to Eq. (3.11). The approximation is $O(\Delta x^2)$ because both the forward and backward approximations are $O(\Delta x)$ and are divided by Δx. The approach used in Eq. (3.12) is useful for deriving reservoir simulation equations that involve spatial heterogeneities in permeability or grid size. Example 3.2 demonstrates the use of finite difference approximations to derivatives.

Example 3.2. Finite Difference Approximations
The analytical solution for radial flow and constant production rate during infinite acting flow was introduced in Chapter 2 (Eq. 2.35). The solution includes the exponential integral, but at sufficiently large times an approximate solution can be used,

$$p_D \approx -\frac{1}{2}\left[\ln\left(\frac{r_D^2}{4t_D}\right) + \gamma \right]$$

where $\gamma = 0.5772$ is Euler's constant.

a. Evaluate the following partial derivatives analytically and then compute them at $t_D = 1.0 \times 10^5$ and $r_D = 1$.

$$\frac{\partial p_D}{\partial t_D}, \frac{\partial p_D}{\partial r_D}, \frac{\partial^2 p_D}{\partial r_D^2}$$

b. By hand (with a pencil and calculator), numerically calculate at $t_D = 1.0 \times 10^5$ and $r_D = 1$, the
   a. forward, backward, and centered approximation of dp/dt using $\Delta t_D = 1.0 \times 10^3$
   b. forward, backward, and centered approximation of dp/dr using $\Delta r_D = 0.01$
   c. centered approximation of d^2p/dr^2 using $\Delta r_D = 0.01$

Solution
a. Analytical derivatives

$$\frac{\partial p_D}{\partial t_D} = \frac{1}{2t_D} \Rightarrow \frac{1}{2 \cdot 10^5} = 5 \times 10^{-6}$$

$$\frac{\partial p_D}{\partial r_D} = -\frac{1}{r_D} \Rightarrow -\frac{1}{1} = -1$$

$$\frac{\partial^2 p_D}{\partial r_D^2} = \frac{1}{r_D^2} \Rightarrow \frac{1}{1^2} = 1$$

b. Difference approximations
First derivative with respect to time,

$$\frac{\partial p_D}{\partial t_D} = \frac{p_D(t_D + \Delta t_D) - p_D(t_D)}{\Delta t_D} = -\frac{1}{2 \cdot 10^3}\left[\ln\left(\frac{r_D^2}{t_D + \Delta t_D}\right) - \ln\left(\frac{r_D^2}{t_D}\right)\right]$$

$$= -\left[\ln\left(\frac{1^2}{10^5 + 10^3}\right) - \ln\left(\frac{1^2}{10^5}\right)\right] = 4.9752e{-}06$$

$$\frac{\partial p_D}{\partial t_D} = \frac{p_D(t_D) - p_D(t_D - \Delta t_D)}{\Delta t_D} = \frac{1}{2 \cdot 10^3}\left[\ln\left(\frac{r_D^2}{t_D}\right) - \ln\left(\frac{r_D^2}{t_D - \Delta t_D}\right)\right]$$

$$= -\left[\ln\left(\frac{1^2}{10^5}\right) - \ln\left(\frac{1^2}{10^5 - 10^3}\right)\right] = 5.0252e{-}06$$

$$\frac{\partial p_D}{\partial t_D} = \frac{p_D(t_D + \Delta t_D) - p_D(t_D - \Delta t_D)}{2\Delta t_D}$$

$$= \frac{1}{4 \cdot 10^3}\left[\ln\left(\frac{r_D^2}{t_D + \Delta t_D}\right) - \ln\left(\frac{r_D^2}{t_D - \Delta t_D}\right)\right]$$

$$= -\left[\ln\left(\frac{1^2}{10^5 + 10^3}\right) - \ln\left(\frac{1^2}{10^5 - 10^3}\right)\right] = 5.0002e{-}06$$

Continued

Example 3.2. Finite Difference Approximations—cont'd

First derivative with respect to radius

$$\frac{\partial p_D}{\partial r_D} = \frac{p_D(r_D + \Delta r_D, t_D) - p_D(r_D, t_D)}{\Delta r_D} = -\frac{1}{2 \cdot 0.01}\left[\ln\left(\frac{(r_D + \Delta r_D)^2}{t_D}\right) - \ln\left(\frac{r_D^2}{t_D}\right)\right]$$

$$= -\frac{1}{2 \cdot 0.01}\left[\ln\left(\frac{(1 + 0.01)^2}{10^5}\right) - \ln\left(\frac{1^2}{10^5}\right)\right] = -0.9950$$

$$\frac{\partial p_D}{\partial r_D} = \frac{p_D(r_D, t_D) - p_D(r_D - \Delta r_D, t_D)}{\Delta r_D} = -\frac{1}{2 \cdot 0.01}\left[\ln\left(\frac{r_D^2}{t_D}\right) - \ln\left(\frac{(r_D - \Delta r_D)^2}{t_D}\right)\right]$$

$$= -\frac{1}{2 \cdot 0.01}\left[\ln\left(\frac{1^2}{10^5}\right) - \ln\left(\frac{(1 - 0.01)^2}{10^5}\right)\right] = -1.0050$$

$$\frac{\partial p_D}{\partial r_D} = \frac{p_D(r_D + \Delta r_D, t_D) - p_D(r_D - \Delta r_D, t_D)}{2\Delta r_D}$$

$$= -\frac{1}{4 \cdot 0.01}\left[\ln\left(\frac{(r_D + \Delta r_D)^2}{t_D}\right) - \ln\left(\frac{(r_D - \Delta r_D)^2}{t_D}\right)\right]$$

$$= -\frac{1}{2 \cdot 0.01}\left[\ln\left(\frac{(1 + 0.01)^2}{10^5}\right) - \ln\left(\frac{(1 - 0.01)^2}{10^5}\right)\right] = -1.00003$$

Second derivative with respect to radius,

$$\frac{\partial^2 p_D}{\partial r_D^2} = \frac{p(r_D + \Delta r_D, t_D) - 2 \cdot p(r_D, t_D) + p(r_D - \Delta r_D, t_D)}{(\Delta r_D)^2}$$

$$= -\frac{1}{2 \cdot (\Delta r_D)^2}\left[\ln\left(\frac{(r_D + \Delta r_D)^2}{t_D}\right) - 2 \cdot \ln\left(\frac{r_D^2}{t_D}\right) + \ln\left(\frac{(r_D - \Delta r_D)^2}{t_D}\right)\right]$$

$$= -\frac{1}{2 \cdot (0.01)^2}\left[\ln\left(\frac{(1 + 0.01)^2}{10^5}\right) - 2\ln\left(\frac{1^2}{10^5}\right) + \ln\left(\frac{(1 - 0.01)^2}{10^5}\right)\right] = 1.00005$$

3.2.5 Generalization to higher-order approximations

All the approximations to the first and second derivatives shown until now are either first- or second-order accurate. It is possible to derive formulas with higher-order accuracy by including more terms in the Taylor series. The additional accuracy, however, generally comes at the cost of more computational effort. Here we present a generalized approach for derivation of these finite differences. The derivative can be approximated as a weighted sum of points,

$$\frac{\partial^k p}{\partial x^k} \approx \frac{1}{(\Delta x)^k} \sum_j \beta_j p_j, \tag{3.13a}$$

where k is the derivative order and j is the *stencil*. For example, $k = 2$ indicates a second-order derivative and the centered difference would include $j = -1, 0, 1$. Each of the stencil pressures is then defined using a Taylor series, e.g.,

$$p_{i-1} = p_i - \Delta x p_{xi} + \frac{1}{2}\Delta x^2 p_{xxi} - \frac{1}{6}\Delta x^3 p_{xxxi} + O(\Delta x^4)$$

$$p_i = p_i \tag{3.13b}$$

$$p_{i+1} = p_i + \Delta x p_{xi} + \frac{1}{2}\Delta x^2 p_{xxi} + \frac{1}{6}\Delta x^3 p_{xxxi} + O(\Delta x^4)$$

where the shorthand notation, p_{xi}, p_{xxi}, and p_{xxxi} refer to the first, second, and third derivatives, respectively. In the Taylor series, the number of terms used will determine the accuracy. Eq. (3.13b) can be substituted into Eq. (3.13a) and then like derivatives can be grouped together, e.g.,

$$\frac{\partial^2 p}{\partial x^2} \approx (\beta_{i-1} + \beta_i + \beta_{i+1})\frac{1}{(\Delta x)^2}p_i + (-\beta_{i-1} + \beta_{i+1})\frac{1}{(\Delta x)}p_{xi}$$

$$+ \left(\frac{1}{2}\beta_{i-1} + \frac{1}{2}\beta_{i+1}\right)p_{xxi} + \left(-\frac{1}{6}\beta_{i-1} + \frac{1}{6}\beta_{i+1}\right)\Delta x p_{xxxi} + O(\Delta x^2).$$

Finally, set the coefficients of $p_{ki} = 1$ (where k is the order of the derivative) and all other coefficients to zero in order to minimize the truncation error. This will lead to a system of equations that can be solved to obtain βs. There should be the same number of linearly independent equations as unknowns (if more equations than unknowns are used they will not be linearly independent).

Table 3.1a and b summarizes finite difference formulas of varying accuracy for the first and second-order derivatives. Formulas for higher derivatives (third, fourth, etc.) also exist but are outside the scope of this text. Example 3.3 demonstrates the derivation of a fourth-order accurate approximation to the second derivative.

TABLE 3.1A Centered finite difference formulas.

Derivative	Accuracy	−2	−1	0	1	2
1	2nd		−1/2	0	1/2	
	4th	1/12	−2/3	0	2/3	−1/12
2	2nd		1	−2	1	
	4th	−1/12	4/3	−5/2	4/3	−1/12

TABLE 3.1B Forward/backward finite difference formulas.

Derivative	Accuracy	−1	0	1	2	3	4	5
1	1st		−1	1				
	2nd		−3/2	2	−1/2			
	3rd		−11/6	3	−3/2	1/3		
	3rd[a]	−1/3	−1/2	1	−1/6			
	4th		−25/12	4	−3	4/3	−1/4	
2	1st		1	−2	1			
	2nd		12	−5	4	−1		
	3rd		35/12	−26/3	19/2	−14/3	11/12	
	4th		15/4	−77/6	107/6	−13	61/12	−5/6

[a]This is an alternate third-order accurate finite difference that is often employed. It is a hybrid of a forward and centered difference.

Example 3.3. Higher-order derivative approximations
Derive finite difference equations for the second derivative with fourth-order accuracy. Use a 5-point stencil with points $i - 2$, $i - 1$, i, $i + 1$, $i + 2$.

Solution
The approximation has the general form:

$$p_{xxi} = \frac{1}{(\Delta x^2)} \sum_{j=i-2}^{i+2} \beta_j p_j = \beta_{i-2} p_{i-2} + \beta_{i-1} p_{i-1} + \beta_i p_i + \beta_{i+1} p_{i+1} + \beta_{i+2} p_{i+2}$$

where each of the points can be described by a Taylor series:

$$p_{i-2} = p_i - (2\Delta x)p_{xi} + \frac{1}{2}(2\Delta x)^2 p_{xxi} - \frac{1}{6}(2\Delta x)^3 p_{xxxi} + \frac{1}{24}(2\Delta x)^4 p_{xxxxi} - \frac{1}{120}(2\Delta x)^5 p_{xxxxxi} + O(\Delta x^6)$$

$$p_{i-1} = p_i - \Delta x p_{xi} + \frac{1}{2}\Delta x^2 p_{xxi} - \frac{1}{6}\Delta x^3 p_{xxxi} + \frac{1}{24}(\Delta x)^4 p_{xxxxi} - \frac{1}{120}(\Delta x)^5 p_{xxxxxi} + O(\Delta x^6)$$

$$p_i = p_i$$

$$p_{i+1} = p_i + \Delta x p_{xi} + \frac{1}{2}\Delta x^2 p_{xxi} + \frac{1}{6}\Delta x^3 p_{xxxi} + \frac{1}{24}(\Delta x)^4 p_{xxxxi} - \frac{1}{120}(\Delta x)^5 p_{xxxxxi} + O(\Delta x^6)$$

$$p_{i+2} = p_i + (2\Delta x)p_{xi} + \frac{1}{2}(2\Delta x)^2 p_{xxi} + \frac{1}{6}(2\Delta x)^3 p_{xxxi} + \frac{1}{24}(2\Delta x)^4 p_{xxxxi} + \frac{1}{120}(2\Delta x)^5 p_{xxxxxi} + O(\Delta x^6)$$

Substitution into the approximation gives:

$$p_{xxi} \approx \frac{1}{(\Delta x^2)} \sum_{j=i-2}^{i+2} \beta_j p_j = \beta_{i-2}\left(p_i - (2\Delta x)p_{xi} + \frac{1}{2}(2\Delta x)^2 p_{xxi} - \frac{1}{6}(2\Delta x)^3 p_{xxxi} + \frac{1}{24}(2\Delta x)^4 p_{xxxxi} - \frac{1}{120}(2\Delta x)^5 p_{xxxxxi}\right)$$

$$+ \beta_{i-1}\left(p_i - \Delta x p_{xi} + \frac{1}{2}\Delta x^2 p_{xxi} - \frac{1}{6}\Delta x^3 p_{xxxi} + \frac{1}{24}(\Delta x)^4 p_{xxxxi} - \frac{1}{120}(\Delta x)^5 p_{xxxxxi}\right)$$

$$+ \beta_i p_i + \beta_{i+1}\left(p_i + \Delta x p_{xi} + \frac{1}{2}\Delta x^2 p_{xxi} + \frac{1}{6}\Delta x^3 P_{xxxi} + \frac{1}{24}(\Delta x)^4 p_{xxxxi} + \frac{1}{120}(\Delta x)^5 p_{xxxxxi}\right)$$

$$+ \beta_{i+2}\left(p_i + (2\Delta x)p_{xi} + \frac{1}{2}(2\Delta x)^2 p_{xxi} + \frac{1}{6}(2\Delta x)^3 p_{xxxi} + \frac{1}{24}(2\Delta x)^4 p_{xxxxi} + \frac{1}{120}(2\Delta x)^5 p_{xxxxxi}\right) + O(\Delta x^4)$$

Grouping like terms,

$$p_{xxi} = (\beta_{i-2} + \beta_{i-1} + \beta_i + \beta_{i+1} + \beta_{i+2})\frac{1}{\Delta x^2}p_i + (-2\beta_{i-2} - \beta_{i-1} + \beta_{i+1} + 2\beta_{i+2})\frac{1}{\Delta x}p_{xi}$$

$$+ \left(2\beta_{i-2} + \frac{1}{2}\beta_{i-1} + \frac{1}{2}\beta_{i+1} + 2\beta_{i+2}\right)p_{xxi} + \left(-\frac{8}{6}\beta_{i-2} - \frac{1}{6}\beta_{i-1} + \frac{1}{6}\beta_{i+1} + \frac{8}{6}\beta_{i+2}\right)\Delta x p_{xxxi}$$

$$+ \left(\frac{16}{24}\beta_{i-2} + \frac{1}{24}\beta_{i-1} + \frac{1}{24}\beta_{i+1} + \frac{16}{24}\beta_{i+2}\right)\Delta x^2 p_{xxxxi}$$

$$+ \left(-\frac{32}{120}\beta_{i-2} - \frac{1}{120}\beta_{i-1} + \frac{1}{120}\beta_{i+1} + \frac{32}{120}\beta_{i+2}\right)\Delta x^3 p_{xxxxxi} + O(\Delta x^4)$$

This gives a system of equations,

$$\begin{bmatrix} 1 & 1 & 1 & 1 & 1 \\ -2 & -1 & 0 & 1 & 2 \\ 2 & 0.5 & 0 & 0.5 & 2 \\ -\frac{8}{6} & -\frac{1}{6} & 0 & \frac{1}{6} & \frac{8}{6} \\ \frac{16}{24} & \frac{1}{24} & 0 & \frac{1}{24} & \frac{16}{24} \\ -\frac{32}{120} & -\frac{1}{120} & 0 & \frac{1}{120} & \frac{32}{120} \end{bmatrix} \begin{bmatrix} \beta_{i-2} \\ \beta_{i-1} \\ \beta_i \\ \beta_{i+1} \\ \beta_{i+2} \end{bmatrix} = \begin{bmatrix} 0 \\ 0 \\ 1 \\ 0 \\ 0 \\ 0 \end{bmatrix}$$

Continued

Example 3.3. Higher-order derivative approximations—cont'd

The system appears to be overdetermined because there are six equations and five unknowns; however, there are only five linearly independent equations. Solution gives

$$\beta_{i-2} = -1/12; \quad \beta_{i-1} = 4/3; \quad \beta_i = -5/2; \quad \beta_{i+1} = 4/3; \quad \beta_{i+2} = -1/12$$

Therefore, the fourth-order second derivative approximation is:

$$p_{xxi} = \frac{-p_{i-2} + 16p_{i-1} - 30\,p_i + 16p_{i+1} - p_{i+2}}{12\,(\Delta x)^2} + O(\Delta x^4)$$

3.3 Discretization of the parabolic diffusivity (heat) equation

The PDE describing 1D single-phase, slightly compressible flow in a homogenous porous medium without sources or sinks was derived in Chapter 2.

$$\frac{\partial p}{\partial t} = \alpha \frac{\partial^2 p}{\partial x^2} \tag{3.14}$$

The PDE is well posed if one initial condition and two boundary conditions are included, provided at least one of those initial or boundary conditions defines pressure.

Finite differencing is one numerical technique to find approximate solutions to these PDEs. The goal is to transform the PDE into algebraic equations using approximations for the partial derivatives. This can be accomplished by spatially discretizing the reservoir into discrete *grid blocks* (also referred to as *cells* or *elements*) each with length Δx_i and discretizing the time into intervals of Δt. Fig. 3.2 shows a 1D porous medium (e.g., rock core sample) with length L discretized into N equally spaced grid blocks.

Fig. 3.2 also illustrates the grid blocks for the 1D problem and a qualitative solution for a pressure profile, $p(x)$, at some time, t. For the numerical solution, it is assumed the pressure is constant in each block i and over each time interval, Δt; then an algebraic equation is written for each block. In 1D flow, the grid blocks are, in general, connected to two neighbors: $i - 1$ to the left of the block and $i + 1$ to the right. In this book, it is generally assumed that grid points are placed exactly at the center of the grid block and are therefore referred to as *cell-centered* or *block-centered* grids. Reservoir properties and state variables (porosity, pressure, saturation, etc.) are defined at the block centers. The 1D reservoir in Fig. 3.2 is discretized into N equal-sized blocks. The points in the figure refer to the approximate block pressures and the curve represents the true solution to a PDE.

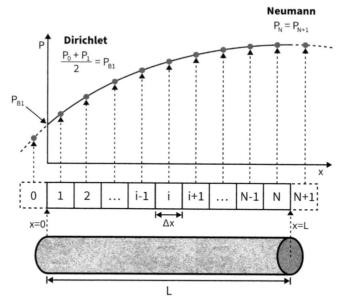

FIGURE 3.2 1D porous medium (core) discretized into N equally spaced grid blocks where $\Delta x = L/N$. *Blocks 0 and N+1 are imaginary/ghost blocks outside the computational domain for handling boundary conditions.*

Using the finite difference approximations derived in Section 3.2, the spatial derivatives can be discretized using a centered difference approximation which was shown to be of $O(\Delta x^2)$ in accuracy based on the truncated terms in the Taylor series,

$$\frac{\partial^2 p}{\partial x^2} = \frac{P_{i-1} - 2P_i + P_{i+1}}{(\Delta x)^2},$$
(3.15a)

where P is used to denote the approximate solution for pressure. Note the important switch from lower-case p (true, exact pressure) to upper-case P (approximate, finite difference pressure).

Likewise, the PDE is solved in discrete increments of time, each separated by an interval Δt. A time interval with a superscript "n" is denoted and the next time level as "$n + 1$". A finite difference approximation in time can be employed. Note that the superscript indicates the time level and the subscript the spatial location. Discretization of the time derivative gives,

$$\frac{\partial p}{\partial t} = \frac{P_i^{n+1} - P_i^n}{\Delta t}.$$
(3.15b)

Eq. (3.15b) is a forward difference approximation if the current time is envisioned to be n and the future time to be $n + 1$, is a backward difference if the current time is $n + 1$ and the previous time is n, or is a centered difference if the current time is $n + 1/2$. The backward (explicit method), forward (implicit method), and centered (Crank-Nicolson method) time difference approximations are the starting points for the numerical solutions described below and throughout this text.

Eqs. (3.15a) and (3.15b) can be substituted directly into the 1D diffusivity Eq. (3.14) as shown in Eq. (3.16),

$$P_i^{n+1} - P_i^n = \eta P_{i-1} - 2\eta P_i + \eta P_{i+1}, \tag{3.16}$$

where the dimensionless diffusivity constant (sometimes called the Fourier number), η, is defined as,

$$\eta = \frac{\alpha \Delta t}{(\Delta x)^2}. \tag{3.17}$$

Each term in Eq. (3.16) has units of pressure. It is often more convenient to write the equations in units of flow rate (e.g., ft^3/day or RB/day); after all, a primary interest of the reservoir engineer is to determine the *rate* of oil, gas, and water produced. Multiplying Eq. (3.16) by A and c_t,

$$Ac_t\left(P_i^{n+1} - P_i^n\right) = TP_{i-1} - 2TP_i + TP_{i+1}, \tag{3.18}$$

where the accumulation, A, and transmissibility, T, are defined by,

$$A = \frac{a \Delta x \phi}{\Delta t}; \quad T = Ac_t \eta = \frac{ka}{\mu \Delta x}, \tag{3.19}$$

where c_t is the total compressibility defined by Eq. (1.18), and $a = \Delta y \Delta z$ is the cross-sectional area perpendicular to the flow. Eqs. (3.16) and (3.18) are equivalent, but Eq. (3.18) will be used throughout this text.

Given an initial condition (pressures of all blocks at time n), Eq. (3.18) can be solved for P_i^{n+1} in all blocks with $i = 1$ to N. However, it must be determined at which time level (n or $n + 1$) the right-hand side pressures of Eq. (3.18) should be evaluated. Additionally, in blocks $i = 1$ and $i = N$, a pressure in a nonexistent ghost block, $P_{i-1} = P_0$ and $P_{i+1} = P_{N+1}$, respectively, must be evaluated using boundary conditions.

3.4 Boundary and initial conditions

Boundary conditions must be applied to solve any PDE. For the 1D parabolic diffusivity equation, two boundary conditions (one at $x = 0$ and the other at $x = L$) in addition to one initial condition are needed to define the problem. There are many different possibilities for boundary conditions, but the most common that are applied to flow in reservoir simulation are Dirichlet (constant pressure) and Neumann (constant rate).

3.4.1 Dirichlet boundary condition

A Dirichlet boundary condition is one of constant pressure. Consider Fig. 3.2, which represents grids in a 1D finite difference reservoir simulator. At $x = 0$, the pressure is specified as a constant, $p = P_{B1}$, and is an example of a Dirichlet boundary condition. For cell-centered grids, boundary pressure is not defined for any grid point in the reservoir because the grid point is at the center of the block while the boundary conditions are imposed at the edges (i.e., $x = 0$ or $x = L$). A crude approximation is to specify the pressure of the left ghost block as being equal to the specified boundary condition (in this example, $P_0 = P_{B1}$). However, this results in first-order error, $O(\Delta x)$, since the center of the ghost block is actually located at $x = -\Delta x/2$ instead of $x = 0$. A better approximation to the boundary condition is to specify the pressure at the edge as an arithmetic average of the pressure in the boundary block (#1) and the pressure in a ghost block (#0) outside the reservoir as shown in Fig. 3.2 (left). Therefore, for a Dirichlet boundary condition at $x = 0$, one can write,

$$P_{B1} = \frac{P_0 + P_1}{2} \quad \Rightarrow \quad P_0 = 2P_{B1} - P_1, \tag{3.20}$$

and an equivalent expression can be written for a Dirichlet condition at $x = L$. This approximation is of error $O(\Delta x^2)$, which is more accurate than using the crude approximation of making $P_0 = P_{B1}$. Eq. (3.18) includes the ghost block pressure (P_0) if $i = 1$; Eq. (3.20) can be substituted into Eq. (3.18) if the boundary is subject to a Dirichlet boundary condition.

3.4.2 Neumann boundary condition

A Neumann boundary condition is one of constant rate (or flux), e.g., $q = q_{B2}$ at $x = L$. Darcy's law states that flux is proportional to pressure gradient,

$$\frac{\partial p}{\partial x} \approx \frac{P_N - P_{N+1}}{\Delta x} = \frac{q_{B2}\mu}{ka} \quad \Rightarrow \quad P_{N+1} = P_N - \frac{1}{T}q_{B2}, \tag{3.21}$$

where a centered finite difference approximation is used for the first derivative. Eq. (3.21) may "look like" a forward or backward difference, but because the flux is specified at the boundary (i.e., the interface between block "N" and "$N + 1$") it is a centered difference approximation with accuracy $O(\Delta x^2)$.

If the Neumann boundary condition is simply a "no flux" condition, i.e., $q_{B2} = 0$, a mirroring technique ($P_N = P_{N+1}$) can be used as shown in Fig. 3.2 ($x = L$). This occurs for a sealed or insulated boundary. Physically, this may represent the beginning of an impermeable zone, like shale, that does not allow for a significant amount of flow.

3.4.3 Robin boundary conditions

Dirichlet and Neumann are the simplest and most common boundary conditions employed in reservoir simulation, but a few other boundary conditions are used in special situations. The *Robin* boundary condition is a combination of the Dirichlet and Neumann conditions,

$$b\frac{dp}{dx} + cp = g(t), \tag{3.22}$$

where b and c are assumed constants here (although they could be functions) and $g(t)$ is potentially a time-dependent function. Eq. (3.22) reduces to a Dirichlet condition when $b = 0$ and $g/c = P_B$ and to a Neumann condition when $c = 0$ and $g/b = -q_B\mu/(ka)$.

For a Robin boundary condition at the $x = L$ boundary, the finite difference approximation can then be written as,

$$b\frac{P_{N+1} - P_N}{\Delta x} + c\frac{P_N + P_{N+1}}{2} = g(t), \tag{3.23}$$

Solving for P_{N+1},

$$P_{N+1} = \left(\frac{2b - c\Delta x}{2b + c\Delta x}\right)P_N + \left(\frac{2\Delta x}{2b + c\Delta x}\right)g(t), \tag{3.24}$$

which can be substituted into the finite difference equation to eliminate the imaginary block pressure P_{N+1}. An equivalent expression can be written for ghost block P_0 if the boundary at $x = 0$ is subject to a Robin boundary condition.

Physically, the Robin boundary condition might refer to a reservoir that has a constant pressure far from the boundary, such as an aquifer, although Dirichlet and Neumann boundary conditions are also used for aquifers. In heat transfer problems, this boundary condition is even more common, representing heat flux from both conduction and convection.

3.5 Solution methods

3.5.1 Explicit solution to the diffusivity equation

The porous medium is discretized into N grid blocks; therefore, there are N equations, each associated with a corresponding block in the discretized reservoir simulator. To solve for pressures in each grid block i (i.e., block $i = 1, 2, ...,$ N) using the explicit method, Eq. (3.16) is written for each grid block,

$$P_1^{n+1} = P_1^n + \frac{T}{Ac_t}\left[P_0^n - 2P_1^n + P_2^n\right]$$

$$P_2^{n+1} = P_2^n + \frac{T}{Ac_t}\left[P_1^n - 2P_2^n + P_3^n\right]$$

$$P_3^{n+1} = P_3^n + \frac{T}{Ac_t}\left[P_2^n - 2P_3^n + P_4^n\right] \tag{3.25a}$$

$$\vdots$$

$$P_N^{n+1} = P_N^n + \frac{T}{Ac_t}\left[P_{N-1}^n - 2P_N^n + P_{N+1}^n\right],$$

where the spatial derivatives on the right hand side of the equation are evaluated at the n time level in the explicit method.

Since initial conditions ($n = 0$) for pressure as a function of position are known, the solution of grid block pressures ($P_1^0, P_2^0, ..., P_N^0$) is known. These initial conditions can be substituted into the above Eq. (3.25a) and the right-hand side can be evaluated. Therefore, new pressures can be computed *explicitly* at the new time level, $n + 1$; the equations are not "coupled" (i.e., each equation is independent and does not depend on the others).

Boundary conditions are used to evaluate pressures outside the boundary (ghost blocks), P_0 and P_{N+1}. For brevity, it is possible to write the algebraic Eq. (3.25a), with boundary conditions, in matrix form. The calculation of the new pressure vector, P^{n+1}, requires matrix/vector multiplication and addition/subtraction, but no "system of equations" must be solved.

$$
\underbrace{\begin{pmatrix} P_1 \\ P_2 \\ P_3 \\ \vdots \\ P_N \end{pmatrix}^{n+1}}_{\vec{P}^{n+1}} = \underbrace{\begin{pmatrix} P_1 \\ P_2 \\ P_3 \\ \vdots \\ P_2 \end{pmatrix}^{n}}_{\vec{P}^{n}} + \underbrace{\begin{pmatrix} Ac_t & & & & \\ & Ac_t & & & \\ & & Ac_t & & \\ & & & Ac_t & \\ & & & & Ac_t \end{pmatrix}^{-1}}_{Ac_t}
$$

$$
\left\{ - \left[\underbrace{\begin{pmatrix} T & -T & 0 & 0 & 0 \\ -T & 2T & -T & 0 & 0 \\ 0 & -T & 2T & -T & 0 \\ 0 & 0 & \ddots & \ddots & \ddots \\ 0 & 0 & 0 & -T & T \end{pmatrix}}_{\mathbf{T}} + \underbrace{\begin{pmatrix} J_1 & 0 & 0 & 0 & 0 \\ 0 & 0 & 0 & 0 & 0 \\ 0 & 0 & 0 & 0 & 0 \\ 0 & 0 & \ddots & \ddots & \ddots \\ 0 & 0 & 0 & & J_N \end{pmatrix}}_{\mathbf{J}} \right] \underbrace{\begin{pmatrix} P_1 \\ P_2 \\ P_3 \\ \vdots \\ P_2 \end{pmatrix}^{n}}_{\vec{P}^{n}} + \underbrace{\begin{pmatrix} Q_1 \\ Q_2 \\ \vdots \\ Q_N \end{pmatrix}}_{\vec{Q}} \right\}
$$

$$
(3.25b)
$$

where, from Eq. (3.22), $J_i = 2T$ and $Q_i = 2TP_{Bi}$ for a Dirichlet and $J_i = 0$ and $Q_i = q_{Bi}$, for a Neumann boundary condition. In matrix notation,

$$
\vec{P}^{n+1} = \vec{P}^{n} + (\mathbf{A}\mathbf{c}_t)^{-1} \left[-(\mathbf{T} + \mathbf{J}) \vec{P}^{n} + \vec{Q} \right] \qquad (3.25c)
$$

where $\mathbf{A}$ is the diagonal accumulation matrix, $\mathbf{c}_t$ is the diagonal compressibility matrix, $\mathbf{T}$ is a tridiagonal transmissibility matrix, $\mathbf{J}$ is a diagonal productivity matrix, $\mathbf{I}$ is the identity matrix, and Q is a source vector. The accumulation and compressibility matrix can be combined into one diagonal matrix; however, in later chapters involving compositional and multiphase flow, **A** alone will be needed in some equations. In summary, the initial, known pressure field P^0 is used to calculate P^1 at the next time level using Eq. (3.25). P^1 is then used to calculate P^2, and so forth until the end of the simulation. Example 3.4 demonstrates the solution of a 4-block reservoir using the explicit method.

Example 3.4. Explicit Method: 1D, single-phase homogeneous flow

A 1D reservoir has reservoir length, L, 4000 ft, width, w, 1000 ft, and thickness, h, of 20 feet and the following reservoir and fluid properties: $\phi = 0.2$, $k = 100$ mD, $\mu_o = 5$ cp, $c_t = 10^{-5}$ psi^{-1}. The reservoir is saturated with oil and is at residual water saturation ($S_w = S_{wr} = 0.10$). The endpoint relative permeability to oil, $k_{ro}^0 = 1.0$. The initial condition is $P = 3000$ psia. The boundary conditions are $P = 1000$ psia at $x = L$ and no flow ($q = 0$) at $x = 0$. Using a time step of $\Delta t = 5.0$ days, determine the pressure field in the reservoir using four uniform blocks and the explicit method.

Solution

The pressure is governed by the 1D diffusivity Eq. (2.16) and can be solved by using finite differences. The algebraic equation for the explicit method is given by Eq. (3.25a).

There is no "block #0" or "block #5". We must use our boundary conditions:

(1) At $x = 0$, we have a no-flow Neumann boundary,

$$\frac{P_0 - P_1}{\Delta x} = 0 \Rightarrow P_0 = P_1$$

(2) At $x = L$, we have a constant pressure boundary $P = P_B = 1000$ psia

$$P_B = \frac{P_4 + P_5}{2} \Rightarrow P_5 = 2P_B - P_4$$

Substitution of the boundary conditions into the algebraic equations gives,

$$P_1^{n+1} = P_1^n + \frac{T}{Ac_t}\left[P_1^n - 2P_1^n + P_2^n\right] = P_1^n + \frac{T}{Ac_t}\left[-P_1^n + P_2^n\right]$$

$$P_2^{n+1} = P_2^n + \frac{T}{Ac_t}\left[P_1^n - 2P_2^n + P_3^n\right]$$

$$P_3^{n+1} = P_3^n + \frac{T}{Ac_t}\left[P_2^n - 2P_3^n + P_4^n\right]$$

$$P_4^{n+1} = P_4^n + \frac{T}{Ac_t}\left[P_3^n - 2P_4^n + (2P_B - P_4^n)\right]$$

$$= P_4^n + \frac{T}{Ac_t}\left[P_3^n - P_4^n\right] + \frac{T}{Ac_t}\left(-2P_B - 2P_4^n\right)$$

$$= P_4^n + \frac{1}{Ac_t}\left(T\left[P_3^n - P_4^n\right] - JP_4^n - Q_4\right)$$

where $J_4 = 2T$ and $Q_4 = 2TP_B$

Calculation of the scalar parameters

$$Ac_t = \frac{(hw)\Delta x \phi c_t}{\Delta t} = \frac{20 \text{ ft} \cdot 1000 \text{ ft} \cdot 1000 \text{ ft} \cdot 0.2 \cdot 1.0 \times 10^{-5} \text{ psi}^{-1}}{5.0 \text{ days}} = 8 \frac{\text{ft}^3}{\text{psi} - \text{day}}$$

$$T = \frac{kk_{ro}^0(hw)}{\mu_o \Delta x} = \frac{100 \text{ mD} \cdot 1 \cdot 20 \text{ ft} \cdot 1000 \text{ ft}}{5 \text{ cp} \cdot 1000 \text{ ft}} = 400 \frac{\text{mD} - \text{ft}}{\text{cp}} = 2.532 \frac{\text{ft}^3}{\text{psi} - \text{day}}$$

$$J_4 = 2T = 800 \frac{\text{mD} - \text{ft}}{\text{cp}} = 5.064 \frac{\text{ft}^3}{\text{psi} - \text{day}}$$

$$Q_4 = 2TP_B = 800,000 \frac{\text{mD} - \text{ft}}{\text{cp}} = 5,064 \frac{\text{ft}^3}{\text{psi} - \text{day}}$$

For the explicit method, the stability requirement must be met,

$$\eta = \frac{\alpha \Delta t}{\Delta x^2} = \frac{kk_{ro}^0 hw \Delta t}{\mu_o \phi c_t \Delta x^2} = \frac{T}{Ac_t} = \frac{2.532 \frac{\text{ft}^3}{\text{psi} - \text{day}}}{8 \frac{\text{ft}^3}{\text{psi} - \text{day}}} = 0.3165 < 0.5$$

Since the dimensionless diffusivity constant is less than 0.5, the explicit method is stable. Stability is discussed in Section 3.6.

(a) Explicit Method

At $t = 0$ days, the pressure is uniform at 3000 psia. $P^0 = [3000 \ 3000 \ 3000 \ 3000]$ psia.

At $t = 5.0$ days $(n = 1)$, we solve explicitly for the pressures

$$P_1^{n+1} = P_1^n + \frac{T}{Ac_t}\left[-P_1^n + P_2^n\right] = 3000 + 0.3165[-3000 + 3000] = 3000 \text{ psia}$$

$$P_2^{n+1} = P_2^n + \frac{T}{Ac_t}\left[P_1^n - 2P_2^n + P_3^n\right] = 3000 + 0.3165[3000 - 2 \cdot 3000 + 3000] = 3000 \text{ psia}$$

$$P_3^{n+1} = P_3^n + \frac{T}{Ac_t}\left[P_2^n - 2P_3^n + P_4^n\right] = 3000 + 0.3165[3000 - 2 \cdot 3000 + 3000] = 3000 \text{ psia}$$

$$P_4^{n+1} = P_4^n + \frac{1}{Ac_t}\left(T\left[P_3^n - P_4^n\right] - JP_4^n + Q_4\right) = 3000$$

$$+ \ 0.3165([3000 - 3000] - 2 \cdot 3000 - 2 \cdot 1000) = 1734 \text{ psia}$$

At $t = 10.0$ days $(n = 2)$

$$P_1^{n+1} = P_1^n + \frac{T}{Ac_t}\left[-P_1^n + P_2^n\right] = 3000 + 0.3165[-3000 + 3000] = 3000 \text{ psia}$$

$$P_2^{n+1} = P_2^n + \frac{T}{Ac_t}\left[P_1^n - 2P_2^n + P_3^n\right] = 3000 + 0.3165[3000 - 2 \cdot 3000 + 3000] = 3000 \text{ psia}$$

$$P_3^{n+1} = P_3^n + \frac{T}{Ac_t}\left[P_2^n - 2P_3^n + P_4^n\right] = 3000 + 0.3165[3000 - 2 \cdot 3000 + 1734] = 2599.3 \text{ psia}$$

$$P_4^{n+1} = P_4^n + \frac{1}{Ac_t}\left(T\left[P_3^n - P_4^n\right] - JP_4^n - Q_4\right) = 3000$$

$$+ \ 0.3165([3000 - 1734] - 2 \cdot 1734 - 2 \cdot 1000) = 1670 \text{ psia}$$

Continued

Example 3.4. Explicit Method: 1D, single-phase homogeneous flow—cont'd

At $t = 15.0$ days ($n = 3$)

$$P_1^{n+1} = P_1^n + \frac{T}{Ac_t}\left[-P_1^n + P_2^n\right] = 3000 + 0.3165[-3000 + 3000] = 3000 \text{ psia}$$

$$P_2^{n+1} = P_2^n + \frac{T}{Ac_t}\left[P_1^n - 2P_2^n + P_3^n\right] = 3000 + 0.3165[3000 - 2\cdot3000 + 2599.3] = 2873 \text{ psia}$$

$$P_3^{n+1} = P_3^n + \frac{T}{Ac_t}\left[P_2^n - 2P_3^n + P_4^n\right] = 3000 + 0.3165[3000 - 2\cdot2599.3 + 1670] = 2432 \text{ psia}$$

$$P_4^{n+1} = P_4^n + \frac{1}{Ac_t}\left(T\left[P_3^n - P_4^n\right] - JP_4^n - Q_4\right) = 3000$$

$$+ 0.3165([3000 - 1670] - 2\cdot1670 - 2\cdot1000) = 1540 \text{ psia}$$

At $t = \infty$, the reservoir comes to equilibrium at $P = 1000$ psia in all blocks.

3.5.2 Implicit solution to the diffusivity equation

To solve for pressures in each grid block i (i.e., block $i = 1, 2, ..., N$) using the implicit method, Eq. (3.18) is written for each grid block,

$$-TP_0^{n+1} + (Ac_t + 2T)P_1^{n+1} - TP_2^{n+1} = Ac_t P_1^n$$
$$-TP_1^{n+1} + (Ac_t + 2T)P_2^{n+1} - TP_3^{n+1} = Ac_t P_2^n$$
$$-TP_2^{n+1} + (Ac_t + 2T)P_3^{n+1} - TP_4^{n+1} = Ac_t P_3^n$$
$$\vdots$$
$$-TP_{N-1}^{n+1} + (Ac_t + 2T)P_N^{n+1} - TP_{N+1}^{n+1} = Ac_t P_N^n \tag{3.26a}$$

where the superscript $n + 1$ on the spatial derivative terms indicates they are evaluated at the $n + 1$ time level. New pressures can be computed *implicitly* at the new time level, $n + 1$; the equations are "coupled" (i.e., each equation is dependent on the others). A system of linear equations must be solved. In expanded matrix form these equations can be written as,

$$\left[\underbrace{\begin{pmatrix} Ac_t & & & & \\ & Ac_t & & & \\ & & Ac_t & & \\ & & & Ac_t & \\ & & & & Ac_t \end{pmatrix}}_{Ac_t} + \underbrace{\begin{pmatrix} T & -T & 0 & 0 & 0 \\ -T & 2T & -T & 0 & 0 \\ 0 & -T & 2T & -T & 0 \\ 0 & 0 & \ddots & \ddots & \ddots \\ 0 & 0 & 0 & -T & T \end{pmatrix}}_{T} + \underbrace{\begin{pmatrix} J_1 & 0 & 0 & 0 & 0 \\ 0 & 0 & 0 & 0 & 0 \\ 0 & 0 & 0 & 0 & 0 \\ 0 & 0 & \ddots & \ddots & \ddots \\ 0 & 0 & 0 & & J_N \end{pmatrix}}_{J}\right]$$

$$\times \underbrace{\begin{pmatrix} P_1 \\ P_2 \\ P_3 \\ \vdots \\ P_N \end{pmatrix}^{n+1}}_{\overrightarrow{P}^{n+1}} = \underbrace{\begin{pmatrix} Ac_t & & & & \\ & Ac_t & & & \\ & & Ac_t & & \\ & & & Ac_t & \\ & & & & Ac_t \end{pmatrix}}_{Ac_t} \underbrace{\begin{pmatrix} P_1 \\ P_2 \\ P_3 \\ \vdots \\ P_N \end{pmatrix}^{n}}_{\overrightarrow{P}^{n}} + \underbrace{\begin{pmatrix} Q_1 \\ \\ \\ \vdots \\ Q_N \end{pmatrix}}_{\overrightarrow{Q}}, \tag{3.26b}$$

or in matrix notation,

$$(\mathbf{Ac}_t + \mathbf{T} + \mathbf{J})\vec{P}^{n+1} = \mathbf{Ac}_t\,\vec{P}^{n} + \vec{Q} \tag{3.26c}$$

where the matrices and vectors in Eq. (3.26c) are the same as those in Eq. (3.25c).

3.5.3 Mixed methods and Crank–Nicolson

The time derivative in the diffusivity equation was approximated as a forward difference in the explicit method and a backward difference in the implicit method. As a result, both methods are only first-order accurate in time but different answers will be obtained when using each of the methods. To improve the accuracy of the solution in time without using smaller timesteps, a weighted average of the two methods can be used. Multiplying the explicit method by θ (a weighting factor between 0 and 1) and the implicit method by $1 - \theta$ gives,

$$\theta\left[\vec{P}^{n+1} = (\mathbf{Ac}_t)^{-1}\left\{(\mathbf{Ac}_t - \mathbf{T} - \mathbf{J})\vec{P}^{n} + \vec{Q}\right\}\right] +$$
$$(1-\theta)\left[(\mathbf{Ac}_t + \mathbf{T} + \mathbf{J})\vec{P}^{n+1} = \mathbf{Ac}_t\,\vec{P}^{n} + \vec{Q}\right] \tag{3.27a}$$

or after rearranging,

$$[(1-\theta)(\mathbf{T}+\mathbf{J}) + \mathbf{Ac}_t]\vec{P}^{n+1} = [\mathbf{Ac}_t - \theta(\mathbf{T}+\mathbf{J})]\vec{P}^{n} + \vec{Q}. \tag{3.27b}$$

Eq. (3.27b) reduces to the implicit method if $\theta = 0$ and the explicit method if $\theta = 1$; it is a mixed method for $0 < \theta < 1$. If $\theta = \frac{1}{2}$, the method is known as the Crank–Nicolson (C–N) scheme and has accuracy $O(\Delta t^2)$. For all mixed methods, a system of equations must still be solved and has slightly more computational work compared to the implicit method, but larger timesteps can be taken in C–N and maintain accuracy.

The additional accuracy obtained by using C–N might be more easily understood if the method is viewed as using a centered difference approximation in time, where the time derivative is centered around the $n + 1/2$ time level.

$$\frac{\partial p^{n+\frac{1}{2}}}{\partial t} \approx \frac{P^{n+1} - P^n}{\Delta t} \tag{3.28}$$

Since this is a centered difference, the time derivative is $O(\Delta t^2)$ in accuracy. The spatial derivative is also at the $n+1/2$ time level; intermediate time levels do not exist but can be approximated as an average of the "n" and "$n + 1$" time intervals.

$$\frac{\partial^2 p^{n+\frac{1}{2}}}{\partial x^2} \approx \frac{P_{i-1}^{n+\frac{1}{2}} - 2P_i^{n+\frac{1}{2}} + P_{i+1}^{n+\frac{1}{2}}}{(\Delta x)^2} \approx \frac{P_{i-1}^{n} - 2P_i^{n} + P_{i+1}^{n}}{2(\Delta x)^2} + \frac{P_{i-1}^{n+1} - 2P_i^{n+1} + P_{i+1}^{n+1}}{2(\Delta x)^2} \tag{3.29}$$

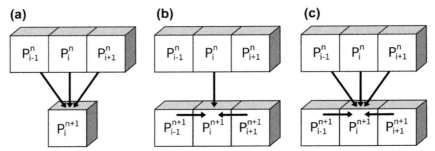

FIGURE 3.3 Schematic of solution method for the (A) explicit method which calculates the pressure in the new time level based on all pressures at the old time level and (B) implicit method which solves for new pressure using old and new (unknown) neighboring pressures, (C) mixed method which is a weighted average of the explicit and implicit equations.

After some algebra, one obtains the C−N Eq. (3.27a,b with $\theta = 1/2$) which are $O(\Delta t^2)$ accuracy in time and $O(\Delta x^2)$ accurate in space. Fig. 3.3 illustrates a comparison of the three methods (explicit, implicit, and mixed). Example 3.5 demonstrates the solution for a 4-block reservoir using the explicit, implicit and C-N methods.

Example 3.5. Implicit and C−N Methods: 1D, single-phase homogeneous flow

Repeat Example 3.4 but use the explicit, implicit and Crank-Nicolson methods.

Solution

The matrix equations, (3.27b), can be used for the explicit ($\theta = 1$), implicit ($\theta = 0$), or Crank−Nicolson ($\theta = 1/2$) methods. The matrices are given by,

$$\mathbf{T} = \begin{pmatrix} T & -T & & \\ -T & 2T & -T & \\ & -T & 2T & -T \\ & & -T & T \end{pmatrix} ; \mathbf{Ac}_t = \begin{pmatrix} Ac_t & & & \\ & Ac_t & & \\ & & Ac_t & \\ & & & Ac_t \end{pmatrix} ;$$

$$\mathbf{J} = \begin{pmatrix} J_1 & & \\ & \ddots & \\ & & J_4 \end{pmatrix} ; \vec{Q} = \begin{pmatrix} Q_1 \\ \vdots \\ Q_4 \end{pmatrix}$$

where $T = 2.532$ ft³/psi-day and $Ac_t = 8.0$ ft³/psi-day were defined in Example 3.4; $J_1 = 0$, $J_4 = 2T = 5.064$ ft³/psi-day, $Q_1 = 0$, and $Q_4 = 2TP_B = 5064$ ft³/day

(1) Explicit solution

The matrix equations for the explicit method can be written

$$\vec{P}^{n+1} = \vec{P}^n + (\mathbf{Ac}_t)^{-1} \left[-(\mathbf{T} + \mathbf{J}) \vec{P}^n + \vec{Q} \right]$$

Plugging in the matrices and vectors and starting with P^0 results in the exact same solution as obtained in Example 3.4. The solution requires matrix-vector multiplication, but no system of linear equations must be solved.

(2) Implicit Solution

The equations for each block can be written as a system of linear equations for the implicit method

$$(\mathbf{Ac}_t + \mathbf{T} + \mathbf{J})\,\vec{P}^{n+1} = \mathbf{Ac}_t\,\vec{P}^n + \vec{Q}$$

The terms on the right $(\mathbf{Ac}_t P^n + Q)$ are vectors and change with time because P^n changes with time. However, the terms on the left in parentheses do not change with time,

$$\mathbf{Ac}_t + \mathbf{T} + \mathbf{J} = \begin{pmatrix} 10.532 & -2.532 & 0 & 0 \\ -2.532 & 13.064 & -2.532 & 0 \\ 0 & -2.532 & 13.064 & -2.532 \\ 0 & 0 & -0.2532 & 15.596 \end{pmatrix} \frac{ft^3}{psi - day}$$

The system of equations must be solved using a linear solver (e.g., Gauss Elimination). Here we use a computer program with a linear algebra package to solve the equations.

At $t = 0$, $P^n = P^0$

At $t = 5$ days, the solution can be found by solving the system of equations:

$$\begin{pmatrix} 10.532 & -2.532 & 0 & 0 \\ -2.532 & 13.064 & -2.532 & 0 \\ 0 & -2.532 & 13.064 & -2.532 \\ 0 & 0 & -0.2532 & 15.596 \end{pmatrix} \begin{pmatrix} P_1^1 \\ P_2^1 \\ P_3^1 \\ P_4^1 \end{pmatrix}$$

$$= \begin{pmatrix} 8 & & & \\ & 8 & & \\ & & 8 & \\ & & & 8 \end{pmatrix} \begin{pmatrix} 3000 \\ 3000 \\ 3000 \\ 3000 \end{pmatrix} + \begin{pmatrix} 0 \\ 0 \\ 0 \\ 2 \cdot 2.532 \cdot 1000 \end{pmatrix}$$

which has the solution $P^1 = [2993.3\ 2972.4\ 2864.5\ 2328.6]$ psia.

At $t = 10$ days, use the solution at the previous time step. Note that the matrix remains the same, but the "right hand side (RHS)" vector changes from the previous time step

Continued

Example 3.5. Implicit and C–N Methods: 1D, single-phase homogeneous flow—cont'd

$$
\begin{pmatrix}
10.532 & -2.532 & 0 & 0 \\
-2.532 & 13.064 & -2.532 & 0 \\
0 & -2.532 & 13.064 & -2.532 \\
0 & 0 & -0.2532 & 15.596
\end{pmatrix}
\begin{pmatrix}
P_1^1 \\
P_2^1 \\
P_3^1 \\
P_4^1
\end{pmatrix}
$$

$$
=
\begin{pmatrix}
8 & & & \\
& 8 & & \\
& & 8 & \\
& & & 8
\end{pmatrix}
\begin{pmatrix}
2993.3 \\
2972.4 \\
2864.5 \\
2328.6
\end{pmatrix}
+
\begin{pmatrix}
0 \\
0 \\
0 \\
2 \cdot 2.532 \cdot 1000
\end{pmatrix}
$$

which has the solution $P^2 = [2975.7; 2920.2; 2699.5; 1957.4]$ psia.
At $t = 15$ days, use the solution at the previous time step. Note that the η matrix remains the same.

$$
\begin{pmatrix}
10.532 & -2.532 & 0 & 0 \\
-2.532 & 13.064 & -2.532 & 0 \\
0 & -2.532 & 13.064 & -2.532 \\
0 & 0 & -0.2532 & 15.596
\end{pmatrix}
\begin{pmatrix}
P_1^1 \\
P_2^1 \\
P_3^1 \\
P_4^1
\end{pmatrix}
$$

$$
=
\begin{pmatrix}
8 & & & \\
& 8 & & \\
& & 8 & \\
& & & 8
\end{pmatrix}
\begin{pmatrix}
2975.7 \\
2920.2 \\
2699.5 \\
1957.4
\end{pmatrix}
+
\begin{pmatrix}
0 \\
0 \\
0 \\
2 \cdot 2.532 \cdot 1000
\end{pmatrix}
$$

which has the solution $P^3 = [2946.0; 2852.2; 2543.4; 1741.6]$ psia.
Eventually, the solution reaches steady state and $P^{ss} = [1000; 1000; 1000; 1000]$ psia.

(3) Crank–Nicolson Solution

The equations for each block can be written as a system of linear equations in the Crank–Nicolson method

$$
[0.5(\mathbf{T} + \mathbf{J}) + \mathbf{Ac}_t]\vec{P}^{n+1} = [\mathbf{Ac}_t - 0.5(\mathbf{T} + \mathbf{J})]\vec{P}^n + \vec{Q}
$$

Again, the terms on the right change with time because P^n changes with time. However, the terms on the left in parentheses do not change with time,

$$
\mathbf{Ac}_t + 0.5(\mathbf{T} + \mathbf{J}) =
\begin{pmatrix}
9.266 & -1.266 & 0 & 0 \\
-1.266 & 10.532 & -1.266 & 0 \\
0 & -1.266 & 10.532 & -1.266 \\
0 & 0 & -1.266 & 11.798
\end{pmatrix}
\frac{ft^3}{psi - day}
$$

$$\mathbf{Ac}_t - 0.5(\mathbf{T} + \mathbf{J}) = \begin{pmatrix} 6.734 & 1.266 & 0 & 0 \\ 1.266 & 5.468 & 1.266 & 0 \\ 0 & 1.266 & 5.468 & 1.266 \\ 0 & 0 & 1.266 & 4.202 \end{pmatrix} \frac{\text{ft}^3}{\text{psi} - \text{day}}$$

Again, the system of equations must be solved using a linear solver (e.g., Gauss Elimination). Here we use a computer program to solve the equations.

At $t = 0$, $P^n = P^0$

At $t = 5$ days, the solution can be found by solving the system of equations:

$$\begin{pmatrix} 9.266 & -1.266 & 0 & 0 \\ -1.266 & 10.532 & -1.266 & 0 \\ 0 & -1.266 & 10.532 & -1.266 \\ 0 & 0 & -1.266 & 11.798 \end{pmatrix} \begin{pmatrix} P_1^1 \\ P_2^1 \\ P_3^1 \\ P_4^1 \end{pmatrix}$$

$$= \begin{pmatrix} 6.734 & 1.266 & 0 & 0 \\ 1.266 & 5.468 & 1.266 & 0 \\ 0 & 1.266 & 5.468 & 1.266 \\ 0 & 0 & 1.266 & 4.202 \end{pmatrix} \begin{pmatrix} 3000 \\ 3000 \\ 3000 \\ 3000 \end{pmatrix} + \begin{pmatrix} 0 \\ 0 \\ 0 \\ 2 \cdot 2.532 \cdot 1000 \end{pmatrix}$$

which has the solution $P^1 = [2998.2; 2987.0; 2893.9; 2130.2]$ psia.

At $t = 10$ days, use the solution at the previous time step. Note that the matrix remains the same, but the "right hand side (RHS)" vector changes from the previous time step

$$\begin{pmatrix} 9.266 & -1.266 & 0 & 0 \\ -1.266 & 10.532 & -1.266 & 0 \\ 0 & -1.266 & 10.532 & -1.266 \\ 0 & 0 & -1.266 & 11.798 \end{pmatrix} \begin{pmatrix} P_1^2 \\ P_2^2 \\ P_3^2 \\ P_4^2 \end{pmatrix}$$

$$= \begin{pmatrix} 6.734 & 1.266 & 0 & 0 \\ 1.266 & 5.468 & 1.266 & 0 \\ 0 & 1.266 & 5.468 & 1.266 \\ 0 & 0 & 1.266 & 4.202 \end{pmatrix} \begin{pmatrix} 2998.2 \\ 2987.0 \\ 2893.9 \\ 2130.2 \end{pmatrix} + \begin{pmatrix} 0 \\ 0 \\ 0 \\ 2 \cdot 2.532 \cdot 1000 \end{pmatrix}$$

which has the solution $P^2 = [2988.9; 2941.2; 2685.9; 1786.7]$ psia.

At $t = 15$ days, use the solution at the previous time step. Note that the matrix remains the same.

Continued

Example 3.5. Implicit and C–N Methods: 1D, single-phase homogeneous flow—cont'd

$$
\begin{pmatrix}
9.266 & -1.266 & 0 & 0 \\
-1.266 & 10.532 & -1.266 & 0 \\
0 & -1.266 & 10.532 & -1.266 \\
0 & 0 & -1.266 & 11.798
\end{pmatrix}
\begin{pmatrix}
P_1^3 \\
P_2^3 \\
P_3^3 \\
P_4^3
\end{pmatrix}
$$

$$
=
\begin{pmatrix}
6.734 & 1.266 & 0 & 0 \\
1.266 & 5.468 & 1.266 & 0 \\
0 & 1.266 & 5.468 & 1.266 \\
0 & 0 & 1.266 & 4.202
\end{pmatrix}
\begin{pmatrix}
2988.9 \\
2941.2 \\
2685.9 \\
1786.7
\end{pmatrix}
+
\begin{pmatrix}
0 \\
0 \\
0 \\
2 \cdot 2.532 \cdot 1000
\end{pmatrix}
$$

which has the solution $P^3 = [2965.7; 2866.4; 2502.3; 1622.3]$ psia. Eventually, the solution reaches steady state and $P^{ss} = [1000; 1000; 1000; 1000]$ psia.

TABLE 3.1 Comparison of numerical solution by hand (Example 3.5) to the analytical

	Method	Block #1 (psia) 500 ft	Block #2 (psia) 1500 ft	Block #3 (psia) 2500 ft	Block #4 (psia) 3500 ft
Initial		3000	3000	3000	3000
n = 1 5 days	Explicit	3000	3000	3000	1734
	Implicit	2993.3	2972.4	2864.5	2328.6
	C-N	2998.2	2987.0	2893.9	2130.2
	Analytical	3000.0	2996.6	2881.2	1940.6
n = 2 10 days	Explicit	3000	3000	2599.3	1670.1
	Implicit	2975.7	2920.2	2699.5	1957.4
	C-N	2988.9	2941.2	2685.9	1786.7
	Analytical	2996.1	2947.4	2635.0	1686.4
n = 3 15 days	Explicit	3000	2873.2	2432	1540.0
	Implicit	2946.0	2852.2	2543.4	1741.6
	C-N	2965.7	2866.4	2502.3	1622.3
	Analytical	2975.6	2860.6	2447.2	1566.5
∞		1000	1000	1000	1000

The solution makes physical sense. The pressure in the reservoir decreases over time as a result of the Dirichlet (constant pressure) boundary condition of $P = 1000$ psia, until the reservoir comes to steady state at 1000 psia. Pressure decreases fastest in block #4 because it is near the Dirichlet boundary and slowest in block #1 because it is far from the Dirichlet boundary.

3.5.4 Linear systems of equations

Comparison of Eqs. (3.26) to (3.25) shows that, unlike the explicit equations, the implicit (and mixed methods) equations are coupled; a system of N simultaneous linear equations are formed. The system of linear equations can be solved using direct or indirect methods. Direct methods attempt to solve the system exactly and are not subject to truncation error but are subject to roundoff error. Examples include Gauss Elimination and LU decomposition, which are theoretically both $O(N^3)$ methods, meaning that increasing number of grids by 10-fold results in a computation time that is 1000-fold larger. Thus, direct methods are often not feasible for very large (e.g., tens of thousands or millions of blocks) systems of equations. The Thomas algorithm is a fast, $O(N)$, direct method, but only applicable to 1D, tridiagonal problems.

Indirect methods (Saad, 2003) begin with a guess value for the solution and iterate until convergence to a predetermined tolerance and are thus subject to truncation error. Iterative methods are usually classified as *stationary* or *Krylov subspace* methods. Stationary methods such as Jacobi, weighted Jacobi, Gauss—Seidel, or successive overrelaxation (SOR) are faster than most direct methods for moderately large systems (thousands of grids). However, they are only guaranteed to converge for certain matrices (e.g., symmetric, positive definite) and are usually slow compared to Krylov subspace methods for the very large systems often employed in reservoir simulation. Krylov subspace methods (Shewchuk, 1994) formulate the system of equations as a minimization problem and find the direction of steepest descent to find the optimum. Examples include the conjugate gradient and GMRES methods. Iterative methods often employ a preconditioner on the matrix which allows for convergence in less iterations.

Advanced simulators may have several linear solver subroutine options that can be called for the purpose of solving the system of equations. The user may choose the method in advance or the simulator can determine the best option based on the size, structure, and condition number of the matrix. Solution to the linear system is often the computational bottleneck of the simulator and the development of new solvers and preconditioners are an ongoing area of research. However, detailed discussion of solvers for linear systems of equations is outside the scope of this text. Linear algebra software libraries, such as LAPACK (Anderson et al., 1999), are available and compatible with most programming languages. Solving a system of equations in these languages can be as simple as calling a built-in linear solver function and sending the $N \times N$ matrix and the $N \times 1$ right-hand side vector. In the example problems and exercises in this text, we utilize such built-in functions even for small (4×4) systems and encourage the reader to do the same. Methods for solving nonlinear equations are discussed in Chapter 6.

3.6 Stability and convergence

The implicit method requires significantly more computational work per timestep than the explicit method because a system of linear equations must be solved. As mentioned previously, a significant percentage of the computation time required in reservoir simulators is attributed to the solution of large systems of linear equations. Implicit methods are *not*, in general, more ac-curate than explicit methods. The order of accuracy is the same for both the implicit and explicit methods since the spatial derivatives in both methods were approximated to second-order accuracy, $O(\Delta x^2)$, and the temporal de-rivatives were first-order accurate, $O(\Delta t)$. Implicit methods must offer some advantage over explicit methods or else they would not be used given the extra computation time. The advantage of the implicit method is that it is more stable and convergent. A method is said to *converge* if the solution becomes more accurate by using smaller step sizes (Δx or Δt). A method is said to be *stable* if errors (e.g., roundoff error) do not propagate in the solution. Stability and convergence are two different numerical issues; however, for linear PDEs, stability and convergence occur under the same criteria (Smith, 1985). One can show using von Neumann stability analysis (Crank and Nicolson, 1947; Charney et al., 1950) that an explicit solution to the 1D homogeneous, diffusivity equation is stable and convergent if and only if the following condition holds.

$$\eta \equiv \frac{\alpha \Delta t}{(\Delta x)^2} \leq \frac{1}{2} \tag{3.30}$$

The solution becomes unstable if the chosen timestep is too large or the grid block size is too small. This is unfortunate because one may want to improve accuracy by reducing Δx or decrease computation time by using a larger Δt. An example of instability for the solution of the 1D diffusivity equation using various η can be seen in Fig. 3.4. For values less than $\frac{1}{2}$, a stable solution is obtained but for values greater than $\frac{1}{2}$ it is unstable.

The implicit and Crank-Nicolson solutions, on the other hand, *is uncon-ditionally stable and convergent,* meaning that any size grid block or any timestep can be used and the solution is guaranteed to be stable. However, small timesteps may still be required to obtain an accurate solution.

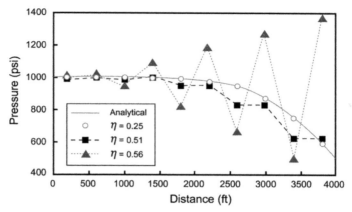

FIGURE 3.4   Numerical stability of a 1D diffusivity equation with Neumann boundary condition on the *left* and Dirichlet boundary condition on the *right*.

3.7 Higher-order approximations

We chose to use a second-order accurate, centered difference approximation for the spatial derivatives and a first-order accurate (explicit/implicit) or a second-order accurate (Crank–Nicolson) approximation for the temporal derivatives. However, additional accuracy can be achieved by substituting approximations of higher-order accuracy as defined in Section 3.2.5. The additional accuracy comes at additional computation cost for the same discretization sizes. In Example 3.6, the equations are derived and the problem in Example 3.4 is solved using fourth-order accurate spatial derivatives.

Example 3.6. Fourth-order accurate problem
Repeat Example 3.4 but use fourth-order accurate spatial derivatives.

Solution
Discretize the PDE using finite differences and a fourth-order accurate spatial derivative,

$$\frac{P_i^{n+1} - P_i^n}{\Delta t} = \frac{\alpha}{\Delta x^2} \left[-\frac{1}{12}P_{i-2}^{n+1} + \frac{4}{3}P_{i-1}^{n+1} - \frac{5}{2}P_i^{n+1} + \frac{4}{3}P_{i+1}^{n+1} - \frac{1}{12}P_{i+2}^{n+1} \right]$$

Multiply through by a constant, Ac_t,

$$Ac_t\left(P_i^{n+1} - P_i^n\right) = T\left[-\frac{1}{12}P_{i-2}^{n+1} + \frac{4}{3}P_{i-1}^{n+1} - \frac{5}{2}P_i^{n+1} + \frac{4}{3}P_{i+1}^{n+1} - \frac{1}{12}P_{i+2}^{n+1} \right]$$

Continued

Example 3.6. Fourth-order accurate problem—cont'd

Writing the algebraic equation for all grid blocks

$$A_1 c_{t,1} \left(P_1^{n+1} - P_1^n \right) = T \left[-\frac{1}{12} P_{-1}^{n+1} + \frac{4}{3} P_0^{n+1} - \frac{5}{2} P_1^{n+1} + \frac{4}{3} P_2^{n+1} - \frac{1}{12} P_3^{n+1} \right]$$

$$A_2 c_{t,2} \left(P_2^{n+1} - P_2^n \right) = T \left[-\frac{1}{12} P_0^{n+1} + \frac{4}{3} P_1^{n+1} - \frac{5}{2} P_2^{n+1} + \frac{4}{3} P_3^{n+1} - \frac{1}{12} P_4^{n+1} \right]$$

$$A_3 c_{t,3} \left(P_3^{n+1} - P_3^n \right) = T \left[-\frac{1}{12} P_1^{n+1} + \frac{4}{3} P_2^{n+1} - \frac{5}{2} P_3^{n+1} + \frac{4}{3} P_4^{n+1} - \frac{1}{12} P_5^{n+1} \right]$$

$$A_4 c_{t,4} \left(P_4^{n+1} - P_4^n \right) = T \left[-\frac{1}{12} P_2^{n+1} + \frac{4}{3} P_3^{n+1} - \frac{5}{2} P_4^{n+1} + \frac{4}{3} P_5^{n+1} - \frac{1}{12} P_6^{n+1} \right]$$

Blocks −1, 0, 5, and 6 are ghost blocks. Use the boundary conditions to replace them

$$\frac{P_0 + P_1}{2} = 0 \Rightarrow P_0 = P_1 \quad \text{(Neumann, reflection)}$$

$$\frac{P_{-1} + P_2}{2} = 0 \Rightarrow P_{-1} = P_2 \quad \text{(Neumann, reflection)}$$

$$\frac{P_4 + P_5}{2} = P_B \Rightarrow P_5 = 2P_B - P_4 \quad \text{(Dirichlet)}$$

$$\frac{P_3 + P_6}{2} = P_B \Rightarrow P_6 = 2P_B - P_3 \quad \text{(Dirichlet)}$$

The above is one simple way to apply the boundary conditions which is only second-order accurate at the boundaries. At $x = 0$, there is a no-flow boundary conditions, so we apply symmetry (reflection technique between blocks 0 and 1 and also blocks −1 and 2). At $x = L$, the pressure is defined at the boundary so the boundary pressure is applied as the arithmetic mean of blocks 4 and 5 as well as blocks 3 and 6.

Substitution of boundary conditions gives,

$$A_1 c_{t,1} \left(P_1^{n+1} - P_1^n \right) = \frac{T}{12} \left[-P_2^{n+1} + 16 P_1^{n+1} - 30 P_1^{n+1} + 16 P_2^{n+1} - P_3^{n+1} \right]$$

$$= \frac{T}{12} \left[-14 P_1^{n+1} + 15 P_2^{n+1} - P_3^{n+1} \right]$$

$$A_2 c_{t,2} \left(P_2^{n+1} - P_2^n \right) = \frac{T}{12} \left[-P_1^{n+1} + 16 P_1^{n+1} - 30 P_2^{n+1} + 16 P_3^{n+1} - P_4^{n+1} \right]$$

$$= \frac{T}{12} \left[15 P_1^{n+1} - 30 P_2^{n+1} + 16 P_3^{n+1} - P_4^{n+1} \right]$$

$$A_3 c_{t,3} \left(P_3^{n+1} - P_3^n \right) = \frac{T}{12} \left[-P_1^{n+1} + 16 P_2^{n+1} - 30 P_3^{n+1} + 16 P_4^{n+1} - \left(2P_B - P_4^{n+1} \right) \right]$$

$$= \frac{T}{12} \left[-P_1^{n+1} + 16 P_2^{n+1} - 30 P_3^{n+1} + 15 P_4^{n+1} \right] - J_{3,4} P_3^{n+1} + Q_3$$

$$A_4 c_{t,4}\left(P_4^{n+1} - P_4^n\right) = \frac{T}{12}\left[-P_2^{n+1} + 16P_3^{n+1} - 30P_4^{n+1} + 16\left(2P_B - P_4^{n+1}\right) - \left(2P_B - P_3^{n+1}\right)\right]$$

$$= \frac{T}{12}\left[-P_2^{n+1} + 15P_3^{n+1} - 14P_4^{n+1} - 32P_4^{n+1} + 2P_3^{n+1} + 30P_B\right]$$

$$= \frac{T}{12}\left[-P_2^{n+1} + 15P_3^{n+1} - 14P_4^{n+1}\right] - \left[J_{4,3}P_4^{n+1} + J_{4,4}P_3^{n+1}\right] + Q_4$$

where,

$$J_{3,4} = -\frac{2}{12}T; \quad Q_3 = -\frac{2}{12}TP_B$$

$$J_{4,3} = -\frac{2}{12}T; \quad J_{4,4} = \frac{32}{12}T; \quad Q_4 = \frac{30}{12}TP_B$$

$$\mathbf{T} = \frac{T}{12}\begin{pmatrix} 14 & -15 & 1 & 0 \\ -15 & 30 & -16 & 1 \\ 1 & -16 & 30 & -15 \\ 0 & 1 & -15 & 14 \end{pmatrix}\frac{\text{scf} - \text{psi}}{\text{day}}; \quad \mathbf{J} = \frac{T}{12}\begin{pmatrix} 0 & 0 & 0 & 0 \\ 0 & 0 & & \\ & & & -2 \\ 0 & 0 & -2 & 32 \end{pmatrix}\frac{\text{scf} - \text{psi}}{\text{day}}$$

Matrices,

$$\vec{Q} = \frac{T}{12}\begin{pmatrix} 0 \\ 0 \\ -2P_B \\ 32P_B - 2P_B \end{pmatrix}\frac{\text{scf}}{\text{day}}; \quad \mathbf{Ac_t} = \begin{pmatrix} A_1 c_{t,1} & & & \\ & A_2 c_{t,2} & & \\ & & A_3 c_{t,3} & \\ & & & A_4 c_{t,4} \end{pmatrix}\frac{\text{scf}}{\text{psi}}$$

where, Ac_t, T, and P_B were determined in Example 3.4.

Solving the problem using the implicit method,

$$(\mathbf{T} + \mathbf{J} + \mathbf{Ac_t})\vec{P}^{n+1} = \mathbf{Ac_t}\vec{P}^n + \mathbf{Q}$$

Using $P^0 = [3000; 3000; 3000; 3000]$ psia, solve system of equations

$$\vec{P}^0 = \begin{pmatrix} 3000 \\ 3000 \\ 3000 \\ 3000 \end{pmatrix}\text{psia}; \quad \vec{P}^1 = \begin{pmatrix} 2996.5 \\ 2979.2 \\ 2868.4 \\ 2258.6 \end{pmatrix}\text{psia}; \quad \vec{P}^2 = \begin{pmatrix} 2982.5 \\ 2927.9 \\ 2690.4 \\ 1888.1 \end{pmatrix}\text{psia};$$

$$\vec{P}^3 = \begin{pmatrix} 2955.1 \\ 2857.2 \\ 2524.6 \\ 1688.0 \end{pmatrix}\text{psia}; \quad \vec{P}^\infty = \begin{pmatrix} 1000 \\ 1000 \\ 1000 \\ 1000 \end{pmatrix}\text{psia}$$

Continued

Example 3.6. Fourth-order accurate problem—cont'd

Comments:

- The T matrix is pentadiagonal, symmetric, positive definite, and diagonally dominant.
- The solution makes physical sense (grid pressures slowly decrease from 3000 psia to 1000 psia and decrease fastest near the $x = L$ boundary).
- A second-order scheme was used at the boundaries. This is likely acceptable when a large number of grids are used. In Exercise 3.7, a computer code is written which can be applied to a large number of grids.

3.8 Pseudocode for 1D, single-phase flow

In real applications, reservoir simulators do not utilize a small number of grid blocks as depicted in Examples 3.4, 3.5, and 3.6 but rather employ thousands, millions, or even billions of grids. Solutions to problems of this size require computers and computer programs.

All of the reservoir (e.g., permeability, porosity, reservoir dimensions), fluid (compressibility, viscosity), and numerical (number of grids) properties as well as boundary conditions should be included in an input file as described in Chapter 1. In the main code, the input file is read into the code using a PREPROCESS subroutine (see Chapter 1 psueodcode), and then dimensionless diffusivity constant is computed. A subroutine (GRID ARRAYS) to compute the matrices and vectors Eq. (3.25), Eq. (3.26), and Eq. (3.27) is called and then the solution is looped through time, each time updating the vector of grid pressures. In the subroutine GRID ARRAYS, we loop through all $i = 1$ to N grids and create the *ith* row of each matrix and vector needed. Only the terms in the tridiagonal entries of the matrix are nonzero (for 1D problems). The first and last grids are subject to boundary conditions to those matrices and vectors need to be updated. The code can be adapted for higher-order approximations.

```
MAIN CODE
CALL PREPROCESS
CALL GRID ARRAYS
WHILE t < t_final
        SET P_old = P
        CALCULATE P using eqn 3.25,3.26, or 3.27
        INCREMENT t = t + Δt
ENDWHILE
PLOT Pressure versus distance and time
```

```
SUBROUTINE GRID ARRAYS
INPUTS: numerical, petrophysical, fluid properties, boundary
conditions
OUTPUTS: T, A, J, Q, ct

Pre-allocate (sparse storage format preferred) T, A, Q, J, ct
CALC COMPRESS using EQN. 1.18
CALC ACCUM using EQN 3.19a
CALC TRANS using EQN 3.19b
FOR i = 1 to # grids (N)
     SET A(i,i) = ACCUM
     SET ct(i,i) = COMPRESS

     IF NOT on left (x=0) boundary
          SET T(i,i-1) = -TRANS
          SET T(i,i) = T(i,i) + TRANS
     ELSEIF Dirichlet boundary condition
          SET J(i,i) = J(i,i) + 2*TRANS
          SET Q(i) = Q(i) + 2*TRANS*PLB
     ELSEIF Neumann boundary condition
          SET Q(i) = boundary rate
     ENDIF

     IF NOT on right (x=L) boundary
          SET T(i,i+1) = -TRANS
          SET T(i,i) = T(i,i) +TRANS
     ELSEIF Dirichlet boundary condition
          SET J(i,i) = J(i,i) + 2*TRANS
          SET Q(i) = Q(i) + 2*TRANS*PRB
     ELSEIF Neumann boundary condition
          SET Q(i) = boundary rate
     ENDIF

ENDFOR
```

3.9 Exercises

Exercise 3.1. Programming numerical derivatives. Write a short computer program to compute the numerical derivatives requested in Example 3.2 but do the calculations for both $10 < \Delta t < 100$ and $0.0001 < \Delta r < 0.001$ and use logarithmic spacing. Also calculate the absolute error and fit a "straight line" function to the curve with the horizontal axis appropriately chosen in each case to match the theoretical accuracy (e.g., Δt, Δt^2, Δr^2, etc.). From the best-fit line, comment on whether the theoretical accuracy of the curves is reached.

Exercise 3.2. Deriving finite difference equations. Derive finite difference equations for the first derivative with third-order accuracy. Use a 4-point stencil with points $i - 1, i, i + 1, i + 2$.

Exercise 3.3. Reverse boundary conditions. Repeat Example 3.4 but change the boundary conditions to no flux at $x = L$ and constant pressure ($P = 4000$ psia) at $x = 0$ and a timestep of 3 days. Determine whether the problem is stable for the explicit method and then solve for grid block pressures for three timesteps using the explicit, implicit, and Crank–Nicolson methods.

Exercise 3.4. Dirichlet boundary conditions. Repeat Exercise 3.3 but change the boundary conditions to constant pressure at both $x = 0$ ($P = 4000$ psia) and $x = L$ ($P = 1000$ psia).

Exercise 3.5. Fourth-order accurate spatial derivatives. Repeat Exercise 3.3 using a fourth-order accurate derivative in space.

Exercise 3.6. Coding a reservoir simulator. Write a computer program (reservoir simulator) to solve flow in a 1D reservoir using a finite difference method that is first-order accurate in time and second-order accurate in space.

The program should allow the user to define properties in an input file like the one introduced in Chapter 1, "Thomas.yml". These properties include reservoir and fluid, numerical (final time or condition to end the time stepping, number of grid blocks and timestep size, and the numerical method, i.e., explicit, implicit, or Crank–Nicolson). The program should also be flexible enough to allow for different boundary conditions, numerical method, etc. Edit the input file as appropriate.

a. Verify your code against the solution obtained in the 4-block Example 3.4 (and/or Exercises 3.3 through 3.5). If they are not in agreement, the mostly likely error is in the formation of the matrices or vectors, so compare these against the solution found in the example/exercises.
b. Use your code to solve Exercise 3.3 but using $N = 10$ grid blocks, $\Delta t = 1$ day, and the explicit method. Make a plot of pressure versus position at three different times (choose 3 times that are interesting; a relatively early time, a middle time, and a time close to steady state).
c. On the same plot as part (b), plot the analytical solution (see Chapter 2) at those same three timesteps to make a comparison.
d. Repeat part (b) and (c) but using the implicit method.
e. Repeat part (b) and (c) but using the Crank–Nicolson method.
f. From the criteria for the stability of the explicit method, find the maximum timestep size for which the explicit method is stable for $N = 10$ grid blocks. Then pick a timestep exactly 0.5 days longer than that maximum and plot the simulation results using the explicit method. Comment on the results.

Exercise 3.7. Fourth-order accurate simulator. Alter your code in Exercise 3.6 to allow for fourth-order accurate derivatives. Verify your solution against Exercise 3.5 and Example 3.6.

Exercise 3.8. Heterogeneities. Derive the finite difference equations for a reservoir that has heterogeneous (spatially dependent) permeability but other properties, including the grid sizes, are homogenous and uniform. The PDE is described by,

$$\frac{\partial p}{\partial t} = \frac{1}{\phi c_t \mu} \frac{\partial}{\partial x}\left(k \frac{\partial p}{\partial x}\right).$$

In the derivation, a permeability between two blocks (interblock permeability) may arise. Use the nomenclature, $k_{i-\frac{1}{2}}, k_{i+\frac{1}{2}}$ for permeability between blocks $i-1$, i and i, $i+1$, respectively. Define an interblock transmissibility as,

$$T_{i-\frac{1}{2}} = \frac{k_{i-\frac{1}{2}}a}{\Delta x \mu}.$$

Exercise 3.9. 2D reservoirs. Derive the matrix equations (analogous to Eq. 3.27b) for a 2D reservoir with homogenous and uniform reservoir and fluid properties using finite differences (second-order space, first-order time) without sources and sinks and negligible gravitational effects. The PDE is described by,

$$\frac{\partial p}{\partial t} = \alpha \left(\frac{\partial^2 p}{\partial x^2} + \frac{\partial^2 p}{\partial y^2}\right).$$

Exercise 3.10. Gravitational effects. Derive the 1D matrix equations to solve for pressure (analogous to Eq. 3.27b) using finite differences (second-order space, first-order time) that include gravitational effects, that is, the reservoir depth, D, varies with position, x. Include any gravitational effects into a vector, G, and define G.

References

Anderson, E., Bai, Z., Bischof, C., Blackford, S., Demmel, J., Dongarra, J., Du Croz, J., Greenbaum, A., Hammarling, S., McKenney, A., Sorensen, D., 1999. LAPACK Users' Guide, third ed. Society for Industrial and Applied Mathematics, Philadelphia, PA, ISBN 0-89871-447-8.

Charney, J.G., Fjørtoft, R., von Neumann, J., 1950. Numerical integration of the barotropic vorticity equation. Tellus 2, 237–254. https://doi.org/10.3402/tellusa.v2i4.8607.

Crank, J., Nicolson, P., 1947. A practical method for numerical evaluation of solutions of partial differential equations of heat conduction type. Proceedings of the Cambridge Philosophical Society 43, 50–67. https://doi.org/10.1007/BF02127704.

Saad, Y., 2003. Iterative Methods for Sparse Linear Systems, second ed. SIAM, ISBN 0-89871-534-2. OCLC 51266114.

Shewchuk, J.R., 1994. An Introduction to the Conjugate Gradient Method without the Agonizing Pain.

Smith, G.D., 1985. Numerical Solution of Partial Differential Equations: Finite Difference Methods. Published in the. United States by Oxford University Press Inc., New York.

Chapter 4

Multidimensional reservoir domains, the control volume approach, and heterogeneities

4.1 Introduction

In Chapter 3, reservoir simulation equations were derived for a 1D, homogenous reservoir by discretizing the domain using cell-centered grids. Continuous partial differential equations described the physical problem and finite difference approximations for the partial derivatives were substituted into these PDEs to describe the numerical problem. A system of coupled (implicit) or uncoupled (explicit) equations were obtained that could be solved to obtain pressure in the reservoir as a function of space and time. An advantage of the finite difference approach is that the order of accuracy can be determined *a priori* because the error is proportional to the leading term truncated in the Taylor series.

In this chapter, balance equations are extended to multidimensions (2D or 3D), gravity is included, and heterogeneities in permeability, porosity, cross-sectional area, grid sizes, and fluid properties are considered. The reservoir is first discretized into blocks of finite size and the mass balance equations imposed on these blocks. PDEs are never directly used in this chapter. This is referred to herein as the *control volume* approach. The resulting algebraic equations are identical to those that would be obtained by discretizing the PDE.

4.2 Gridding and block numbering in multidimensions

1D simulation models are excellent tools for gaining physical insight and verification against analytical models, but modern commercial reservoir simulators are multidimensional. Here, we first describe the grid systems in multidimensional problems before imposing mass balances.

An Introduction to Multiphase, Multicomponent Reservoir Simulation.
https://doi.org/10.1016/B978-0-323-99235-0.00003-8

4.2.1 Grid block indexing in 2D and 3D

Fig. 4.1A and B[1] shows example 2D and 3D domains, respectively, which are discretized into N_x grids in the x-direction, N_y grids in the y-direction, and N_z grids in the z-direction (for 3D). The total number of grids is $N = N_x \times N_y \times N_z$. The reservoir has a length, L, width, w, and thickness, h. The figure depicts a small number of grids, although useful simulators often utilize thousands, millions, or more grids (Dogru et al., 2011).

Blocks are indexed in this text in two different ways. The block can be defined with a j, k, l integer index, where j represents the grid index in the x-direction, k the grid index in the y-direction, and l the grid index in the z-direction. It is also useful to have a single index defining the block and here the integer i is used. i-indexing can be done using different approaches (Abou-Kassem et al., 2006), but in this text, $i = 1$ is defined at the bottom left corner ($j = 1, k = 1, l = 1$) and increases from left to right, then from bottom to top, and finally from front to back as shown in Fig. 4.1. In the right, upper, back corner, $i = N = N_x \times N_y \times N_z$. For 2D systems, one of the dimensions (e.g., N_z) is unity and in a 1D system, two of the dimensions (e.g., N_y and N_z) are unity.

The i-indexing can be mapped from j, k, l-indexing using Eq. (4.1),

$$i = j + (k - 1)N_x + (l - 1)N_xN_y, \qquad (4.1)$$

where Eq. (4.1) can be simplified for 2D ($l = 1$) and 1D ($k = 1, l = 1$) problems. In Fig. 4.1A, the grid outlined by the box is $j = 2, k = 3$ but from Eq. (4.1), it is also $i = 8$. It is often convenient to refer to a block using i-indexing, but refer to the interface between two blocks using j, k, l indexing. For

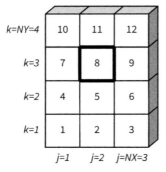

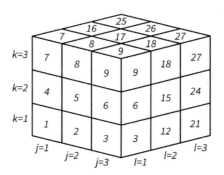

FIGURE 4.1 (A) Example of a 2D grid with 12 total grid blocks; $N_x = 3$ and $N_y = 4$ grids. The numbers inside the grid block use the i-indexing system. Block ($j = 2, k = 3$) or $i = 8$ is highlighted and (B) Example of a 3D grid with 27 total grid blocks; $N_x = 3, N_y = 3, N_z = 3$ grids.

---

1. The Fig. 4.1A is shown upright, but an areal reservoir would be flat with respect to the surface and both figures would be turned on their side.

example, the interface between blocks $j - 1, k, l$ and j, k, l is $j - 1/2, k, l$. Throughout this text, both j, k, l as well as i-indexing are used interchangeably.

4.2.2 Grid dimensions

The dimensions of a grid, i, is given as Δx_i, Δy_i, and Δz_i and the bulk volume of the block as $V_i = \Delta x_i \times \Delta y_i \times \Delta z_i$. If the reservoir can be approximated as a 2D areal problem, $\Delta z_i = h$, the thickness, and for 1D problems, $\Delta z_i = h$ and $\Delta y_i = w$, the width. A simple approach to gridding uses uniform grids as was done in Chapter 3. However, it is often desired to use nonuniform grid block sizes to optimize accuracy and computational speed. Relatively small grid blocks are desired where changes in variables, e.g., pressure, are significant over short distances, e.g., near a well, and larger grid blocks where pressure does not change much with distance, e.g., far from a well. In a reservoir simulator, grid sizes can vary from the order of feet to hundreds or thousands of feet. There are many complex gridding algorithms (Aziz, 1993; Quandalle, 1993) that can be employed. Fig. 4.2 shows an example for a 2D reservoir with three wells.

The figure shows that fine (small) grids are employed near the wells and coarse grids far from the wells as desired. However, fine grids are also used in some places where not needed. Therefore, this gridding algorithm is not ideal, but advanced techniques (Wheeler et al., 1999) are beyond the scope of this text.

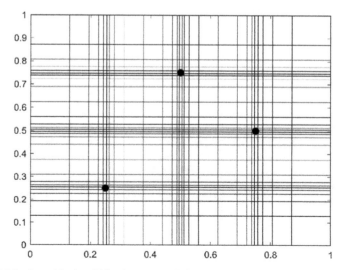

FIGURE 4.2   Logarithmic gridding between wells in a 2D reservoir. The positions on the x- and y- axis are dimensionless.

4.2.3 Irregular geometry and inactive grids

Reservoir simulation is often performed in simple, cuboid geometries that are sections of a larger reservoir (e.g., 5-spot pattern). However, the geometry of naturally occurring reservoirs is irregular and therefore the domain cannot, in general, be accurately described by cuboids. Instead, grids are employed to approximate the reservoir geometry. Fig. 4.3 depicts a finite difference grid on an irregular domain.

There are multiple approaches used in reservoir simulation to solve problems of irregular geometries. *Orthogonal* (those with right angles) grids can be arranged in such a way that the grid system approximates the irregular shape of the reservoir (Fig. 4.3). However, the grid can poorly approximate the geometry at the boundaries unless fine grids are used. Both the finite element method (FEM) and finite volume method (FVM) utilize polygons for grid cells and are not limited to orthogonal grids. The more complex grid shapes allow for better modeling of the geometry at the boundaries and also improve numerical accuracy. However, many simulators continue to use orthogonal grids. Neither the finite element nor the finite volume method is discussed in this text, although cell-centered finite differences can be shown to be subsets of both FEM and FVM methods.

Two methods for implementing orthogonal grids in a noncuboid geometry include: (a) inactive grids and (b) reindexing. Inactive grids are depicted in Fig. 4.4A. In this approach, the domain is modeled by a cuboid geometry, but blocks outside the reservoir are considered "inactive"; they are assigned nonphysical parameters such as zero porosity and/or permeability. Numerically, matrices of size $N \times N$ are still generated, but rows and columns corresponding to the inactive grid contain the nonphysical values. The system of equations may be *singular* or *ill-conditioned* when using inactive grids (more unknowns than unique equations making the matrix equations numerically difficult to solve). However, the numerical value of pressure in the inactive grids is not meaningful and iterative solvers should still converge to the solution in the active grid blocks.

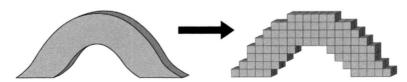

FIGURE 4.3   2D reservoir with irregular geometry and gridding to approximate the geometry.

141	142	143	144	145	146	147	148	149	150	151	152	153	154	155	156	157	158	159	160
121	122	123	124	125	126	127	128	129	130	131	132	133	134	135	136	137	138	139	140
101	102	103	104	105	106	107	108	109	110	111	112	113	114	115	116	117	118	119	120
81	82	83	84	85	86	87	88	89	90	91	92	93	94	95	96	97	98	99	100
61	62	63	64	65	66	67	68	69	70	71	72	73	74	75	76	77	78	79	80
41	42	43	44	45	46	47	48	49	50	51	52	53	54	55	56	57	58	59	60
21	22	23	24	25	26	27	28	29	30	31	32	33	34	35	36	37	38	39	40
1	2	3	4	5	6	7	8	9	10	11	12	13	14	15	16	17	18	19	20

(a)

							78	79	80	81	82	83							
				68	69	70	71	72	73	74	75	76	77						
			56	57	58	59	60	61	62	63	64	65	66	67					
			42	43	44	45	46	47	48	49	50	51	52	53	54	55			
		31	32	33	34	35	36					37	38	39	40	41			
		21	22	23	24	25						26	27	28	29	30			
	11	12	13	14	15							16	17	18	19	20			
1	2	3	4	5											6	7	8	9	10

(b)

FIGURE 4.4 Irregular reservoir using orthogonal, finite difference grids with (A) inactive grids and (B) renumbering. Grids in white are not part of the reservoir.

The bookkeeping in using "inactive grids" is simple because the grid indexing follows the simple geometric pattern shown in Fig. 4.4A and Eq. (4.1). This approach is slightly more computationally demanding (than the reindexing approach) because additional, unnecessary equations/grids are included in the model. The alternative is to reindex the grids as shown in Fig. 4.4B, in such a way that only active blocks are assigned a block number. As a result, the number of equations is smaller than the inactive blocks method; however, the connectivity between blocks is not as simple as described by Eq. (4.1). The connectivity of blocks must be determined and saved in a reference list. In this text, we will generally use "inactive grids" for simplicity.

4.3 Single-phase flow in multidimensions and the control volume approach

The general diffusivity equation for single-phase (α) flow in 3D (Cartesian), with anisotropic, heterogeneous permeability, gravitational effects, and sources/sinks can be expanded from the general form (Eq. 2.10a),

$$\phi c_t \frac{\partial p}{\partial t} = \frac{1}{\rho_\alpha} \frac{\partial}{\partial x} \left(\frac{\rho_\alpha k_x k_{r,\alpha}^0}{\mu_\alpha} \left(\frac{\partial p}{\partial x} - \rho_\alpha g \frac{\partial D}{\partial x} \right) \right) + \frac{1}{\rho_\alpha} \frac{\partial}{\partial y} \left(\frac{\rho_\alpha k_y k_{r,\alpha}^0}{\mu_\alpha} \left(\frac{\partial p}{\partial y} - \rho_\alpha g \frac{\partial D}{\partial y} \right) \right)$$
$$+ \frac{1}{\rho_\alpha} \frac{\partial}{\partial z} \left(\frac{\rho_\alpha k_z k_{r,\alpha}^0}{\mu_\alpha} \left(\frac{\partial p}{\partial z} - \rho_\alpha g \frac{\partial D}{\partial z} \right) \right) + B_\alpha \tilde{q}_{sc}$$

(4.2)

Endpoint relative permeability is included because only one flowing phase is assumed here (multiphase flow is covered in Chapter 9). Eq. (4.2) can be discretized using finite differences using the approaches in Chapter 3. An alternative is to use the *control-volume approach* to develop mass balance equations for each block. The same N algebraic equations result in both approaches.

In the control volume approach, the reservoir domain is first discretized into grid blocks of finite size and then mass balance equations are imposed on each grid. The grid blocks are the control volume for the balance. Fig. 4.5 depicts these control volume grids in 2D.

A mass balance (in − out + generation − consumption = accumulation) on each phase α can be imposed on each control volume (grid) as was done in Chapter 2. Mass flux, $\rho_\alpha u_\alpha$, enters or exits the block from six faces (in 3D); mass is generated/consumed by wells and is accumulated by changes in density, porosity, and saturation,

$$V_i \left[(\phi \rho_\alpha S_\alpha)_i^{n+1} - (\phi \rho_\alpha S_\alpha)_i^n \right] = \left(\rho_\alpha u_{\alpha,x} \Delta y \Delta z \big|_{j-\frac{1}{2},k,l} + \rho_\alpha u_{\alpha,x} \Delta y \Delta z \big|_{j+\frac{1}{2},k,l} \right) \Delta t$$

$$+ \left(\rho_\alpha u_{\alpha,y} \Delta x \Delta z \big|_{j,k-\frac{1}{2},l} + \rho_\alpha u_{\alpha,y} \Delta x \Delta z \big|_{j,k+\frac{1}{2},l} \right) \Delta t$$

$$+ \left(\rho_\alpha u_{\alpha,z} \Delta x \Delta y \big|_{j,k,l-\frac{1}{2}} + \rho_\alpha u_{\alpha,z} \Delta x \Delta y \big|_{j,k,l+\frac{1}{2}} \right) \Delta t$$

$$+ \rho_\alpha \tilde{q}_{\alpha,i} \Delta x_i \Delta y_i \Delta z_i \Delta t,$$

(4.3)

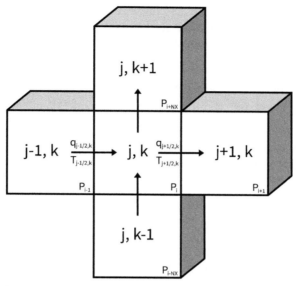

FIGURE 4.5 2D reservoir discretized into grid blocks. Each block can be taken as a control volume.

where interblock (e.g., $j - 1/2, k, l$) fluxes are between two adjacent (e.g., $j - 1, k$, l and j, k, l) blocks. Phase velocities can be positive or negative depending on the direction of flow. In Eq. (4.3), i-indexing is used for grids (e.g., accumulation and source terms) and j, k, l-indexing is used for interfaces. In the following sections each term in the mass balance equation is derived and expanded without discretizing the PDE (Eq. 4.2).

4.3.1 Accumulation

The accumulation terms of Eq. (4.3) can be simplified by first recognizing that the time difference can be expanded,

$$
(\phi \rho_\alpha S_\alpha)_i^{n+1} - (\phi \rho_\alpha S_\alpha)_i^n = \phi_i^{n+1} \rho_{\alpha,i}^{n+1} \left(S_{\alpha,i}^{n+1} - S_{\alpha,i}^n \right) + S_{\alpha,i}^n \rho_{\alpha,i}^{n+1} \left(\phi_i^{n+1} - \phi_i^n \right)
$$
$$
+ \phi_i^n S_{\alpha,i}^n \left(\rho_{\alpha,i}^{n+1} - \rho_{\alpha,i}^n \right).
$$

(4.4)

The porosity and density (for slightly compressible fluids) differences in time can be simplified by using a first-order Taylor Series approximation,

$$
\phi_i^{n+1} - \phi_i^n = \phi^0 c_f \left(P_i^{n+1} - P^0 \right) - \phi^0 c_f \left(P_i^n - P^0 \right) = \phi^0 c_f \left(P_i^{n+1} - P_i^n \right) \quad (4.5a)
$$

$$
\rho_i^{n+1} - \rho_i^n = \rho^0 c_\alpha \left(P_i^{n+1} - P^0 \right) - \rho^0 c_\alpha \left(P_i^n - P^0 \right) = \rho^0 c_\alpha \left(P_i^{n+1} - P_i^n \right), \quad (4.5b)
$$

where ρ^0 and ϕ^0 are the density and porosity, respectively, at some reference pressure, P^0. Analogous equations for the density change for a compressible gas can be derived using a real gas law. Eqs. 4.5a and 4.5b can be substituted into Eq. (4.4) to give,

$$(\phi\rho_\alpha S_\alpha)_i^{n+1} - (\phi\rho_\alpha S_\alpha)_i^n = \phi_i \rho_{\alpha,i}\left(S_{\alpha,i}^{n+1} - S_{\alpha,i}^n\right) + \phi_i S_{\alpha,i}^n \rho_{\alpha,i}\left(c_{f,i} + c_{\alpha,i}\right)\left(P_i^{n+1} - P_i^n\right).$$

(4.6)

In Eq. (4.6), an approximation was made that, for a slightly compressible fluid, the porosity and density could be treated as constants and the superscripts for time index were removed. The pressure/time-dependence on these variables was accounted for in Eq. (4.5).

4.3.2 Flux terms

There are six flux terms (one for each face of the block) in a three-dimensional balance. Darcy's law can be substituted for velocities in Eq. (4.3). For example, in the x-direction at the interface of blocks $j-1, k, l$ and j, k, l,

$$q_{\alpha,j-\frac{1}{2},k,l} = \left(u_{\alpha,x}\Delta y \Delta z\right)_{j-\frac{1}{2},k,l} = (T_\alpha B_\alpha)_{j-\frac{1}{2},k,l}\left[(P_{i-1} - P_i) - (\rho g)_{\alpha,j-\frac{1}{2},k,l}(D_{i-1} - D_i)\right]$$

(4.7)

where the phase (Eq. 4.8a) and total (Eq. 4.8b) interblock transmissibility between the two blocks,

$$T_{\alpha,j-\frac{1}{2},k,l} = \left.\frac{k_x k_{r,\alpha}^0 a}{\mu_\alpha B_\alpha \Delta x}\right|_{j-\frac{1}{2},k,l}$$

(4.8a)

$$T_{j-\frac{1}{2},k,l} = \sum_{\alpha=1}^{N_p} B_{\alpha,i} T_{\alpha,j-\frac{1}{2},k,l} = \sum_{\alpha=1}^{N_p} B_{\alpha,i}\left(\frac{k_x k_{r,\alpha}^0 a}{\mu_\alpha B_\alpha \Delta x}\right)_{j-\frac{1}{2},k,l}$$

(4.8b)

and D_i is the depth of block i. Similar expressions can be written for the other faces of the block. The transmissibility terms include permeability in the direction of flow, which may be anisotropic ($k_x \neq k_y \neq k_z$). Spatially dependent heterogeneities of properties in the interblock transmissibility are discussed in Section 4.5.

In Eq. (4.8a) an interblock phase (T_α) transmissibility is defined, which includes an interblock formation volume factor in the denominator. The units of this interblock transmissibility are standard conditions (e.g., scf/psi-day). In Eq. (4.8b), the total (T) transmissibility is the *interblock* phase transmissibility multiplied by the *block* formation volume factor, summed over all phases. If only one phase is flowing the summation is not required. The units of the total transmissibility are reservoir conditions (e.g., ft³/psi-day). Formation volume factor of the phase appears in both the numerator and denominator of Eq. (4.8b), but the block and interblock formation volume factors are not, in general, equal because they are pressure dependent. However, for slightly compressible fluids they can be approximated as equal (this approximation was made indirectly in Chapter 3). For compressible fluids, such as gases, the difference between the two formation volume factors may not be negligible.

4.3.3 Sources and sinks (wells)

Mass can be generated or consumed in a grid block through wells and, therefore, are implemented as sources (injection wells) or sinks (producer wells). Usually, the well rate is measured at surface conditions (sc) instead of reservoir conditions (rc) and formation volume factor is needed for the conversion,

$$\widetilde{q}_{\alpha,i}\Delta x_i \Delta y_i \Delta z_i = Q^{rc}_{\alpha,i} = B_{\alpha,i}Q^{sc}_{\alpha,i}. \tag{4.9}$$

4.3.4 Single-phase flow

Eqs. (4.6), (4.7), and (4.9) can be substituted into the phase balance, Eq. (4.3) to give,

$$B_{\alpha,i}\Delta T_\alpha \Delta P - B_{\alpha,i}\Delta(\rho_\alpha g T_\alpha)\Delta D + B_{\alpha,i}Q^{sc}_i = A_i\left[\Delta_t S_\alpha + S^n_{\alpha,i}(c_{f,i} + c_{\alpha,i})\Delta_t P\right] \tag{4.10}$$

where $A_i = V_i\phi_i/\Delta t$ and total, T (Eq. 4.8b), not phase, transmissibility is used. The shorthand notation is defined as,

$$\Delta T_\alpha \Delta U = T_{\alpha,j-\frac{1}{2},k,l}(U_i - U_{i-1}) + T_{\alpha,j+\frac{1}{2},k,l}(U_i - U_{i+1}) + T_{\alpha,j,k-\frac{1}{2},l}(U_i - U_{i-NX})$$

$$+ T_{\alpha,j,k+\frac{1}{2},l}(U_i - U_{i+NX}) + T_{\alpha,j,k,l-\frac{1}{2}}(U_i - U_{i-NX*NY})$$

$$+ T_{\alpha,j,k,l+\frac{1}{2}}(U_i - U_{i+NX*NY}), \tag{4.11a}$$

$$\Delta_t U = U^{n+1}_i - U^n_i, \tag{4.11b}$$

where U is any spatially or time-dependent variable, e.g., P, D or S_α. All phases can be summed together to give,

$$\Delta T \Delta P - \left(\sum_{\alpha=1}^{Np} B_\alpha \Delta(\rho_\alpha g T_\alpha)\right)\Delta D + \sum_{\alpha=1}^{N_p} B_{\alpha,i}Q^{sc}_{\alpha,i} = A_i c_{t,i}\Delta_t P \tag{4.12}$$

and c_t is the saturated-weighted total compressibility, defined in Chapter 1 (Eq. 1.18). Time differences in saturation in Eq. (4.9) are eliminated in the pressure equation because saturations of all phases must sum to unity. If there is only one flowing phase, the summation of phases on the left-hand side of Eq. (4.12) is not necessary because production rates are only finite for the flowing phase (i.e., those above residual saturation). For simplicity, the phase density of the flowing phase is assumed constant in the gravity terms.

In Eq. (4.12), the time level for pressure in the flow terms was not specified; pressures are evaluated at the n time level for the explicit method, $n + 1$ time level for the implicit method, and an intermediate time level for mixed methods as defined in Chapter 3. In matrix form Eq. (4.12) can be written for all blocks,

$$\vec{P}^{n+1} = \vec{P}^{n} + (\mathbf{Ac}_t)^{-1}\left[-(\mathbf{T}+\mathbf{J})\,\vec{P}^{n} + \vec{Q} + \vec{G}\right] \qquad (4.13a)$$

$$(\mathbf{T}+\mathbf{J}+\mathbf{Ac}_t)\,\vec{P}^{n+1} = \mathbf{Ac}_t\,\vec{P}^{n} + \vec{Q} + \vec{G} \qquad (4.13b)$$

$$[(1-\theta)(\mathbf{T}+\mathbf{J}) + \mathbf{Ac}_t]\,\vec{P}^{n+1} = [\mathbf{Ac}_t - \theta(\mathbf{T}+\mathbf{J})]\,\vec{P}^{n} + \vec{Q} + \vec{G} \qquad (4.13c)$$

where Eq. (4.13a) is the explicit method, Eq. (4.13b) the implicit method, and Eq. (4.13c) a mixed method. In the mixed method, $0 < \theta < 1$; $\theta = 1$ reduces to the explicit method, $\theta = 0$ reduces to the implicit method, and $\theta = 1/2$ is the Crank–Nicolson method. The matrix $\mathbf{J}$ is used for boundary conditions and constant bottomhole pressure wells (discussed in Chapter 5). The transmissibility matrix is tridiagonal in 1D, pentadiagonal in 2D, and heptadiagonal in 3D. Importantly, the phase transmissibility, $\mathbf{T}_\alpha$, is symmetric but $\mathbf{T}$ is only strictly symmetric for the case that formation volume factor is uniform. Section 4.7 explains each of the arrays in Eq. (4.13) in more detail.

4.4 Wells, boundary conditions, and initial conditions

4.4.1 Constant rate wells

Wells and well models are discussed in detail in Chapter 5. Here, we consider only wells, injectors or producers, in which the rate is specified (in standard conditions). The grid block, i, that contains the well is included as a source term, $Q_i = B_{\alpha,i}Q_i^{sc}$, where the specified standard rate is converted to reservoir conditions using the formation volume factor. The convention used here is that injector wells have a positive rate and producers a negative rate. For our purposes, the location of the well in the grid does not affect the value of Q_i; however, good practice is to place the grids in such a way that the wells are near the grid centers (Ding et al., 1998).

4.4.2 Neumann boundary conditions

Neumann conditions are those of constant specified rate on a boundary and can be positive (into the reservoir), negative (out of the reservoir), or zero (a no-flux condition). Therefore, Neumann boundary conditions can be treated similarly to constant rate wells. The specified rate at the boundary can be directly included into the source vector, Q_i, where i refers to the block with the specified boundary condition. Since boundary fluxes are generally specified at reservoir conditions, they do not need to be converted from standard conditions using formation volume factors. A common Neumann boundary condition is one of zero flux (sealed boundary), in which case $Q_i = 0$.

If the specified rate is defined across several blocks or an entire boundary (in 2D or 3D) the flux must be distributed among those grids. Any approach used to distribute the flux is an approximation. One simple method is to distribute evenly between the blocks, but this can be inaccurate if the transmissibility varies significantly along the boundary blocks because fluid preferentially flows through the blocks of highest transmissibility. A more accurate approach is to weight the fluxes by the block transmissibility,

$$Q_i \approx \frac{T_i}{\sum\limits_{i\text{bound}} T_{i\text{bound}}} Q_{\text{total}}, \tag{4.14}$$

where the summation is performed across all blocks on the boundary. Eq. (4.14) (weighting the flow rates by transmissibility), although approximate, usually results in acceptable accuracy. For wells that are perforated by multiple blocks, the well rate can be distributed using a weighting of well productivity (discussed in Chapter 5) indices instead of transmissibilities.

The flux at the boundary edge is defined by Darcy's law; for example, at the left boundary,

$$q_{\frac{1}{2},k,l} = T_{\frac{1}{2},k,l}\left(P_{0,k,l} - P_{1,k,l}\right) = 0, \tag{4.15}$$

where block 0, k, l is outside the domain and therefore a ghost block. The flux is defined as zero in Eq. (4.15), even for a finite flux boundary condition, because the specified rate was already included in Q. Setting the interblock transmissibility to zero forces the expression (Eq. 4.15) to be true and therefore interblock transmissibilities at Neumann boundaries are zero and included as such in the Transmissibility matrix, T. This can be shown to be equivalent to the "reflection technique" presented in Chapter 3.

4.4.3 Dirichlet conditions

Dirichlet (constant pressure) boundary conditions were implemented in Chapter 3 by approximating the boundary pressure as the arithmetic average of the pressures of the imaginary block and boundary block (e.g., $P_B = (P_{0,k,l} + P_{1,k,l})/2$). Equivalently, one can rewrite Darcy's law at the boundary using the boundary pressure,

$$q_{j-\frac{1}{2},k,l} = T_{j-\frac{1}{2},k,l}\left(P_{0,k,l} - P_{1,k,l}\right) = T_{j,k,l}(2P_B - P_i - P_i) = 2T_i(P_B - P_i) = J_i(P_B - P_i), \tag{4.16}$$

where the "2" arises because the distance from the boundary edge to the center of the block equals to half of the length. The interblock transmissibility is then equal to twice the transmissibility of the block, i, and the result is that the transmissibility matrix includes a "$2T_i$" in the main diagonal. Mathematically, a diagonal matrix, $\mathbf{J}$, is used which includes $+2T_i$ in the main diagonal for any block that has a constant pressure boundary condition. In the source vector, $Q_i = +2T_i P_B$. Constant bottomhole pressure wells are similar to constant pressure boundary conditions and are thus mathematically treated similarly as discussed in Chapter 5.

4.4.4 Corner blocks

2D domains have four boundary faces and 3D domains have six. Blocks that are adjacent to more than one boundary are referred to here as "corner blocks." It may seem like blocks on corners in 2D and 3D domains need to be handled differently than other boundary blocks. However, corner blocks are no different than other boundary blocks; they are just subject to more than one boundary condition. For the purpose of developing a code, extra conditional statements are not required for these corner blocks as shown in the pseudocode provided at the end of this chapter. Example 4.1 demonstrates the block equations for a 2D reservoir with various boundary conditions.

Example 4.1. Mass balance equations
  Write the algebraic mass balance equations for a heterogeneous, anisotropic 2D reservoir with $N_x = 3$, $N_y = 3$ grids. The boundary conditions are $P = P_B$ at $x = L$ and no flow on the other boundaries. Only one phase is flowing.

Solution
The mass balance for any block i with four neighboring blocks can be written as,

$$T_{j-\frac{1}{2},k}(P_{i-1} - P_i) + T_{j+\frac{1}{2},k}(P_{i+1} - P_i) + T_{j,k-\frac{1}{2}}(P_{i-NX} - P_i)$$

$$+ T_{j,k+\frac{1}{2}}(P_{i+NX} - P_i) = A_i c_{t,i}\left(P_i^{n+1} - P_i^n\right).$$

  The equation can be rewritten for each of the nine blocks, boundary conditions substituted where appropriate, and then rearranged to group terms of the same block pressure.

Block #i = 1 (j = 1, k = 1).
There is a no flow, Neumann boundary condition, on the $x = 0$ and $y = 0$ boundaries, so there is no contribution to flux in those directions.

$$\cancel{T_{\frac{1}{2},1}\left(P_{0,1} - P_1\right)} + T_{\frac{3}{2},1}\left(P_2 - P_1\right) + \cancel{T_{1,\frac{1}{2}}\left(P_{1,0} - P_1\right)} + T_{1,\frac{3}{2}}\left(P_4 - P_1\right) = A_1 c_{t,1}\left(P_1^{n+1} - P_1^n\right)$$

Example 4.1. Mass balance equations—cont'd

Rearranging and grouping like terms

$$\left(T_{\frac{3}{2},1} + T_{1,\frac{3}{2}}\right)P_1 - T_{\frac{3}{2},1}P_2 - T_{1,\frac{3}{2}}P_4 = -\left(A_1 c_{t,1}\right)\left(P_1^{n+1} - P_1^n\right)$$

Block #i = 2 (j = 2, k = 1).
There is a no flow, Neumann boundary condition, on the $y = 0$ boundary (so there is no contribution to flux in that direction).

$$T_{\frac{3}{2},1}\left(P_1 - P_2\right) + T_{\frac{5}{2},1}\left(P_3 - P_2\right) + T_{2,\frac{1}{2}}\left(\cancel{P_{2,0} - P_2}\right) + T_{2,\frac{3}{2}}\left(P_5 - P_2\right) = A_2 c_{t,2}\left(P_2^{n+1} - P_2^n\right)$$

Rearranging and grouping like terms

$$-T_{\frac{3}{2},1}P_1 + \left(T_{\frac{3}{2},1} + T_{\frac{5}{2},1} + T_{2,\frac{3}{2}}\right)P_2 - T_{\frac{5}{2},1}P_3 - T_{2,\frac{3}{2}}P_5 = -\left(A_2 c_{t,2}\right)\left(P_2^{n+1} - P_2^n\right)$$

Block #i = 3 (j = 3, k = 1).
There is a no flow, Neumann boundary condition, on the $y = 0$ boundary (so there is no contribution to flux in that direction). On the $x = L$ boundary, there is a constant pressure, Dirichlet, boundary condition, P_B.

$$T_{\frac{5}{2},1}\left(P_2 - P_3\right) + \underbrace{T_{\frac{7}{2},1}\left(P_{NX+1,1} - P_3\right)}_{2T_{x,3}(P_B - P_3)} + T_{3,\frac{1}{2}}\left(\cancel{P_{3,0} - P_3}\right) + T_{3,\frac{3}{2}}\left(P_6 - P_3\right) = A_3 c_{t,3}\left(P_3^{n+1} - P_3^n\right)$$

Rearranging and grouping like terms

$$-T_{\frac{5}{2},1}P_2 + \left(T_{\frac{5}{2},1} + T_{3,\frac{3}{2}} + J_3\right)P_3 - T_{3,\frac{3}{2}}P_6 = -\left(A_3 c_{t,3}\right)\left(P_3^{n+1} - P_3^n\right) + Q_3$$

where $J_3 = 2T_{x,3}$ and $Q_3 = 2T_{x,3}P_B$ and $T_{x,3}$ is the x-direction transmissibility of block #3.

Block #i = 4 (j = 1, k = 2).
There is a no flow, Neumann boundary condition, on the left boundary (so there is no contribution to flux in that direction).

$$T_{1,\frac{3}{2},2}\left(\cancel{P_{0,2} - P_4}\right) + T_{\frac{3}{2},2}\left(P_5 - P_4\right) + T_{1,\frac{3}{2}}\left(P_1 - P_4\right) + T_{1,\frac{5}{2}}\left(P_7 - P_4\right) = A_4 c_{t,4}\left(P_4^{n+1} - P_4^n\right)$$

Continued

Example 4.1. Mass balance equations—cont'd

Rearranging and grouping like terms

$$-T_{1,\frac{3}{2}}P_1 + \left(T_{\frac{3}{2},2} + T_{1,\frac{3}{2}} + T_{1,\frac{5}{2}} \right) P_4 - T_{\frac{3}{2},2}P_5 - T_{1,\frac{5}{2}}P_7 = -(A_4 c_{t,4})\left(P_4^{n+1} - P_4^n\right)$$

Block #i = 5 (j = 2, k = 2).
There are no boundary conditions applied since block #5 is an interior block.

$$T_{\frac{3}{2},2}(P_4 - P_5) + T_{\frac{5}{2},2}(P_6 - P_5) + T_{2,\frac{3}{2}}(P_2 - P_5) + T_{2,\frac{5}{2}}(P_8 - P_5) = A_5 c_{t,5}\left(P_5^{n+1} - P_5^n\right)$$

Rearranging and grouping like terms

$$-T_{2,\frac{3}{2}}P_2 - T_{\frac{3}{2},2}P_4 + \left(T_{2,\frac{3}{2}} + T_{\frac{3}{2},2} + T_{\frac{5}{2},2} + T_{2,\frac{5}{2}} \right) P_5 - T_{\frac{5}{2},2}P_6 - T_{2,\frac{5}{2}}P_8$$

$$= -(A_5 c_{t,5})\left(P_5^{n+1} - P_5^n\right)$$

Block #i = 6 (j = 3, k = 2).
On the right boundary, there is a constant pressure, Dirichlet, boundary condition, P_B.

$$T_{3,\frac{3}{2}}(P_3 - P_6) + T_{\frac{5}{2},1}(P_5 - P_6) + \underbrace{T_{\frac{7}{2},1}\left(P_{NX+1,1} - P_6\right)}_{2T_{x,6}(P_B-P_6)} + T_{3,\frac{5}{2}}(P_9 - P_6) = A_6 c_{t,6}\left(P_6^{n+1} - P_6^n\right)$$

Rearranging and grouping like terms

$$-T_{3,\frac{3}{2}}P_3 - T_{\frac{5}{2},1}P_5 + \left(T_{3,\frac{3}{2}} + T_{\frac{5}{2},1} + T_{3,\frac{5}{2}} + J_6 \right) P_6 - T_{3,\frac{5}{2}}P_9 = -(A_6 c_{t,6})\left(P_6^{n+1} - P_6^n\right) + Q_6$$

where $J_6 = 2T_{x,6}$ and $Q_6 = 2T_{x,6}P_B$ and $T_{x,6}$ is the x-direction transmissibility of block #6.

Block #i = 7 (j = 1, k = 3).
There is a no flow, Neumann boundary condition, on both the left and top boundaries, so there is no contribution to flux in those directions.

$$T_{1,\frac{5}{2}}(P_4 - P_7) + T_{\frac{1}{2},3}\left(P_{0,3} - P_7\right) + T_{\frac{3}{2},3}(P_8 - P_7) + T_{1,\frac{7}{2}}\left(P_{1,NY+1} - P_7\right) = A_7 c_{t,7}\left(P_7^{n+1} - P_7^n\right)$$

Rearranging and grouping like terms

$$-T_{1,\frac{5}{2}}P_4 + \left(T_{1,\frac{5}{2}} + T_{\frac{3}{2},3} \right) P_7 - T_{\frac{3}{2},3}P_8 = -(A_7 c_{t,7})\left(P_7^{n+1} - P_7^n\right)$$

Block #i = 8 (j = 2, k = 3).
There is a no flow, Neumann boundary condition, on the (y = w) boundary (so there is no contribution to flux in that direction).

Example 4.1. Mass balance equations—cont'd

$$T_{2,\frac{5}{2}}\left(P_5 - P_8\right) + T_{\frac{3}{2},3}\left(P_7 - P_8\right) + T_{\frac{5}{2},3}\left(P_9 - P_8\right) + T_{2,\frac{7}{2}}\left(\cancel{P_{2,NT+1} - P_8}\right) = A_8 c_{t,8}\left(P_8^{n+1} - P_8^n\right)$$

Rearranging and grouping like terms

$$-T_{2,\frac{5}{2}}P_5 - T_{\frac{3}{2},3}P_7 + \left(T_{2,\frac{5}{2}} + T_{\frac{3}{2},3} + T_{\frac{5}{2},3}\right)P_8 - T_{\frac{5}{2},3}P_9 = -\left(A_8 c_{t,8}\right)\left(P_8^{n+1} - P_8^n\right)$$

Block #i = 9 (*j* = 3, *k* = 3).
There is a no flow, Neumann boundary condition, on the *y* = *w* boundary (so there is no contribution to flux in that direction). On the *x* = *L* boundary, there is a constant pressure, Dirichlet, boundary condition, P_B.

$$T_{3,\frac{5}{2}}\left(P_6 - P_9\right) + T_{\frac{5}{2},3}\left(P_8 - P_9\right) + T_{\frac{7}{2},3}\left(\underbrace{P_{4,3} - P_9}_{2T_{x,9}(P_B - P_9)}\right) + T_{3,\frac{7}{2}}\left(\cancel{P_{2,NT+1} - P_9}\right) = A_9 c_{t,9}\left(P_9^{n+1} - P_9^n\right)$$

Rearranging and grouping like terms

$$-T_{3,\frac{5}{2}}P_6 - T_{\frac{5}{2},3}P_8 + \left(T_{3,\frac{5}{2}} + T_{\frac{5}{2},3} + J_9\right)P_9 = -\left(A_9 c_{t,9}\right)\left(P_9^{n+1} - P_9^n\right) + Q_9$$

where $J_9 = 2T_{x,9}$ and $Q_9 = 2T_{x,9}P_B$ and $T_{x,9}$ is the *x*-direction transmissibility of block #9.

There are nine algebraic equations and nine unknown block pressures. In the equations, the time index was not specified on the spatial differences, which is dependent on the numerical scheme (explicit, implicit, etc.). The interblock transmissibilities and block accumulation terms can be computed from reservoir/fluid properties and is done so in Example 4.2. The nine equations can also be arranged in matrices and vectors; these are shown in Example 4.3.

4.4.5 Initial conditions

Initial pressure of reservoir blocks is determined by using the depth of the grid blocks, density of the fluid, and a reference pressure at a reference depth (e.g., surface or water-oil contact line) as was described in Chapter 1.

4.5 Reservoir heterogeneities

In most practical applications, heterogeneities in permeability, porosity, grid size, fluid properties, etc., must be accounted for in the reservoir simulator. It is common to employ variable grid block sizes in reservoir simulation because it allows for large blocks in regions of the reservoir that have small changes in pressure and small grids where more accuracy is needed or pressure changes significantly over short distances (e.g., near a well, a fracture, or a shock front). Fig. 4.6A depicts a 1D reservoir with variable grid block sizes and permeability.

Other reservoir and fluid variables in Darcy's law can vary spatially. For example, the width (Δy) and/or thickness (Δz) of the reservoir can vary with position (even in 1D simulations), meaning that the cross-sectional area varies along the length of the reservoir (if Δy or Δz varies significantly, it is recommended that a 2D or 3D simulator is used). If the changes are small, interblock areas can be used in the simulation.

Flow occurs from one block (e.g., $j - 1, k, l$) to another block (e.g., j, k, l). Therefore, if those two blocks have different permeability, grid size, cross-sectional area, fluid properties, etc., then the interblock phase transmissibility should be some sort of mean between the two blocks,

$$T_{\alpha,j+\frac{1}{2},k,l} = \left(\frac{k_{r,\alpha}^0}{\mu_\alpha B_\alpha}\right)_{j+\frac{1}{2},k,l} \left(\frac{k_x a}{\Delta x}\right)_{j+\frac{1}{2},k,l} \tag{4.17}$$

In Eq. (4.17), endpoint relatively permeability of the one flowing phase, α, is used but in later chapters it is shown the equation can be used for multiphase flow where relative permeability is a function of saturation. Interblock fluid-dependent properties ($k_{r,\alpha}^0$, μ_a) are generally treated differently than geometric properties (k, Δx, Δy, Δz) and thus are separated. In Eq. (4.17), transmissibility is in the x-direction, so x-directional permeability is used along with Δx as the distance between blocks and $\Delta y \Delta z$ the cross-sectional area. For flow in the y- and z-direction, Eq. (4.17) must be adjusted accordingly.

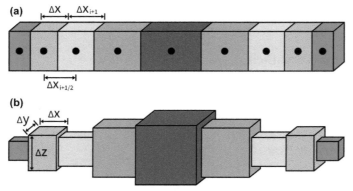

FIGURE 4.6 (A) 1D reservoir discretized into N nonuniform, cell-centered blocks. (B) 1D reservoir discretized blocks of varying cross-sectional area.

4.5.1 Fluid properties

Viscosity and formation volume factor may be assumed a constant or be given as a function of pressure as discussed in Chapter 1. For slightly compressible fluids (e.g. aqueous, oleic phases), viscosity and formation volume factor are weak functions of pressure. Pressure-dependent properties may vary in both space and time; they can be evaluated in each grid block using the known pressure in the previous timestep, $\mu_{i,} = f(P_i^n)$. Endpoint relative permeability is usually assumed to be constant, although could vary spatially if the rock type or wettability varied in the reservoir or with time if the wettability changed. For single-phase flow problems, an arithmetic mean for the interblock fluid properties is often an acceptable approximation. For multiphase flow problems, a technique referred to as *upwinding* is used to evaluate the interblock fluid properties, but this is left for discussion in later chapters.

4.5.2 Geometric properties

Geometric properties (e.g., k, Δx, Δy, Δz) can vary significantly between two adjacent grid blocks, even orders of magnitude in the case of permeability. Equations for interblock geometric properties are derived by assuming steady-state flow between two adjacent blocks. Interblock transmissibility can be written as follows by assuming the flow rate, q, is constant through the two adjacent blocks,

$$\frac{1}{T_{\alpha,j+1/2,k,l}} = \frac{\left(P_{j,k,l} - P_{j+1,k,l}\right)}{q_{\alpha,j+1/2,k,l}} = \frac{\left(P_{j,k,l} - P_{j+1/2,k,l}\right)}{q_{\alpha,j+1/2,k,l}} + \frac{\left(P_{j+1/2,k,l} - P_{j+1,k,l}\right)}{q_{\alpha,j+1/2,k,l}} = \frac{2}{T_{\alpha,j,k,l}} + \frac{2}{T_{\alpha,j+1,k,l}}$$

(4.18)

where $P_{j+1/2,k,l}$ is the pressure at the interface between the two adjacent blocks and the total pressure drop between the two blocks is the sum of the pressure drop from the center of block i (j,k,l) to the interface and the interface to block $i+1$ $(j+1,k,l)$. Eq. (4.18) is a harmonic mean. Assuming the interblock fluid properties can be determined separately, Eq. (4.18) can be simplified to give a harmonic mean of geometric properties,

$$\left(\frac{k_x \Delta y \Delta z}{\Delta x}\right)_{j+\frac{1}{2},k,l} = 2\left[\frac{1}{\left(\frac{k_x \Delta y \Delta z}{\Delta x}\right)_{j,k,l}} + \frac{1}{\left(\frac{k_x \Delta y \Delta z}{\Delta x}\right)_{j+1,k,l}}\right]^{-1}$$

$$= \frac{2(k_x \Delta y \Delta z)_{j,k,l}(k_x \Delta y \Delta z)_{j+1,k,l}}{(k_x \Delta y \Delta z)_{j,k,l}\Delta x_{j+1,k,l} + (k_x \Delta y \Delta z)_{j+1,k,l}\Delta x_{j,k,l}},$$

(4.19)

where Eq. (4.19) was derived for flow in the x-direction between blocks i, $i+1$, but equivalent expressions can be used for the other interblock transmissibilities. It can be further simplified if the block cross-sectional areas and/or the grid sizes are uniform.

The harmonic mean for interblock transmissibility was derived mathematically but it also makes sense physically. Consider an extreme case, where a relatively high permeability block (e.g., a sandstone with 100 mD) is adjacent to an impermeable block (e.g., shale with 0 mD) and grids are of uniform size. A simple arithmetic mean suggests that the effective permeability is $(100 + 0)/2 = 50$ mD and fluid flows, relatively easily, between the two blocks. In reality, no flow would occur at all between the blocks because the shale block is impermeable. In the example of a block of 100 mD sandstone adjacent to 0 mD shale, the harmonic mean (Eq. 4.19) of permeability is 0 mD, just as one would expect. The harmonic mean of permeability is exact for 1D, steady-state problems. For 2D and 3D problems, the harmonic mean can underestimate the effective permeability or transmissibility. Advanced approaches for upscaling permeability have been developed (Christie, 1996; Holden & Nielsen, 2000; Durlofsky, 2005), but this is beyond the scope of this chapter. Upscaling is also commonly used to coarsen grids when fine grids are provided in order to improve computational efficiency. Example 4.2 demonstrates the calculation of interblock transmissibilities for a 2D, heterogeneous and anisotropic reservoir.

Example 4.2. Interblock transmissibilities

A heterogeneous 2D reservoir ($L = 3000$ ft, $w = 3000$ ft, $h = 20$ ft) is saturated with oil at residual water saturation. The endpoint relative permeability and viscosity are constant and are 1.0 and 5 cp, respectively. Formation volume factor of oil is uniform throughout the reservoir. The total compressibility is $1E{-}5$ psi^{-1}. The reservoir is discretized into $N_x = 3$, $N_y = 3$, $N_z = 1$ nonuniform grids, $\Delta x = [750\ 1000\ 1250]$ ft and $\Delta y = [750\ 1000\ 1250]$ ft. The permeability is anisotropic, $k_y = 0.5 \times k_x$, $k_z = 0.1 \times k_x$ and both x-direction permeability and porosity are heterogeneous as given below. The depth is uniform.

Calculate the 12 interblock transmissibilities (6 in the x-direction and 6 in the y-direction) and 9 block accumulation terms.

$$k_x = [50 \quad 100 \quad 200 \quad 100 \quad 150 \quad 250 \quad 150 \quad 200 \quad 300]\ \text{mD}$$

$$\phi = [0.15 \quad 0.18 \quad 0.20 \quad 0.17 \quad 0.20 \quad 0.22 \quad 0.22 \quad 0.25 \quad 0.26]$$

Solution

Calculate the six x-direction interblock transmissibilities.

$$T_{j+1/2,k} = B_{\alpha,i} T_{\alpha,j+1/2,k} = B_{\alpha,i} \left(\frac{k_{r,\alpha}^0}{\mu_\alpha B_\alpha}\right)_{j+1/2,k} \frac{2(k_x \Delta y \Delta z)_{j,k}(k_x \Delta y \Delta z)_{j+1,k}}{(k_x \Delta y \Delta z)_{j,k}\Delta x_{j+1,k} + (k_x \Delta y \Delta z)_{j+1,k}\Delta x_{j,k}}$$

$$= \left(\frac{k_{r,\alpha}^0}{\mu_\alpha}\right)_{j+1/2,k} \frac{2\Delta z(k_x \Delta y)_{j,k}(k_x \Delta y)_{j+1,k}}{(k_x \Delta y)_{j,k}\Delta x_{j+1,k} + (k_x \Delta y)_{j+1,k}\Delta x_{j,k}}$$

Example 4.2. Interblock transmissibilities—cont'd

Note that block formation volume factors are equal to interblock values because they are stated as uniform in the problem.

$$T_{3/2,1} = \left(\frac{k_{r,\alpha}^0}{\mu_\alpha}\right) \frac{2\Delta z(k_x\Delta y)_{1,1}(k_x\Delta y)_{2,1}}{(k_x\Delta y)_{1,1}\Delta x_{2,1} + (k_x\Delta y)_{2,1}\Delta x_{1,1}} = \left(\frac{1}{5}\right) \frac{2\cdot 20(50\cdot 750)(100\cdot 750)}{(50\cdot 750)1000 + (100\cdot 750)750}$$

$$= 240 \frac{mD - ft}{cp} = 1.5192 \frac{ft^3}{psi - day}$$

$$T_{5/2,1} = \left(\frac{k_{r,\alpha}^0}{\mu_\alpha}\right) \frac{2\Delta z(k_x\Delta y)_{2,1}(k_x\Delta y)_{3,1}}{(k_x\Delta y)_{2,1}\Delta x_{3,1} + (k_x\Delta y)_{3,1}\Delta x_{2,1}} = \left(\frac{1}{5}\right) \frac{2\cdot 20(100\cdot 750)(200\cdot 750)}{(100\cdot 750)1250 + (200\cdot 750)1000}$$

$$= 369 \frac{mD - ft}{cp} = 2.3372 \frac{ft^3}{psi - day}$$

$$T_{3/2,2} = \left(\frac{k_{r,\alpha}^0}{\mu_\alpha}\right) \frac{2\Delta z(k_x\Delta y)_{1,2}(k_x\Delta y)_{2,2}}{(k_x\Delta y)_{1,2}\Delta x_{2,2} + (k_x\Delta y)_{2,2}\Delta x_{1,2}} = \left(\frac{1}{5}\right) \frac{2\cdot 20(100\cdot 1000)(150\cdot 1000)}{(100\cdot 1000)1000 + (150\cdot 1000)750}$$

$$= 564.7 \frac{mD - ft}{cp} = 3.5746 \frac{ft^3}{psi - day}$$

$$T_{5/2,2} = \left(\frac{k_{r,\alpha}^0}{\mu_\alpha}\right) \frac{2\Delta z(k_x\Delta y)_{2,2}(k_x\Delta y)_{3,2}}{(k_x\Delta y)_{2,2}\Delta x_{3,2} + (k_x\Delta y)_{3,2}\Delta x_{2,2}} = \left(\frac{1}{5}\right) \frac{2\cdot 20(150\cdot 1000)(250\cdot 1000)}{(150\cdot 1000)1250 + (250\cdot 1000)1000}$$

$$= 685.7 \frac{mD - ft}{cp} = 4.3406 \frac{ft^3}{psi - day}$$

$$T_{3/2,3} = \left(\frac{k_{r,\alpha}^0}{\mu_\alpha}\right) \frac{2\Delta z(k_x\Delta y)_{1,3}(k_x\Delta y)_{2,3}}{(k_x\Delta y)_{1,3}\Delta x_{2,3} + (k_x\Delta y)_{2,3}\Delta x_{1,3}} = \left(\frac{1}{5}\right) \frac{2\cdot 20(150\cdot 1250)(200\cdot 1250)}{(150\cdot 1250)1000 + (200\cdot 1250)750}$$

$$= 1000 \frac{mD - ft}{cp} = 6.33 \frac{ft^3}{psi - day}$$

$$T_{5/2,3} = \left(\frac{k_{r,\alpha}^0}{\mu_\alpha}\right) \frac{2\Delta z(k_x\Delta y)_{2,3}(k_x\Delta y)_{3,3}}{(k_x\Delta y)_{2,3}\Delta x_{3,3} + (k_x\Delta y)_{3,3}\Delta x_{2,3}}$$

$$= \left(\frac{1}{5}\right) \frac{2\cdot 20(200\cdot 1250)(300\cdot 1250)}{(200\cdot 1250)1250 + (300\cdot 1250)1000}$$

$$= 1090.9 \frac{mD - ft}{cp} = 6.9055 \frac{ft^3}{psi - day}$$

Continued

Example 4.2. Interblock transmissibilities—cont'd

Calculate the six y-direction transmissibilities

$$T_{j,k+1/2} = B_{\alpha,i} T_{\alpha,j,k+1/2} = B_{\alpha,i} \left(\frac{k_{r,\alpha}^0}{\mu_\alpha B_\alpha}\right)_{j,k+\frac{1}{2}} \frac{2(k_y \Delta x \Delta z)_{j,k}(k_y \Delta x \Delta z)_{j,k+1}}{(k_y \Delta x \Delta z)_{j,k}\Delta y_{j,k+1} + (k_y \Delta x \Delta z)_{j,k+1}\Delta y_{j,k}}$$

$$= \left(\frac{k_{r,\alpha}^0}{\mu_\alpha}\right)_{j,k+\frac{1}{2}} \frac{2\Delta z(k_y \Delta x)_{j,k}(k_y \Delta x)_{j,k+1}}{(k_y \Delta x)_{j,k}\Delta y_{j,k+1} + (k_y \Delta x)_{j,k+1}\Delta y_{j,k}}$$

$$T_{1,3/2} = \left(\frac{k_{r,\alpha}^0}{\mu_\alpha}\right)_{1,\frac{3}{2}} \frac{2\Delta z(k_y \Delta x)_{1,1}(k_y \Delta x)_{1,2}}{(k_y \Delta x)_{1,1}\Delta y_{1,2} + (k_y \Delta x)_{1,2}\Delta y_{1,1}} = \left(\frac{1}{5}\right) \frac{2 \cdot 20(25 \cdot 750)(50 \cdot 750)}{(25 \cdot 750)1000 + (50 \cdot 750)750}$$

$$= 120\frac{mD - cp}{ft} = 0.7596\frac{ft^3}{psi - day}$$

$$T_{2,3/2} = \left(\frac{k_{r,\alpha}^0}{\mu_\alpha}\right)_{2,\frac{3}{2}} \frac{2\Delta z(k_y \Delta x)_{2,1}(k_y \Delta x)_{2,2}}{(k_y \Delta x)_{2,1}\Delta y_{2,2} + (k_y \Delta x)_{2,2}\Delta y_{2,1}} = \left(\frac{1}{5}\right) \frac{2 \cdot 20(50 \cdot 1000)(75 \cdot 1000)}{(50 \cdot 1000)1000 + (75 \cdot 1000)750}$$

$$= 282.4\frac{mD - cp}{ft} = 1.7873\frac{ft^3}{psi - day}$$

$$T_{3,3/2} = \left(\frac{k_{r,\alpha}^0}{\mu_\alpha}\right)_{3,\frac{3}{2}} \frac{2\Delta z(k_y \Delta x)_{3,1}(k_y \Delta x)_{3,2}}{(k_y \Delta x)_{3,1}\Delta y_{3,2} + (k_y \Delta x)_{3,2}\Delta y_{3,1}} = \left(\frac{1}{5}\right) \frac{2 \cdot 20(100 \cdot 1250)(125 \cdot 1250)}{(100 \cdot 1250)1000 + (125 \cdot 1250)750}$$

$$= 645.2\frac{mD - cp}{ft} = 4.0839\frac{ft^3}{psi - day}$$

$$T_{1,5/2} = \left(\frac{k_{r,\alpha}^0}{\mu_\alpha}\right)_{1,\frac{5}{2}} \frac{2\Delta z(k_y \Delta x)_{1,2}(k_y \Delta x)_{1,3}}{(k_y \Delta x)_{1,2}\Delta y_{1,3} + (k_y \Delta x)_{1,3}\Delta y_{1,2}} = \left(\frac{1}{5}\right) \frac{2 \cdot 20(50 \cdot 750)(75 \cdot 750)}{(50 \cdot 750)1250 + (75 \cdot 750)1000}$$

$$= 163.6\frac{mD - cp}{ft} = 1.0358\frac{ft^3}{psi - day}$$

$$T_{2,5/2} = \left(\frac{k_{r,\alpha}^0}{\mu_\alpha}\right)_{2,\frac{5}{2}} \frac{2\Delta z(k_y \Delta x)_{2,2}(k_y \Delta x)_{2,3}}{(k_y \Delta x)_{2,2}\Delta y_{2,3} + (k_y \Delta x)_{2,3}\Delta y_{2,2}} = \left(\frac{1}{5}\right) \frac{2 \cdot 20(75 \cdot 1000)(100 \cdot 1000)}{(75 \cdot 1000)1250 + (100 \cdot 1000)1000}$$

$$= 309.7\frac{mD - cp}{ft} = 1.9603\frac{ft^3}{psi - day}$$

$$T_{3,5/2} = \left(\frac{k_{r,\alpha}^0}{\mu_\alpha}\right)_{3,\frac{5}{2}} \frac{2\Delta z(k_y \Delta x)_{3,2}(k_y \Delta x)_{3,3}}{(k_y \Delta x)_{3,2}\Delta y_{3,3} + (k_y \Delta x)_{3,3}\Delta y_{3,2}} = \left(\frac{1}{5}\right) \frac{2 \cdot 20(125 \cdot 1250)(150 \cdot 1250)}{(125 \cdot 1250)1250 + (150 \cdot 1250)1000}$$

$$= 612.2\frac{mD - cp}{ft} = 3.8755\frac{ft^3}{psi - day}$$

Calculate the nine block accumulation terms

$$A_i = \frac{\Delta x_i \Delta y_i \Delta z \phi_i}{\Delta t}$$

Example 4.2. Interblock transmissibilities—cont'd

$$A_1 = \frac{\Delta x_1 \Delta y_1 \Delta z \phi_1}{\Delta t} = \frac{750 \cdot 750 \cdot 20 \cdot 0.15}{5} = 3.375 \times 10^5 \frac{\text{ft}^3}{\text{day}}$$

$$A_2 = \frac{\Delta x_2 \Delta y_2 \Delta z \phi_2}{\Delta t} = \frac{1000 \cdot 750 \cdot 20 \cdot 0.18}{5} = 5.40 \times 10^5 \frac{\text{ft}^3}{\text{day}}$$

$$A_3 = \frac{\Delta x_3 \Delta y_3 \Delta z \phi_3}{\Delta t} = \frac{1250 \cdot 750 \cdot 20 \cdot 0.20}{5} = 7.50 \times 10^5 \frac{\text{ft}^3}{\text{day}}$$

$$A_4 = \frac{\Delta x_4 \Delta y_4 \Delta z \phi_4}{\Delta t} = \frac{750 \cdot 1000 \cdot 20 \cdot 0.17}{5} = 5.10 \times 10^5 \frac{\text{ft}^3}{\text{day}}$$

$$A_5 = \frac{\Delta x_5 \Delta y_5 \Delta z \phi_5}{\Delta t} = \frac{1000 \cdot 1000 \cdot 20 \cdot 0.20}{5} = 8.0 \times 10^5 \frac{\text{ft}^3}{\text{day}}$$

$$A_6 = \frac{\Delta x_6 \Delta y_6 \Delta z \phi_6}{\Delta t} = \frac{1250 \cdot 1000 \cdot 20 \cdot 0.22}{5} = 11.0 \times 10^5 \frac{\text{ft}^3}{\text{day}}$$

$$A_7 = \frac{\Delta x_7 \Delta y_7 \Delta z \phi_7}{\Delta t} = \frac{750 \cdot 1000 \cdot 20 \cdot 0.22}{5} = 8.25 \times 10^5 \frac{\text{ft}^3}{\text{day}}$$

$$A_8 = \frac{\Delta x_8 \Delta y_8 \Delta z \phi_8}{\Delta t} = \frac{1000 \cdot 1250 \cdot 20 \cdot 0.25}{5} = 12.50 \times 10^5 \frac{\text{ft}^3}{\text{day}}$$

$$A_9 = \frac{\Delta x_9 \Delta y_9 \Delta z \phi_9}{\Delta t} = \frac{1250 \cdot 1250 \cdot 20 \cdot 0.26}{5} = 16.25 \times 10^5 \frac{\text{ft}^3}{\text{day}}$$

4.5.3 Accumulation terms

Porosity, compressibility, and block volume may also vary with grid blocks. These variables are included in the accumulation term of the mass balance (i.e., **A**) and interblock values are not needed. The diagonal elements of the accumulation matrix can be computed as follows,

$$A_{i,i} = \frac{V_i \phi_i}{\Delta t} \tag{4.20}$$

4.6 Matrix arrays

4.6.1 Accumulation and compressibility

The accumulation, **A**, and compressibility, c_t, arrays are both $N \times N$ diagonal arrays with each diagonal element corresponding to the respective block. $A_{i,i}$ is given by Eq. (4.20) and the compressibilities, $c_{t,i,i}$ are a saturation-weighted sum of fluid and formation compressibilities (Eq. 1.18).

4.6.2 Transmissibility

The transmissibility, **T**, matrix is tridiagonal, pentadiagonal, and heptadiagonal in 1D, 2D, and 3D, respectively. The $T_{i,j}$ element of the matrix is the negative of interblock transmissibility between blocks i and j and thus $T_{i,j}$ is only nonzero if the two blocks are adjacent (or $i{=}j$). For a sufficiently large number of grids, most entries of T are zero and thus the matrix is sparse. The main diagonal, $T_{i,i}$, is the absolute value of the sum of the off-diagonals. For homogeneous, isotropic media, all the interblock transmissibilities are the same and equal to the block transmissibility. If the permeability is anisotropic and/or heterogeneous, the transmissibility would not be constant. On a given row, i, each off-diagonal term would be unique and the main diagonal would be negative the sum of the off-diagonal terms. The **T** matrix, shown below, depicts the pentadiagonal matrix for a homogeneous 9-block ($N_x{=}3$, $N_y{=}3$) reservoir.

$$
T = \begin{pmatrix}
2T & -T & & -T & & & & & \\
-T & 3T & -T & & -T & & & & \\
& -T & 2T & & & -T & & & \\
-T & & & 3T & -T & & -T & & \\
& -T & & -T & 4T & -T & & -T & \\
& & -T & & -T & 3T & & & -T \\
& & & -T & & & 2T & -T & \\
& & & & -T & & -T & 3T & -T \\
& & & & & -T & & -T & 2T
\end{pmatrix}
$$

The number of columns between the far-left diagonal and the far-right diagonal is referred to as the *bandwidth*. A higher bandwidth can sometimes require additional computer memory and increase the computation time for the linear solver. In the indexing system proposed in this text, the bandwidth equals the number of blocks in the x-direction (N_x), but different indexing systems can be used to reduce (or increase) bandwidth. It is, however, highly recommended to use *sparse storage formatting* in code development, which only saves the nonzero entries.

4.6.3 Source terms

For constant rate wells or constant rate boundary conditions, the rate (at reservoir conditions) is directly substituted into the source vector, Q_i, if the well or boundary is associated with that block. The value of Q_i is positive for injectors (or flow is into the boundary) and negative for producers (or flow out of the boundary).

For constant pressure (Dirichlet) boundaries, a diagonal $\mathbf{J}$ matrix is used with $J_{i,i} = +2T_i$, where T_i is the transmissibility of that boundary block. The source vector is $Q_i = +2T_iP_B$ where P_B is the value of the constant pressure at the boundary of block i. If the block is subject to more than one boundary condition (i.e. corner block), Q_i is a summation of those boundary conditions.

4.6.4 Gravity

Reservoirs are usually not horizontal and therefore gravity effects must be included. Fig. 4.7 shows a reservoir with varying depth.

FIGURE 4.7   Reservoir with nonuniform depths.

Gravity was included in the balance equations through Darcy's law and a gravity vector, G, was included in the matrix equations,

$$\overrightarrow{G} = \rho_\alpha g \mathbf{T} \overrightarrow{D}; \quad \overrightarrow{D} = \begin{bmatrix} D_1 \\ D_2 \\ \vdots \\ D_N \end{bmatrix}, \tag{4.21}$$

where D is a vector of depths and D_i is the depth of the center of grid i. Note that in Eq. (4.21), the fluid density has been treated as a constant. If the density is taken to be function of pressure, interblock densities (or formation volume factors) should be used.

Once the arrays are created, they can be used in Eq. (4.13) to solve the system of algebraic equations implicitly, explicitly, or with a mixed method. Example 4.3 demonstrates the solution for a 2D reservoir with uniform depth, and Example 4.4 demonstrates the solution for a 2D reservoir with varying depth.

Example 4.3. Solution to pressure field in 2D

Create the arrays and solve the pressure field implicitly for three timesteps for the 2D, 9-block system in Exercises 4.1 and 4.2. The boundary condition at x = L is P = 1000 psia. The initial condition is P = 3000 psia and the depth is uniform.

Continued

Example 4.3. Solution to pressure field in 2D—cont'd

Solution

The transmissibility matrix is pentadiagonal,

$$
T =
\begin{pmatrix}
\sum |T_1| & -T_{3/2,1} & & -T_{1,3/2} & & & & & \\
-T_{3/2,1} & \sum |T_2| & -T_{5/2,1} & & -T_{2,3/2} & & & & \\
& -T_{5/2,1} & \sum |T_3| & & & -T_{3,3/2} & & & \\
-T_{1,3/2} & & & \sum |T_4| & -T_{3/2,2} & & -T_{1,5/2} & & \\
& -T_{2,3/2} & & -T_{3/2,2} & \sum |T_5| & -T_{5/2,2} & & -T_{2,5/2} & \\
& & -T_{3,3/2} & & -T_{5/2,2} & \sum |T_6| & & & -T_{3,5/2} \\
& & & -T_{1,5/2} & & & \sum |T_7| & -T_{3/2,3} & \\
& & & & -T_{2,5/2} & & -T_{3/2,3} & \sum |T_8| & -T_{5/2,3} \\
& & & & & -T_{3,5/2} & & -T_{5/2,3} & \sum |T_9|
\end{pmatrix}
=
$$

Where the main diagonal is the absolute value of the sum of off-diagonal terms. Using the values obtained in Example 4.2,

$$
T =
\begin{bmatrix}
2.2788 & -1.5192 & & -0.7596 & & & & & \\
-1.5192 & 5.6437 & -2.3372 & & -1.7873 & & & & \\
& -2.3372 & 6.4211 & & & -4.0839 & & & \\
-0.7596 & & & 5.3700 & -3.5746 & & -1.0358 & & \\
& -1.7873 & & -3.5746 & 11.6627 & -4.3406 & & -1.9603 & \\
& & -4.0839 & & -4.3406 & 12.3000 & & & -3.8755 \\
& & & -1.0358 & & & 7.3658 & -6.3300 & \\
& & & & -1.9603 & & -6.3300 & 15.1957 & -6.9055 \\
& & & & & -3.8755 & & -6.9055 & 10.7810
\end{bmatrix}
\frac{ft^3}{psi - day}
$$

Example 4.3. Solution to pressure field in 2D—cont'd

The accumulation matrix is diagonal and the matrix elements were computed in Example 4.2,

$$
\mathbf{Ac}_t =
\begin{pmatrix}
A_1 c_{t,1} & & & & & & & & \\
& A_2 c_{t,2} & & & & & & & \\
& & A_3 c_{t,3} & & & & & & \\
& & & A_4 c_{t,4} & & & & & \\
& & & & A_5 c_{t,5} & & & & \\
& & & & & A_6 c_{t,6} & & & \\
& & & & & & A_7 c_{t,7} & & \\
& & & & & & & A_8 c_{t,8} & \\
& & & & & & & & A_9 c_{t,9}
\end{pmatrix}
$$

$$
=
\begin{pmatrix}
3.38 & & & & & & & & \\
& 5.40 & & & & & & & \\
& & 7.50 & & & & & & \\
& & & 5.10 & & & & & \\
& & & & 8.0 & & & & \\
& & & & & 11.0 & & & \\
& & & & & & 8.25 & & \\
& & & & & & & 12.50 & \\
& & & & & & & & 16.25
\end{pmatrix}
\frac{ft^3}{psi - day}
$$

Dirichlet boundary conditions on the $x = L$ boundary are included in the J matrix and Q vector

$$
\mathbf{J} =
\begin{pmatrix}
2T_{x,3} & & \\
& 2T_{x,6} & \\
& & 2T_{x,9}
\end{pmatrix}
=
\begin{pmatrix}
6.08 \\
10.13 \\
15.19
\end{pmatrix}
\frac{ft^3}{psi - day}
$$

Continued

Example 4.3. Solution to pressure field in 2D—cont'd

$$\vec{Q} = \begin{pmatrix} 0 \\ 0 \\ 2T_{x,3}P_B \\ 0 \\ 0 \\ 2T_{x,6}P_B \\ 0 \\ 0 \\ 2T_{x,9}P_B \end{pmatrix} = \begin{pmatrix} 0 \\ 0 \\ 6076.8 \\ 0 \\ 0 \\ 10128.1 \\ 0 \\ 0 \\ 15192.0 \end{pmatrix} \frac{ft^3}{day}$$

Solving the problem using the implicit method,

$$(\mathbf{T} + \mathbf{J} + \mathbf{Ac}_t)\vec{P}^{n+1} = \mathbf{Ac}_t\vec{P}^n + \vec{Q}$$

$$\vec{P}^0 = \begin{pmatrix} 3000 \\ 3000 \\ 3000 \\ 3000 \\ 3000 \\ 3000 \\ 3000 \\ 3000 \\ 3000 \end{pmatrix} \text{psia; } \vec{P}^1 = \begin{pmatrix} 2927.4 \\ 2779.9 \\ 2196.4 \\ 2899.6 \\ 2752.8 \\ 2166.7 \\ 2891.6 \\ 2749.0 \\ 2162.9 \end{pmatrix} \text{psia; } \vec{P}^2 = \begin{pmatrix} 2791.9 \\ 2521.9 \\ 1781.7 \\ 2730.2 \\ 2480.7 \\ 1759.6 \\ 2711,8 \\ 2474.5 \\ 1758.4 \end{pmatrix} \text{psia;}$$

$$\vec{P}^3 = \begin{pmatrix} 2620.9 \\ 2285.1 \\ 1554.0 \\ 2532.9 \\ 2240.4 \\ 1541.7 \\ 2505.8 \\ 2232.9 \\ 1543.2 \end{pmatrix} \text{psia; } \vec{P}^{\infty} = \begin{pmatrix} 1000 \\ 1000 \\ 1000 \\ 1000 \\ 1000 \\ 1000 \\ 1000 \\ 1000 \\ 1000 \end{pmatrix} \text{psia}$$

Example 4.4. Solution to pressure field in 2D with gravity

Calculate the pressure field at the first three timesteps for the reservoir described in Examples 4.1–4.3 if the reservoir depth is not uniform, but is defined by the vector given below. The density of the fluid is 53 lb_m/ft^3 and $P = 3000$ psia at a depth of 6400 ft. The initial pressure is in static equilibrium.

$$D = [6400 \quad 6000 \quad 5600 \quad 6000 \quad 5800 \quad 5400 \quad 5600 \quad 5400 \quad 5200] ft$$

Solution

First compute the initial pressure using the grid depths, D, and

$$P^0 = P_{ref} + \rho_o g(D - D_{ref})$$

Then compute the gravity, G, vector.

$$G = \rho_o g \mathbf{T} \vec{D}$$

where the **T** matrix was defined in Example 4.3. The solution can be found by solving the system of equations with the gravity vector (**T**, **A**, **J**, c_t, and Q are all unchanged from Example 4.3).

$$(\mathbf{T} + \mathbf{J} + \mathbf{A}c_t) \vec{P}^{n+1} = \mathbf{A}c_t \vec{P}^n + \vec{Q} + \vec{G}$$

$$
\vec{G} = \begin{pmatrix} 335.5 \\ 252.0 \\ -43.5 \\ 303.8 \\ 532.9 \\ -654.3 \\ 313.5 \\ -246.2 \\ -793.6 \end{pmatrix} \frac{ft^3}{day}; \quad
\vec{P}^0 = \begin{pmatrix} 3000 \\ 2852.8 \\ 2705.6 \\ 2852.8 \\ 2779.2 \\ 2631.9 \\ 2705.6 \\ 2531.9 \\ 2558.3 \end{pmatrix} psia; \quad
\vec{P}^1 = \begin{pmatrix} 2939.5 \\ 2668.3 \\ 2027.0 \\ 2771.2 \\ 2577.1 \\ 1952.4 \\ 2620.2 \\ 2434.8 \\ 1903.0 \end{pmatrix} psia;
$$

Continued

Example 4.4. Solution to pressure field in 2D with gravity—cont'd

$$
\vec{P}^2 = \begin{pmatrix} 2827.1 \\ 2453.3 \\ 1679.6 \\ 2633.8 \\ 2356.8 \\ 1620.9 \\ 2478.4 \\ 2218.5 \\ 1585.0 \end{pmatrix} \text{psia}; \quad
\vec{P}^3 = \begin{pmatrix} 2685.8 \\ 2257.1 \\ 1490.2 \\ 2473.9 \\ 2161.7 \\ 1443.6 \\ 2315.3 \\ 2027.4 \\ 1415.1 \end{pmatrix} \text{psia}; \quad
\vec{P}^\infty = \begin{pmatrix} 1365.0 \\ 1208.7 \\ 1037.3 \\ 1235.8 \\ 1159.1 \\ 1005.2 \\ 1112.5 \\ 1042.8 \\ 981.6 \end{pmatrix} \text{psia}
$$

The initial pressure is in static equilbrium; however the steady state solution is not because the boundary conditions are not in static equilibrium (P=1000 psia at x=L, regardless of depth). As a result the pressure in block #9 (981.6 psia) is lower than the boundary pressure at steady state.

4.7 Pseudocode for single-phase flow in multidimensions

Here, we present pseudocode for modeling single phase, multidimensional flow in a reservoir. As usual, it is recommended to create several simple subroutines/functions/modules that complete a very specific task. Although there is no single *correct* way to code this problem, a straightforward approach is presented below. It is recommended to code *bottom up*; that is, create the most basic subroutines first and then create the subroutines that call them. The order in which subroutines are presented below is the recommended, but not required, order that they should be developed. It is also highly recommended that each subroutine is tested and validated upon completion.

4.7.1 Preprocessing

The preprocessing subroutine presented in Chapter 1 for initializing a reservoir can be adapted for the problem of single-phase flow and thus is not shown here. Input reservoir, fluid, petrophysical, well, and numerical data are read in from data files or spreadsheets, grids and grid locations are assigned, and well-containing grids are identified. Initial grid pressures are determined using the grid depths as described in Chapter 1. Desired plots, such as the initial pressure/saturation field, well locations, petrophysical properties, etc. can be made.

4.7.2 Interblock transmissibility

Any two integers (i_1, i_2) corresponding to two (usually adjacent) grid blocks are used as inputs along with reservoir, fluid, petrophysical, and numerical properties of all blocks. Blocks i_1 and i_2 can also be the same block, the interblock transmissibility of which would just be the block transmissibility. The direction of connection (x, y, or z) should be provided or determined in the subroutine from the indexing scheme. First, interblock geometric properties are computed using Eq. (4.19); the coder should be careful to use the correct directional values of permeability, grid size, etc., for anisotropic systems. Second, the interblock fluid properties can be computed. A method for computation of interblock fluid properties in multiphase flow using upwinding will be presented in Chapter 9. The interblock transmissibility of a phase can be computed using Eq. (4.17).

4.7.3 Well Arrays

A vector, Q, can be created where the ith element of Q corresponds to the constant rate of the well in the block. The well can be an injector ($+Q$) or producer ($-Q$). In the code, we loop through all the wells and determine which grid block, i, the well resides. If the well is a constant rate well, then it is added to the source vector. If there are multiple wells in the grid block, the rates can be summed, but it is highly recommended to use smaller grids so as to have no more than one well per grid. Constant bottomhole pressure wells are discussed in Chapter 5. The Q vector should also include boundary conditions as decribed in section 4.6.3 but here we propose to add the boundary conditions in a separate *Grid Arrays* subroutine.

4.7.4 Grid Arrays

The matrices and vectors used for solving pressure (e.g., $\mathbf{A}$, $\mathbf{c_t}$, $\mathbf{T}$, Q, and G) are created by looping through all grid blocks. For a grid, i, the main diagonals of $\mathbf{A}$, $\mathbf{c_t}$ are computed using Eq. (4.20). Interblock transmissibilities are also computed between the block and each of its neighbors by calling the *Interblock* subroutine. "if" statements are needed to determine if the block has a neighbor and if not it is on a boundary, and Q and $\mathbf{J}$ arrays must be updated. Importantly, the vector Q was already created for wells and boundary conditions should be added to the source vector (not replace it). The gravity vector, G, can be computed using Eq. (4.21).

4.7.5 Main code

In the main code, the *Well Arrays* and *Grid Arrays* subroutine are called to compute all the arrays ($\mathbf{T}$, $\mathbf{A}$, $\mathbf{J}$, $\mathbf{c_t}$, Q, G). A loop through time is created and at each timestep pressure is computed implicitly, explicitly, or mixed by solving the system of equations using Eq. (4.13). The time is then updated and the loop continues until the final time is reached or some other condition such as small pressure changes. Pressures can be replaced in each timestep or saved for making plots in the postprocessing code. Saving results at every timestep can be memory intensive; an alternative is to save a subset of times, e.g., every tenth or hundredth of timesteps. It should be noted that the arrays are created only once (outside the loop) assuming they do not change with time. However, if the arrays do change with pressure or time, they must be updated. This is discussed in Chapter 6 (nonlinearities) and Chapter 9 (multiphase flow).

4.7.6 Postprocessing

After the code is successfully run and free of errors, the user will want to analyze and visualize the results. For example, time series plots of average reservoir pressure can be created. Spatially dependent properties, such as reservoir pressure, can be plotted in 2D or 3D at specified times or dimensionless times.

MAIN CODE

```
CALL PREPROCESS
CALL WELL ARRAYS
CALL GRID ARRAYS
WHILE t < t_final
      SET P_old = P
      CALCULATE P using eqn 4.13
      INCREMENT t by Δt
ENDWHILE
CALL POSTPROCESS
```

SUBROUTINE INTERBLOCK
```
INPUTS: i₁, i₂, direction, reservoir, petrophysical, fluid,
numerical properties
OUTPUTS: T_Half (interblock transmissibility)

CALCULATE interblock geometric properties using eqn 4.19
CALCULATE interblock fluid properties
CALCULATE interblock T_Half using eqn 4.17
```

SUBROUTINE WELL ARRAYS
```
INPUTS: numerical, petrophysical, fluid, well properties
OUTPUTS: Q

FOR J = 1 to # wells
    FOR I = 1 to # perforated blocks of well J
        IF well(J) = constant rate well
            CALCULATE Q(i) using eqn. 4.14 and Qwell(J)
        ENDIF
    ENDFOR
ENDFOR
```

SUBROUTINE GRID ARRAYS
```
INPUTS: numerical, petrophysical, fluid properties and boundary
conditions
OUTPUTS: T, A, J, Q, G
```

```
Pre-allocate (sparse storage format preferred) T, A, Q, J, P_wf

FOR i = 1 to # grids (N = Nx*Ny*Nz)
     SET c_t(i,i) using eqn. 1.18
     SET A(i,i) = Δx(i)* Δy(i)* Δz(i)* φ(i)/ Δt

     IF NOT on left (x=0) boundary
          SET T(i,i-1) = -INTERBLOCK(i,i-1,'x')
          SET T(i,i) = T(i,i) - T(i,i-1)
     ELSEIF Dirichlet boundary condition
          SET Tx = INTERBLOCK(i,i,'x')
          SET J(i,i) = J(i,i) + 2*Tx
          SET Q(i) = Q(i) + 2*Tx*P_LB
     ELSEIF Neumann boundary condition
          CALCULATE Q(i) using eqn 4.14
     ENDIF

     IF NOT on right (x=L) boundary
          SET T(i,i+1) = -INTERBLOCK(i,i+1,'x')
          SET T(i,i) = T(i,i) - T(i,i+1)
     ELSEIF Dirichlet boundary condition
          SET Tx = INTERBLOCK(i,i,'x')
          SET J(i,i) = J(i,i) + 2*Tx
          SET Q(i) = Q(i) + 2*Tx*P_RB
     ELSEIF Neumann boundary condition
          CALCULATE Q(i) using eqn 4.14
     ENDIF

     IF NOT on bottom (y=0) boundary
          SET T(i,i-NX) = -INTERBLOCK(i,i-NX,'y')
          SET T(i,i) = T(i,i) - T(i,i-NX)
     ELSEIF Dirichlet boundary condition
          SET Ty = INTERBLOCK(i,i,'y')
          SET J(i,i) = J(i,i) + 2*Ty
          SET Q(i) = Q(i) + 2*Ty*P_BB
     ELSEIF Neumann boundary condition
          CALCULATE Q(i) using eqn 4.14
     ENDIF

     IF NOT on top (y=w) boundary
          SET T(i,i+NX) = -INTERBLOCK(i,i+NX,'y')
          SET T(i,i) = T(i,i) - T(i,i+NX)
     ELSEIF Dirichlet boundary condition
          SET Ty = INTERBLOCK(i,i,'y')
          SET J(i,i) = J(i,i) + 2*Ty
          SET Q(i) = Q(i) + 2*Ty*P_TB
     ELSEIF Neumann boundary condition
          CALCULATE Q(i) using eqn 4.14
     ENDIF
```

```
      IF NOT on front (z=0) boundary
           SET T(i,i-NX*NY) = -INTERBLOCK(i,i-NX*NY,'z')
           SET T(i,i) = T(i,i) - T(i,i-NX*NY)
      ELSEIF Dirichlet boundary condition
           SET Tz = INTERBLOCK(i,i,'z')
           SET J(i,i) = J(i,i) + 2*Tz
           SET Q(i) = Q(i) + 2*Tz*P_FB
      ELSEIF Neumann boundary condition
           CALCULATE Q(i) using eqn 4.14
      ENDIF

      IF NOT on back (z=h) boundary
           SET T(i,i+NX*NY) = -INTERBLOCK(i,i+NX*NY,'z')
           SET T(i,i) = T(i,i) - T(i,i+NX*NY)
      ELSEIF Dirichlet boundary condition
           SET Tz = INTERBLOCK(i,i,'z')
           SET J(i,i) = J(i,i) + 2*Tz
           SET Q(i) = Q(i) + 2*Tz*P_BaB
      ELSEIF Neumann boundary condition
           CALCULATE Q(i) using eqn 4.14
      ENDIF

  ENDFOR
  CALCULATE G using eqn 4.21


SUBROUTINE POSTPROCESS
INPUTS: reservoir, fluid, petrophysical, well, numerical, solution
properties
OUTPUTS: well data, plots

CALCULATE average reservoir pressure at all times
PLOT average reservoir pressure versus time
PLOT (contour, surface, etc.) pressure field of all blocks at desired
times
VALIDATE against analytical solution, if applicable
```

4.8 Exercises

Exercise 4.1. **Expansion of terms.** Show that the accumulation term can be expanded to give the right-hand side of Eq. (4.6).

Exercise 4.2. **Heterogeneities.** Repeat Example 3.4 in Chapter 3 using the implicit method but with heterogeneous permeability, porosity, and grid sizes as defined below,

$$k = \begin{bmatrix} 50 & 75 & 100 & 150 \end{bmatrix}$$

$$\phi = \begin{bmatrix} 0.15 & 0.18 & 0.20 & 0.23 \end{bmatrix}$$

$$\Delta x = \begin{bmatrix} 500 & 1000 & 1000 & 1500 \end{bmatrix}$$

Exercise 4.3. Gravity. Repeat Exercise 4.2 but for the case where the reservoir depth is not uniform. The grid depths (ft) are given below

$$D = \begin{bmatrix} 6420 & 6147 & 5783 & 5328 \end{bmatrix}$$

The initial pressure, p_{init}, in block #1 is 3000 psia and other blocks are in hydrostatic equilibrium. The density of the oleic phase is 50 lb_m/ft^3.

Exercise 4.4. 2D reservoir with irregular geometry. Repeat the 2D Example (4.3) for three timesteps, but (1) use no-flow boundary conditions on all boundaries, (2) add a constant rate producer well of 5000 ft^3/day (reservoir conditions) in block #5, and (3) block #1 is inactive (outside the reservoir boundaries). You do not need to recompute any transmissibilities or accumulation terms that are unchanged from the example problem. Calculate the average reservoir pressure at each time and compare to the analytical solution,

$$\bar{p}(t) = p_{init} - \frac{q}{V_p c_t} t,$$

where the average pressure is the pore volume-weighted sum of block pressures,

$$\bar{p} = \frac{1}{V_p} \sum_{i=1}^{N} p_i \Delta x_i \Delta y_i \Delta z_i \phi_i.$$

Exercise 4.5. 3D reservoir. A 3D shallow aquifer ($L = 2000$ ft, $w = 2000$ ft, $h = 200$ ft) has $N_x = 2$, $N_y = 2$, $N_z = 2$ uniform grids and the following homogenous properties: $\phi = 0.2$, $k_x = k_y = 10 k_z = 100$ mD, $\mu_w = 1$ cp, $B_w = 1$ RB/STB, $c_t = 1 \times 10^{-6}$ psi^{-1}, $\rho_w = 64$ lb_m/ft^3. A vertical injector well is perforated in block #1 and a vertical producer well in block #8. Both have a constant rate of 10,000 ft^3/day. The depth at the center of the bottom layer is $D = 200$ feet and the boundary conditions are no flow on all boundaries. Solve the pressure field implicitly for the first three timesteps and use a timestep of $\Delta t = 1$ day.

Exercise 4.6. Reservoir simulator development. Adapt the computer code (reservoir simulator) developed in Exercise 3.6 that reads in data from an input file and solves for pressure versus time in a 2D reservoir for single-phase flow. The simulator should allow for heterogeneities in permeability and porosity, anisotropy in permeability, and nonuniform grids. It should also allow for both

Dirichlet and Neumann boundary conditions, constant rate wells, and include gravitational effects. Hint: use the pseudocode as a guide and break the specific tasks into subroutines or functions. Validate your code against Examples 4.3, 4.4, and any other exercises and examples. For additional complexity, expand your simulator to 3D.

Exercise 4.7. Project. Use the reservoir simulator developed in Exercise 4.6 to solve for grid pressures versus time in the synthetic Thomas oilfield, first introduced in Exercise 1.8. The boundary conditions are no-flow on all boundaries and a constant rate producer of 10,000 STB/day is located in the reservoir center. The oil viscosity is 5 cp. Assume that $S_w = S_{wr}$ everywhere in the reservoir, but use the initial oleic phase pressures computed in Exercise 1.8. Continue the simulation until $t_D = 2.0$, where,

$$t_{D,pss} = \frac{\pi k k_{r,\alpha}^0 t}{\mu \phi c_t L w},$$

where average values of permeability and porosity can be used. Plot the average reservoir pressure versus time and compare to the analytical solution introduced in Exercise 4.4. Plot the pressure field in the reservoir at $t_D = 0.05$, 0.1, 0.2, and 2.

References

Abou-Kassem, J.H., Farouq Ali, S.M., Islam, M.R., 2006. Petroleum Reservoir Simulation: A Basic Approach. Gulf Publishing Co., Houston, Texas, USA.

Aziz, K., 1993. Reservoir simulation grids: opportunities and problems. Journal of Petroleum Technology 45 (07), 658−663.

Christie, M.A., 1996. Upscaling for reservoir simulation. Journal of Petroleum Technology 48 (11), 1004−1010.

Ding, Y., Renard, G., Weill, L., 1998. Representation of wells in numerical reservoir simulation. SPE Reservoir Evaluation and Engineering 1 (01), 18−23.

Dogru, A.H., Fung, L.S., Middya, U., Al-Shaalan, T.M., Byer, T., Hoy, H., Hahn, W.A., Al-Zamel, N., Pita, J., Hemanthkumar, K., Mezghani, M., Al-Mana, A., Tan, J., Dreiman, W., Fugl, A., Al-Baiz, A., February 2011. New Frontiers in Large Scale Reservoir Simulation. Paper presented at the. SPE Reservoir Simulation Symposium, The Woodlands, Texas, USA. https://doi.org/10.2118/142297-MS.

Durlofsky, L.J., 2005. Upscaling and gridding of fine scale geological models for flow simulation. 8th International forum on reservoir simulation Iles Borromees. Stresa, Italy 2024. Citeseer.

Holden, L., Nielsen, B.F., 2000. Global upscaling of permeability in heterogeneous reservoirs; the output least squares (ols) method. Transport in Porous Media 40 (2), 115−143.

Quandalle, P., 1993. Eighth SPE comparative solution project: gridding techniques in reservoir simulation. In: SPE Symposium on Reservoir Simulation. OnePetro.

Wheeler, M.F., et al., 1999. A parallel multiblock/multidomain approach for reservoir simulation. In: Proceedings of the 15th Reservoir Simulation Symposium. SPE Richardson, TX. No. 51884.

Chapter 5

Radial flow, wells, and well models

5.1 Introduction

Fluids are produced from reservoirs through wells and during secondary or tertiary recovery; fluids may also be injected to increase the reservoir pressure and displace fluids in place. Wells, producers or injectors, may be operated at either a constant rate or a constant bottomhole pressure (BHP). When flow is modeled in the near-well region, cylindrical coordinates are used because the wells are cylindrical in geometry. For modeling the entire reservoir or large sections of the reservoir, the well radii are very small (inches) compared to the reservoir domain (thousands of feet or more) and grid blocks. The reservoir is then modeled using Cartesian coordinates and wells are included as point sources or sinks, as was briefly introduced in Chapter 4. In this chapter, (1) numerical techniques for modeling radial flow near wells and (2) well models for direct implementation into the matrix equations to solve for pressure in the reservoir are developed for Cartesian coordinates.

5.2 Radial flow equations and analytical solutions

Wellbores are cylindrical pipes that are relatively small in diameter (e.g., 4–8 in.) compared to the size of the reservoir. Because of their geometry, a cylindrical coordinate system is used to describe mathematically radial fluid flow near a well (Fig. 5.1).

The partial differential equations that describe radial flow around a well were derived in Chapter 2 (Eq. 2.18). For 1D flow with constant, homogenous fluid and rock properties and negligible gravity, the equation reduced to Eq. (2.19),

$$\frac{\partial p}{\partial t} = \frac{\alpha}{r} \frac{\partial}{\partial r} \left(r \frac{\partial p}{\partial r} \right), \tag{5.1}$$

An Introduction to Multiphase, Multicomponent Reservoir Simulation.
https://doi.org/10.1016/B978-0-323-99235-0.00005-1

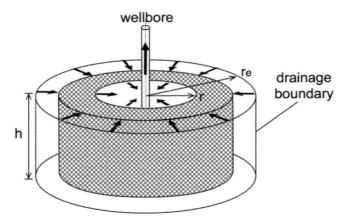

FIGURE 5.1 Wellbore in the center of a cylindrical domain.

where α is the diffusivity constant. The equation has analytical solutions under certain initial and boundary conditions, a few of which were presented in Chapter 2. For 1D, radial flow the boundary conditions are defined at the wellbore, $r = r_w$, and far from the well (i.e., the external drainage radius, r_e, or at $r = \infty$).

Production wells can be operated at constant rate ($q = q_{wf}$ at $r = r_w$) or constant bottomhole pressure, BHP ($P = P_{wf}$ at $r = r_w$). During the early stages of primary production, operators may attempt to maintain constant well flowing rate (Fig. 5.2A). Production of fluids at the surface results in fluid expansion and reservoir/well pressure decreases with time (Fig. 5.2B). Once the well bottomhole pressure reaches a technical limit (e.g., bubble point pressure), the bottomhole pressure is usually constrained to that pressure, P_{lim}. However, wells may be operated at constant BHP from the beginning of production.

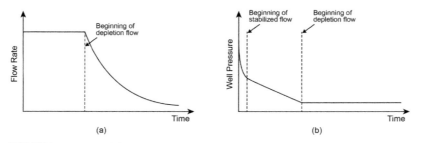

FIGURE 5.2 (A) Well flow rate versus time and (B) well bottomhole pressure versus time.

Wells theoretically undergo four stages of flow during primary recovery: (a) *infinite acting* or *transient*, (b) *transitional*, (c) *stabilized* (steady state or pseudo-steady state), and (d) *depletion* flow (Walsh and Lake, 2003). The first three stages are during the constant rate period and the fourth stage (depletion flow) occurs when the well is operated at constant bottomhole pressure.

At very early times, the well is infinite acting or transient, meaning that the outer *drainage boundary*, r_e, has been impacted by the well. The solution for a constant rate well during infinite acting flow was given in Chapter 2 (Eq. 2.27). If the well is operated at constant BHP, the flow is still infinite acting at early time but the solution is different. *Transitional flow* follows infinite acting flow and as the name indicates is short lived and a transition to the stabilized flow period. This flow stage is generally not well described mathematically. *Stabilized* flow refers to the constant rate period, following transitional flow, in which the drainage boundary is impacted by the well. Stabilized flow is further categorized by two limits, *pseudo-steady* (also known as *semi-steady* or *quasi-steady*) *state flow* and *steady state flow*—or some hybrid between the two limits. In pseudo-steady-state (pss) flow, the boundary is completely sealed and the reservoir pressure decreases linearly with time at all radii, *r*. In steady-state flow, the boundary is completely open, flow into the boundary equals flow produced from the well, and the reservoir pressure is constant with time.

During *depletion* flow, production rate of the well declines with time and approaches zero rate, usually at constant BHP. It can be shown that for a homogenous reservoir during primary recovery, the rate should decline exponentially but many wells follow more complicated behavior such as hyperbolic decline (Arps, 1945) or other empirical models (Ilk et al., 2008; Valko, 2009; Duong, 2011). At some time, the well (or reservoir) production rate is no longer large enough to be economical. The well may then be shut-in or secondary/tertiary methods pursued by converting the well into an injector or drilling new wells.

5.3 Numerical solutions to the radial diffusivity equation

A numerical approach is required to solve the diffusivity equation in cylindrical coordinates if reservoir properties are not homogenous, the problem is multidimensional, boundary conditions are complex, etc.

5.3.1 Gridding

For near-well problems, uniform or nonuniform grids can be employed for the *z*- and *θ*-direction. However, uniform grid sizes are rarely employed in the *r*-direction because the fluid velocity accelerates near the well and as a result larger changes in pressure are observed. Fig. 5.3 illustrates gridding in a cylindrical coordinate system.

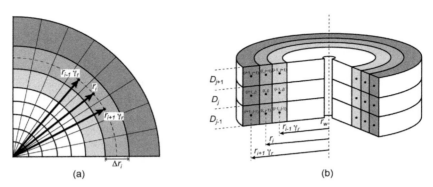

FIGURE 5.3 Schematic of geometric grids in radial flow problems. (A) 2D view and (B) 3D view.

Grids are often spaced using a geometric progression, given by the formulae,

$$r_{i+1} = r_i \gamma_r; \quad \gamma_r = \left(\frac{r_e}{r_w} \right)^{1/N},$$ (5.1a)

where r_e and r_w are the drainage (outer) and well radius, respectively, and the radius of the first grid, $i = 1$, is given as,

$$r_1 = r_w \frac{\ln(\gamma_r)}{1 - \left(1/\gamma_r \right)}.$$ (5.1b)

5.3.2 Discretization

The diffusivity equation in cylindrical coordinates can be discretized directly using a forward/backward difference for the time derivative and centered differences for spatial derivatives. However, Aziz and Settari (1979) note that there are two volumes of interest, a discretized volume, C_{vi}, which can be derived by considering a discretized element in cylindrical coordinates and the actual volume of the block, V_i, which can be found by subtracting the concentric cylinders,

$$C_{vi} = \Delta \theta_i \Delta z_i r_i \left(r_{j+\frac{1}{2},k,l} - r_{j-\frac{1}{2},k,l} \right),$$ (5.2a)

$$V_i = 0.5\Delta\theta_i\Delta z_i r_{j+\frac{1}{2},k,l}^2 - 0.5\Delta\theta_i\Delta z_i r_{j-\frac{1}{2},k,l}^2 = 0.5\Delta\theta_i\Delta z_i \left(r_{j+\frac{1}{2},k,l}^2 - r_{j-\frac{1}{2},k,l}^2 \right),$$

(5.2b)

The equations are only equivalent in the limit that Δr approaches zero, but are not the same in general. An interblock radius can be defined to make the two volumes equal, but then the numerical solution does not give the exact solution for steady-state flow.

An alternative is to perform a variable transformation (Aziz and Settari, 1979) by defining $\rho = r^2$ and substituting into the PDE. This results in the volumes being equal and the numerical solution produces the exact solution for steady-state flow. The discretization and derivation of transmissibilities and block volumes can be found in several texts (Aziz and Settari, 1979; Ertekin et al., 2001; Abou-Kassem et al., 2013) and the final results are summarized here. As was the case in Cartesian coordinates, a system of linear equations results,

$$(\mathbf{T} + \mathbf{J} + \mathbf{Ac}_t)\overrightarrow{P}^{n+1} = \mathbf{Ac}_t\overrightarrow{P}^n + \overrightarrow{Q} + \overrightarrow{G},$$

(5.3)

where the interblock phase transmissibilities for the r, θ, and z-directions are defined as,

$$T_{\alpha,j+\frac{1}{2},k,l} = \left(\frac{k_{r,\alpha}^0}{\mu_\alpha B_\alpha}\right)_{j+\frac{1}{2},k,l} \frac{1}{\left(\frac{1}{k_R\Delta z\Delta\theta}\right)_{j,k,l} \ln\left(\frac{r_{i+1}-r_i}{r_i \ln(r_{i+1}/r_i)}\right) + \left(\frac{1}{k_R\Delta z\Delta\theta}\right)_{j+1,k,l} \ln\left(\frac{r_{i+1}\ln(r_{i+1}/r_i)}{(r_{i+1}-r_i)}\right)},$$

(5.4a)

$$T_{\alpha,j,k+\frac{1}{2},l} = \left(\frac{k_{r,\alpha}^0}{\mu_\alpha B_\alpha}\right)_{j,k+\frac{1}{2},l} \frac{2}{\left(\frac{\Delta\theta}{k_\theta\Delta z}\right)_{j,k,l} + \left(\frac{\Delta\theta}{k_\theta\Delta z}\right)_{j,k+1,l}} \left(\frac{r_{i+1}-r_i}{r_i-r_{i-1}}\right) \frac{\ln r_i/r_{i-1}}{\ln r_{i+1}/r_i},$$

(5.4b)

$$T_{\alpha,j,k,l+\frac{1}{2}} = \left(\frac{k_r^0}{\mu_\alpha B_\alpha}\right)_{j,k,l+\frac{1}{2}} \frac{1}{\left(\frac{\Delta z}{k_z\Delta\theta}\right)_{j,k,l} + \left(\frac{\Delta z}{k_z\Delta\theta}\right)_{j,k,l+1}} \frac{r_{i+1}^2-r_i^2}{\ln\left(r_{i+1}^2/r_i^2\right)} - \frac{r_i^2-r_{i-1}^2}{\ln\left(r_i^2/r_{i-1}^2\right)},$$

(5.4c)

where k_R, k_θ, k_z, are the anisotropic permeabilities in the r-, θ-, and z- directions, respectively. The positions of the cell-centered radii are given in Eq. (5.1). $\Delta\theta$ and Δz are the finite lengths of the block in the θ- and z- direction as depicted in Figure 5.4. Equations 5.4A, 5.4B, and 5.4C are written for adjacent blocks forward in space but equivalent expressions can be used for blocks adjacent backward in space. If the radii are spaced using a geometric progression and all other properties are homogeneous, Eq. (5.4) can be simplified,

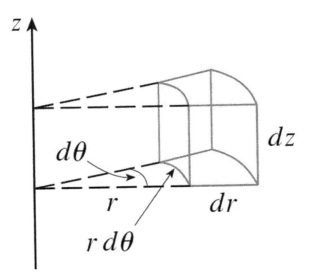

FIGURE 5.4   Schematic of a discretized element in cylindrical coordinates.

$$T_{\alpha,j+\frac{1}{2},k,l} = \frac{k^0_{r,\alpha}}{\mu_\alpha B_\alpha} \frac{\Delta z \Delta \theta}{\ln\left(\frac{r_{i+1}}{r_i}\right)} = \frac{k^0_{r,\alpha}}{\mu_\alpha B_\alpha} \frac{k_R \Delta z \Delta \theta}{\ln(\gamma_r)} \tag{5.5a}$$

$$T_{\alpha,j,k+\frac{1}{2},l} = \frac{k^0_{r,\alpha}}{\mu_\alpha B_\alpha} \frac{2 k_\theta \Delta z}{\Delta \theta} \gamma_r \tag{5.5b}$$

$$T_{\alpha,j,k,l+\frac{1}{2}} = \frac{k^0_r}{\mu_\alpha B_\alpha} \frac{k_z \Delta \theta}{\Delta z} \frac{r^2_{i+1} + r^2_{i-1}}{\ln(\gamma^2_r)} \tag{5.5c}$$

and the overall interblock transmissibility for use in Eq. (5.3) is the phase interblock transmissibility (Eqs. 5.4 and 5.5) weighted by the phase formation volume factor of the block ($T = B_{\alpha,i}T_\alpha$). The accumulation term is given by,

$$A_i = \frac{\phi \Delta \theta_i \Delta z_i}{2 \Delta t} \left(\frac{r^2_{i+1} - r^2_i}{\ln\left(r^2_{i+1}/r^2_i\right)} - \frac{r^2_i - r^2_{i-1}}{\ln\left(r^2_i/r^2_{i-1}\right)} \right) = \frac{\phi \Delta \theta_i \Delta z_i}{2 \Delta t} \left(\frac{r^2_{i+1} - 2r^2_i + r^2_{i-1}}{2 \ln(\gamma_r)} \right) \tag{5.6}$$

where the right-hand side of Eq. (5.6) assumes geometric grid spacing. The source terms in Eq. (5.3) usually come from boundary conditions, since the boundary is the well and can be included using the methods described in Chapters 3 and 4. Example 5.1 demonstrates the solution for 1D radial flow.

Example 5.1. 1D radial flow problem.

Solve the pressure field around a wellbore that has a production rate in reservoir conditions of $q_w = 5000$ ft^3/day. The wellbore radius, r_w, is 0.25 ft and the drainage radius, r_e, is 1000 ft. The initial pressure, p_{init}, is 3000 psia and the reservoir is sealed at the drainage radius. The reservoir has a constant permeability, k, of 50 mD, constant porosity, ϕ, of 0.20, and reservoir thickness, h, of 100 ft. The fluid has a viscosity, μ_o, of 5 cp, and total compressibility, $c_t = 10^{-5}$ psi^{-1}. Flow is single phase with an endpoint relative permeability of 1.0. Use a geometric progression of cell-centered grids and solve for the reservoir pressure for the first three timesteps using $N = 4$ grids and $\Delta t = 1$ day. Assume that formation volume factor is uniform.

Solution

First, compute the constant (γ_r) for the geometric progression of radii,

$$\gamma_r = \left(\frac{r_e}{r_w}\right)^{1/N} = \left(\frac{1000}{0.25}\right)^{1/4} = 7.9527$$

Next, calculate the location of radii of cell-centered grid blocks

$$r_1 = r_w \frac{\ln(\gamma_r)}{1 - \left(\frac{1}{\gamma_r}\right)} = 0.25 \frac{\ln(7.9527)}{1 - \left(\frac{1}{7.9527}\right)} = 0.5929 \text{ ft}$$

$$r_2 = \gamma_r r_1 = 4.715 \text{ ft}$$

$$r_3 = \gamma_r r_2 = 37.50 \text{ ft}$$

$$r_4 = \gamma_r r_3 = 298.23 \text{ ft}$$

$$r_5 = \gamma_r r_4 = 2371.74 \text{ ft}$$

$$r_0 = \frac{r_1}{\gamma_r} = \frac{0.5929 \text{ ft}}{7.9527} = 0.07455 \text{ ft}$$

Note that r_0 and r_5 are outside the reservoir domain but are used in the calculation of the **A** matrix.

Compute interblock transmissibilities and the accumulation terms. All properties are homogeneous and uniform. Using Eq. (5.5a),

$$T_{i+\frac{1}{2}} = B_{\alpha,i} T_{\alpha,i+\frac{1}{2}} = B_{\alpha,i} \left(\frac{k_{r,\alpha}^0}{\mu_\alpha B_\alpha}\right) \frac{k\Delta\theta\Delta z}{\ln(\gamma_r)} = \frac{2\pi k_{r,o}^0 kh}{\mu_o \ln(\gamma_r)} = \frac{2\pi \cdot 1 \cdot (50 \text{ mD})(100\text{ft})}{(5 \text{ cp})\ln(7.9527)}$$

$$= 3030 \frac{\text{mD} - \text{ft}}{\text{cp}} = 19.18 \frac{\text{ft}^3}{\text{psi} - \text{day}}$$

Continued

Example 5.1. 1D radial flow problem.—cont'd

The boundary conditions at the well and drainage radius are both Neumann, $T_{1/2} = T_{9/2} = 0$.

The accumulation matrix can be computed from Eq. (5.6),

$$A_1 = \frac{\varphi \pi h}{\Delta t} \frac{r_2^2 - 2r_1^2 + r_0^2}{2 \ln(\gamma_r)} = \frac{0.2 \cdot \pi (100 \text{ft})}{1 \text{day}} \frac{(4.715)^2 - 2(0.593)^2 + (0.0746)^2}{2 \ln(7.9527)}$$

$$= 3.26 \times 10^2 \frac{\text{ft}^3}{\text{day}}$$

$$A_2 = \frac{\varphi \pi h}{\Delta t} \frac{r_3^2 - 2r_2^2 + r_1^2}{2 \ln(\gamma_r)} = \frac{0.2 \cdot \pi (100 \text{ft})}{1 \text{day}} \frac{(37.50)^2 - 2(4.715)^2 + (0.593)^2}{2 \ln(7.9527)}$$

$$= 2.064 \times 10^4 \frac{\text{ft}^3}{\text{day}}$$

$$A_3 = \frac{\varphi \pi h}{\Delta t} \frac{r_4^2 - 2r_3^2 + r_2^2}{2 \ln(\gamma_r)} = \frac{0.2 \cdot \pi (100 \text{ft})}{1 \text{day}} \frac{(298.23)^2 - 2(37.50)^2 + (4.715)^2}{2 \ln(7.9527)}$$

$$= 1.305 \times 10^6 \frac{\text{ft}^3}{\text{day}}$$

$$A_4 = \frac{\varphi \pi h}{\Delta t} \frac{r_5^2 - 2r_4^2 + r_3^2}{2 \ln(\gamma_r)} = \frac{0.2 \cdot \pi (100 \text{ft})}{1 \text{day}} \frac{(2371.74)^2 - 2(298.23)^2 + (37.50)^2}{2 \ln(7.9527)}$$

$$= 8.255 \times 10^7 \frac{\text{ft}^3}{\text{day}}$$

Build the arrays

$$\mathbf{T} = \begin{pmatrix} T_{1/2} + T_{3/2} & -T_{3/2} & 0 & 0 \\ -T_{3/2} & T_{3/2} + T_{5/2} & -T_{5/2} & 0 \\ 0 & -T_{5/2} & T_{5/2} + T_{7/2} & -T_{7/2} \\ 0 & 0 & -T_{7/2} & T_{7/2} + T_{9/2} \end{pmatrix}$$

$$= \begin{pmatrix} 19.18 & -19.18 & 0 & 0 \\ -19.18 & 38.36 & -19.18 & 0 \\ 0 & -19.18 & 38.36 & -19.18 \\ 0 & 0 & -19.18 & 19.18 \end{pmatrix} \frac{\text{ft}^3}{\text{day} - \text{psi}}$$

Example 5.1. 1D radial flow problem.—cont'd

$$A = \begin{pmatrix} 3.26 \times 10^2 & & & \\ & 2.06 \times 10^4 & & \\ & & 1.305 \times 10^6 & \\ & & & 8.255 \times 10^7 \end{pmatrix} \frac{ft^3}{day};$$

$$c_t = \begin{pmatrix} 1 & & & \\ & 1 & & \\ & & 1 & \\ & & & 1 \end{pmatrix} \times 10^{-5} psi^{-1}$$

$$\vec{P}^0 = \begin{pmatrix} 3000 \\ 3000 \\ 3000 \\ 3000 \end{pmatrix} psia; \quad \vec{Q} = \begin{pmatrix} -5000 \\ 0 \\ 0 \\ 0 \end{pmatrix} \frac{ft^3}{day}$$

The solution is given by,

$$(T + J + Ac_t)\vec{P}^{n+1} = Ac_t\vec{P}^n + \vec{Q} + \vec{G}$$

where the **J** matrix is zero because there are no constant pressure boundary conditions and there is no gravity term.

The solution for the first three timesteps is,

$P^1 = [2328.8\ 2589.4\ 2845.5\ 2996.5]$ psia
$P^2 = [2257.2\ 2517.9\ 2777.7\ 2991.5]$ psia
$P^3 = [2225.7\ 2486.4\ 2746.7\ 2986]$ psia

5.4 Wells and well models in Cartesian grids

Large-scale simulation utilizes a domain that is orders of magnitude larger than the well radius and often contain many wells, spaced far apart. In these models, a Cartesian coordinate system is employed and wells are included in individual grid blocks as sources and sinks.

5.4.1 Well constraints

Wells can be producers or injectors and can be operated at constant rate or at constant bottomhole pressure. Constant rate wells are directly substituted into the source vector (units of reservoir conditions, e.g., ft³/day) of the matrix equations as was done in Chapter 4. Constant BHP wells are included using a well model. In the simulator, the pressure in each grid block is approximated with a constant, average pressure, P_i. If a constant BHP well resides in the block, one might naively assume that the block pressure could be specified as being equal to the well pressure, P_{wf}, but this is incorrect. In fact, it is the difference between the well pressure and the average grid block pressure that provides the driving force to inject or produce fluid. If the well pressure is lower than the average block pressure, the driving force is toward the well, and therefore the well is a producer. If the well pressure is higher than the block pressure, fluid injects into the reservoir.

A well model is needed that relates the well rate and bottomhole pressure. The well model can be used to determine the production/injection rate and substituted into the algebraic mass balance (matrix) equations. Conversely, the well pressure can be computed for a constant-rate well using the well model. We follow the derivation of Peaceman (1978) to derive the well model.

5.4.2 Steady-state radial flow around a well

The analytical solution for steady-state flow in cylindrical coordinates is,

$$P(r) = P_{ref} - \frac{q_{wf}\mu_\alpha}{2\pi k k^0_{r,\alpha} h} \ln\left(\frac{r}{r_{ref}}\right), \tag{5.7}$$

where P_{ref} is the pressure at a reference radius, r_{ref}, and q_{wf} is at reservoir conditions. Fig. 5.5 shows a schematic of the radial pressure profile outward from the well.

The average pressure of block i is P_i, and in cell-centered finite difference simulation a single pressure is assigned to the grid. Since pressure increases or decreases monotonically with radius, the average pressure, P_i, must lie somewhere in between the well radius, r_w, and edge of the block $r = \Delta x/2$. The radius at which $P(r) = P_i$ is referred to as the equivalent radius, r_{eq}. Using r_w as the reference radius (where $P = P_{wf}$) and using the fact that $P = P_i$ at $r = r_{eq}$,

$$P_i = P(r_w) - \frac{q_{wf}\mu}{2\pi k k^0_{r,\alpha} h} \ln\left(\frac{r_{eq}}{r_w}\right) = P_{wf} - \frac{q_{wf}\mu}{2\pi k k^0_{r,\alpha} h}\left[\ln\left(\frac{r_{eq}}{r_w}\right) + s\right], \tag{5.8}$$

where the dimensionless skin factor, s, accounts for an additional (e.g., due to formation damage, s is positive) or less (due to stimulation, s is negative) pressure drop, $P_{wf} - P(r_w)$, near the well.

Rearrangement of Eq. (5.8) yields a well model, i.e., a relationship for well flow rate versus *drawdown pressure* (the difference between reservoir block pressure and wellbore pressure),

$$q_{wf} = J^w_{i,\alpha}\left(P_{wf} - P_i\right) \tag{5.9}$$

where $J_{i,\alpha}$ is a productivity index of phase α of the well in block i defined as,

$$J_{i,\alpha} = \frac{2\pi k_H k^0_{r\alpha} h}{\mu_\alpha\left[\ln\left(\frac{r_{eq}}{r_w}\right) + s\right]}, \tag{5.10}$$

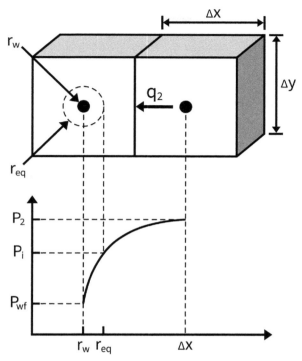

FIGURE 5.5 Schematic of radial pressure distribution from center of block "i" ($r = r_w$) to center of block #2 ($r = \Delta x$). The well radius is exaggerated in the figure.

and the skin, s, and radius, r_w, are properties of the well, and all reservoir/fluid properties are for phase α in block i. The endpoint relative permeability is used because only one phase is assumed flowing. In Chapter 9, the productivity index can be extended to multiple flowing phases using saturation dependent relative permeability. The thickness, h, refers to the thickness of the block which can be Δz, Δy, or Δx depending on the direction of the well. If the medium is anisotropic, the permeability, k_H, is the geometric mean of the directional block permeability.

$$k_H = \sqrt{k_x k_y} \tag{5.11}$$

Eq. (5.11) assumes the well is orthogonal to the x-y plane, i.e., a vertical well, but can be ammended for horizontal wells. The use of Eqs. (5.9) and (5.10) requires knowledge of the equivalent radius, i.e., the distance away from the well that the pressure equals the average grid block pressure.

5.4.3 Mass balance on the well-residing grid block

Consider a block i, which in 2D is connected to four adjacent blocks as shown in Fig. 5.6.

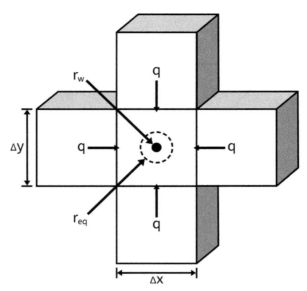

FIGURE 5.6 Grid block with a well connected to four adjacent grids. At the equivalent radius r_{eq} the pressure $P(r)$ is equal to the average pressure of the block, P_i.

Steady-state flow is assumed (no accumulation of mass with time), and a mass balance requires that flow in/out of grid block i from the four-neighboring blocks plus flow in or out from the well yields,

$$q_{wf} + (q_1 + q_2 + q_3 + q_4) = 0, \tag{5.12}$$

i.e., any fluid produced from the block equals the net flow into the block. Using Darcy's law for flow from each of the four adjacent blocks and assuming isotropic, homogenous permeability and square ($\Delta x = \Delta y$) grids,

$$q_{wf} = \frac{-kk_{r\alpha}^o h}{\mu_\alpha}(P_1 + P_2 + P_3 + P_4 - 4P_i). \tag{5.13}$$

Block centers are spaced Δx apart. The pressure varies radially from the well ($r = r_w$) to the center of the adjacent block ($r = \Delta x$) as shown in Fig. 5.5.

Recall Eq. (5.7), the steady-state analytical solution for pressure surrounding the well. Using a reference radius of r_{eq} (where $P = P_i$) and recognizing that the center of the adjacent block is Δx away from the well (Peaceman, 1978), equations for the adjacent block pressures can be used,

$$P_1 = P_2 = P_3 = P_4 = P_i - \frac{q_{wf}\mu_\alpha}{2\pi kk_{r\alpha}^o h}\ln\left(\frac{\Delta x}{r_{eq}}\right). \tag{5.14}$$

Substituting the radial pressure Eq. (5.14) into the mass balance Eq. (5.13) and simplifying gives,

$$1 = \frac{2}{\pi} \ln\left(\frac{\Delta x}{r_{eq}}\right) \tag{5.15}$$

Finally, Peaceman's Correction (1978) is obtained by rearranging Eq. (5.15) for the equivalent radius (r_{eq}):

$$r_{eq} = \Delta x e^{-\frac{\pi}{2}} \approx 0.2078 \Delta x \tag{5.16}$$

Eq. (5.16) suggests that $r_{eq,}$ the radius at which the pressure is equal to the average block pressure, is about 20% of the grid block size. Peaceman (1978) derived this correction in three different ways (numerically and exact in addition to the method shown here). Although a simple approximation to the flow field around the well, Peaceman's correction is surprisingly accurate and it, or simple variations, is implemented in most reservoir simulators through substitution into Eq. (5.9) and (5.10).

5.4.4 Extension to horizontal wells and anisotropy

Many modern wells are horizontal, that is, they traverse horizontally through the pay zone of the reservoir. They are often more effective than vertical wells because they contact more surface area of the reservoir. Mathematically, horizontal wells have a few differences with vertical wells as they traverse multiple grid blocks. Vertical wells can also traverse multiple blocks as well if the z-direction has more than one layer. Fig. 5.7 shows a horizontal well crossing three blocks in a 2D reservoir with one layer of thickness.

Peaceman's correction was derived for a vertical well in a 2D reservoir that is isotropic and homogeneous and the simulation grid uses square grid blocks ($\Delta x = \Delta y$). Peaceman (1983) and other authors (Van Poolen et al., 1986) have

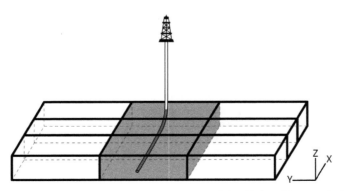

FIGURE 5.7   Schematic of horizontal well traversing multiple grids in a 2D reservoir simulation model.

suggested extensions and improvements. Peaceman's model for anisotropy is given by,

$$r_{eq} = 0.28 \frac{\left(\sqrt{k_2/k_1} \, (\Delta w_1)^2 + \sqrt{k_1/k_2} \, (\Delta w_2)^2 \right)^{1/2}}{\left(k_2/k_1 \right)^{1/4} + \left(k_1/k_2 \right)^{1/4}} \tag{5.17}$$

$$J_{i,\alpha} = \frac{2\pi \Delta w_3 k_{r,\alpha}^0 \sqrt{k_1 k_2}}{\mu_\alpha \left[\ln\left(\frac{r_{eq}}{r_w} \right) + s \right]} \tag{5.18}$$

where w is a proxy for the x-, y-, or z-direction; subscript 3 refers to the direction of the well and subscripts 1 and 2 are the orthogonal directions. For example, for a vertical well $\Delta w_3 = \Delta z$ and Δw_1 and Δw_2 are Δx and Δy, respectively, and k_1 and k_2 are k_x and k_y, respectively. Eq. (5.17) reduces to Peaceman's original Eq. (5.16) if the grids are square and permeability is isotropic.

Wells that traverse multiple grid blocks (such as the horizontal well depicted in Fig. 5.7) are usually assumed to have uniform bottomhole pressure along the length of the well (Havlena and Odeh, 1963; Jackson et al., 2011). If the well is operated at constant rate conditions, the rates must be distributed among all blocks the well passes through. An approximation is to distribute the rates uniformly or by weighting productivity indices. A more accurate approach involves imposing a bottomhole pressure constraint and then iterating to match the specified rate condition. The constrained bottomhole pressure will change every timestep, but the value in the previous timestep is good initial guess.

It is extremely important to calculate the productivity index carefully, especially in code development. For example, one could easily erroneously switch the directional permeability or grid size in Eqs. (5.17) and (5.18). While some errors created in coding are obvious (nonphysical results are obtained), calculation of the productivity index can go unnoticed and lead to poor predictions. Example 5.2 demonstrates the calculation of productivity index.

Example 5.2. Calculation of the productivity index.
For the reservoir and fluid properties described in Examples 4.2/4.3 in Chapter 4, calculate the productivity indices for a
  a. vertical well perforated in block #1. The wellbore radius is 0.25 ft and $s = 2$.
  b. horizontal well perforated in blocks #8 and #9. The wellbore radius is 0.25 ft and $s = 0$.

Example 5.2. Calculation of the productivity index.—cont'd

Solution

The general equation for a productivity index is given by Eqs. (5.17) and (5.18).

a. *Vertical well*

$$r_{eq} = 0.28 \frac{\left(\sqrt{k_{y,1}/k_{x,1}} \, (\Delta x_1)^2 + \sqrt{k_{x,1}/k_{y,1}} \, (\Delta y_1)^2 \right)^{1/2}}{\left(k_{y,1}/k_{x,1} \right)^{1/4} + \left(k_{x,1}/k_{y,1} \right)^{1/4}}$$

$$= 0.28 \frac{\left(\sqrt{25/50} \, (750)^2 + \sqrt{50/25} \, (750)^2 \right)^{1/2}}{\left(25/50 \right)^{1/4} + \left(50/25 \right)^{1/4}} = 150.7 \text{ft}$$

$$J_1 = \frac{2\pi h k_{r,\alpha}^0 \sqrt{k_x k_y}}{\mu_o \left[\ln\left(\frac{r_{eq}}{r_w} \right) + s \right]} = \frac{2\pi \cdot 20 \cdot 1.0 \sqrt{50 \cdot 25}}{5 \left[\ln\left(\frac{150.7}{0.25} \right) + 2 \right]}$$

$$= 105.77 \frac{\text{mD} - \text{ft}}{\text{cp}} = 0.6695 \frac{\text{ft}^3}{\text{psi} - \text{day}}$$

a. *Horizontal well*

The horizontal well is drilled in the x-direction in blocks 8 and 9. It is orthogonal to the y-z plane.

Block #8

$$r_{eq} = 0.28 \frac{\left(\sqrt{k_{z,8}/k_{y,8}} \, (\Delta y_8)^2 + \sqrt{k_{y,8}/k_{z,8}} \, (\Delta z_8)^2 \right)^{1/2}}{\left(k_{z,8}/k_{y,8} \right)^{1/4} + \left(k_{y,8}/k_{z,8} \right)^{1/4}}$$

$$= 0.28 \frac{\left(\sqrt{0.2} \, (1250)^2 + \sqrt{5} \, (20)^2 \right)^{1/2}}{(0.2)^{1/4} + (5)^{1/4}} = 108.2 \text{ ft}$$

$$J_8 = \frac{2\pi \Delta x_8 k_{r,\alpha}^0 \sqrt{k_{y,8} k_{z,8}}}{\mu_o \left[\ln\left(\frac{r_{eq}}{r_w} \right) + s \right]} = \frac{2\pi \cdot 1000 \cdot 1.0 \sqrt{100 \cdot 20}}{5 \left[\ln\left(\frac{108.2}{0.25} \right) + 0 \right]}$$

$$= 9258.0 \frac{\text{mD} - \text{ft}}{\text{cp}} = 58.6 \frac{\text{ft}^3}{\text{psi} - \text{day}}$$

Continued

Example 5.2. Calculation of the productivity index.—cont'd

Block #9

$$r_{eq} = 0.28 \frac{\left(\sqrt{k_{z,9}/k_{y,9}} \, (\Delta y_9)^2 + \sqrt{k_{y,9}/k_{z,9}} \, (\Delta z_9)^2\right)^{1/2}}{\left(k_{z,9}/k_{y,9}\right)^{1/4} + \left(k_{y,9}/k_{z,9}\right)^{1/4}}$$

$$= .28 \frac{\left(\sqrt{0.2} \, (1250)^2 + \sqrt{5} \, (20)^2\right)^{1/2}}{(0.2)^{1/4} + (5)^{1/4}} = 108.2 \text{ ft}$$

$$J_9 = \frac{2\pi \Delta x_9 k_{r,\alpha}^0 \sqrt{k_{y,9} k_{z,9}}}{\mu_o \left[\ln\left(\frac{r_{eq}}{r_w}\right) + s\right]} = \frac{2\pi \cdot 1250 \cdot 1.0 \sqrt{150 \cdot 30}}{5\left[\ln\left(\frac{108.2}{0.25}\right) + 0\right]} = 17,358 \frac{mD - ft}{cp}$$

$$= 109.875 \frac{ft^3}{psi - day}$$

Note that in Example 4.3, there was a constant pressure boundary condition adjacent to block #9 which led to a J_9 and $Q_9 = J_9 P_{wf}$. Those entries in J and Q would sum with the values for the constant BHP well in block #9. In the upcoming Example 5.4, however, we only use no-flow boundary conditions.

5.5 Inclusion of the well model into the matrix equations

Recall the multidimensional mass balance for a grid block, i, from Chapter 4,

$$\Delta T_\alpha \Delta P + \Delta T_\alpha \Delta D + B_{\alpha,i} Q_{\alpha,i}^{sc} + J_{\alpha,i}\left(P_{wf,i} - P_i\right) = A_i c_{t,i}\left(P_i^{n+1} - P_i^n\right), \quad (5.19)$$

where a term has been added to account for a possible constant BHP well in addition to a constant rate source. If no constant BHP well resides (or is perforated) in the block, $J_{\alpha,i} = 0$. Eq. (5.19) can be written for each block in the reservoir simulation model which leads to the usual system of linear equations, Eq. (4.13).

The matrices T and A are defined in previous chapters and are unaffected by wells. The diagonal matrix, J, was introduced to account for Dirichlet boundary conditions. Constant BHP wells are handled similarly to these boundary conditions. The diagonal entry $J_{i,i}$ includes the productivity index of the well *if and only if* a constant BHP pressure well resides and is perforated, in block i. The source vector also includes a term, $J_i P_{wf}$, in entry Q_i if a well exists in that block. Consider, for example, a constant bottomhole pressure well in block #3; the J matrix and Q vector would appear as follows,

$$J = \begin{bmatrix} 0 & & & \\ & 0 & & \\ & & J_3 & \\ & & & \ddots \end{bmatrix} ; \quad \vec{Q} = \begin{bmatrix} 0 \\ 0 \\ J_3 P_{wf} \\ \vdots \end{bmatrix} \qquad (5.20)$$

The productivity index, J_i, is always positive regardless of whether the well is a producer or injector. The source vector, Q, also includes any constant rate wells and boundary conditions. It is mathematically possible to have multiple wells and boundary conditions applied to the same block, in which case they are summed. However, multiple wells in the same grid are not advised and should be avoided if possible. Examples 5.3 and 5.4 demonstrate the inclusion of constant BHP wells in a 1D and 2D simulator, respectively.

Example 5.3. 1D reservoir with wells.
Consider the reservoir and fluid properties in Examples 3.4 and 3.5 from Chapter 3 but with no-flow boundary conditions on both edges. The reservoir has an initial pressure = 3000 psia, a constant rate injector (q_{wf} = 2000 ft³/day) in block #1, and a constant BHP well (P_{wf} = 1000 psia) in block #4. Both wells have a productivity index of 2.384 ft³/psi-day. Calculate,
a. J matrix and Q vector,
b. pressure for the first three timesteps using the implicit method, and
c. the bottomhole pressure of well #1 and the production rate (STB/day) of well #2 after the first timestep. Assume the formation volume factor is 1.2 RB/STB.

Solution
The T and A matrices are exactly the same as Example 3.5 because the reservoir and fluid properties are the same

$$T = \begin{pmatrix} 2.532 & -2.532 & & \\ -2.532 & 5.064 & -2.532 & \\ & -2.532 & 5.064 & -2.532 \\ & & -2.532 & 2.532 \end{pmatrix} \frac{ft^3}{psi-day};$$

$$Ac_t = \begin{pmatrix} 8 & & & \\ & 8 & & \\ & & 8 & \\ & & & 8 \end{pmatrix} \frac{ft^3}{psi-day};$$

where T = 2.532 ft³/psi-day and Ac_t = 8 ft³/psi-day were defined in Example 3.4.

Continued

Example 5.3. 1D reservoir with wells.—cont'd

a. The J matrix and Q vector are given by

$$
\mathbf{J} = \begin{pmatrix} 0 & & \\ & & \\ & & J_4 \end{pmatrix} = \begin{pmatrix} 0 & & \\ & & \\ & & 2.384 \end{pmatrix} \frac{ft^3}{psi - day}
$$

$$
\vec{Q} = \begin{pmatrix} Q_1 \\ \\ J_4 P_{wf} \end{pmatrix} = \begin{pmatrix} 2000 \\ \\ 2384 \end{pmatrix} \frac{ft^3}{day}
$$

Note that the productivity index of well #1 is not included in the J matrix because it is a constant rate well.

b. The simultaneous block equations can be posed as a system of linear equations for the implicit method

$$
(\mathbf{A}c_t + \mathbf{T} + \mathbf{J})\vec{P}^{n+1} = \mathbf{A}c_t \vec{P}^n + \vec{Q} + \vec{G}
$$

where the gravity vector is the null vector. Solving the equations gives,

$$
\vec{P}^0 = \begin{pmatrix} 3000 \\ 3000 \\ 3000 \\ 3000 \end{pmatrix} psia; \quad \vec{P}^1 = \begin{pmatrix} 3195.8 \\ 3024.5 \\ 2930.6 \\ 2617.3 \end{pmatrix} psia; \quad \vec{P}^2 = \begin{pmatrix} 3351.3 \\ 3052.9 \\ 2844.3 \\ 2363.3 \end{pmatrix} psia;
$$

$$
\vec{P}^3 = \begin{pmatrix} 3475.7 \\ 3078.7 \\ 2763.0 \\ 2190.1 \end{pmatrix} psia; \quad \vec{P}^\infty = \begin{pmatrix} 4208.7 \\ 3418.9 \\ 2629.0 \\ 1839.0 \end{pmatrix} psia
$$

c. The well pressure for well #1 is given by,

$$
P_{wf,1} = P_1 + \frac{Q_1}{J_1} = 3195.8 + \frac{2000}{2.384} = 4034.7 psia
$$

The well production rate for well #4 is given by,

$$
Q_4 = J_4 (P_{wf,4} - P_4) = 2.384(1000 - 2617.3)
$$

$$
= -3855.6 \frac{ft^3}{day} \cdot \frac{RB}{5.615 ft^3} \cdot \frac{STB}{1.2 RB} = -572.2 \frac{STB}{day}
$$

Example 5.4. 2D reservoir with wells.
Repeat Example 4.3 (2D heterogeneous reservoir) but with no-flow conditions on all boundaries. There is a constant BHP vertical well in block #1 ($P_{wf} = 4000$ psia) and constant BHP horizontal well perforated in blocks #8 and #9 ($P_{wf} = 2000$ psia). The wellbore radius is 0.25 ft.

Solution
The **T** and **A** matrices from Example 4.3 are unchanged, but both **J** and Q are different because (1) the boundary conditions are different (now no-flow) and (2) there are two wells. The productivity indexes of both wells were computed in Example 5.2.

$$\mathbf{J} = \begin{pmatrix} J_1 & & \\ & & \\ & & \\ & J_8 & \\ & & J_9 \end{pmatrix} = \begin{pmatrix} 0.6695 & & \\ & & \\ & & \\ & 58.6 & \\ & & 109.9 \end{pmatrix} \frac{ft^3}{psi - day}$$

$$\vec{Q} = \begin{pmatrix} J_1 P_{wf,1} \\ \\ J_8 P_{wf,8} \\ J_9 P_{wf,9} \end{pmatrix} = \begin{pmatrix} 2678 \\ \\ 117201 \\ 219753 \end{pmatrix} \frac{ft^3}{day}$$

Continued

Example 5.4. 2D reservoir with wells.—cont'd

Solving the problem using the implicit method,

$$(\mathbf{T} + \mathbf{J} + \mathbf{Ac}_t)\vec{P}^{n+1} = \mathbf{Ac}_t\,\vec{P}^n + \vec{Q} + \vec{G}$$

$$\vec{P}^0 = \begin{pmatrix} 3000 \\ 3000 \\ 3000 \\ 3000 \\ 3000 \\ 3000 \\ 3000 \\ 3000 \\ 3000 \end{pmatrix} \text{psia;} \quad \vec{P}^1 = \begin{pmatrix} 3092.8 \\ 2980.2 \\ 2945.5 \\ 2930.9 \\ 2870.0 \\ 2825.4 \\ 2682.0 \\ 2226.9 \\ 2153.5 \end{pmatrix} \text{psia;} \quad \vec{P}^2 = \begin{pmatrix} 3121.1 \\ 2939.1 \\ 2869.2 \\ 2836.2 \\ 2752.1 \\ 2689.1 \\ 2450.8 \\ 2086.4 \\ 2042.1 \end{pmatrix} \text{psia;}$$

$$\vec{P}^3 = \begin{pmatrix} 3112.2 \\ 2885.7 \\ 2790.2 \\ 2742.8 \\ 2657.1 \\ 2590.3 \\ 2308.5 \\ 2052.0 \\ 2024.3 \end{pmatrix} \text{psia;} \quad \vec{P}^\infty = \begin{pmatrix} 2662.6 \\ 2303.4 \\ 2184.6 \\ 2202.4 \\ 2153.2 \\ 2116.6 \\ 2034.8 \\ 2007.5 \\ 2004.2 \end{pmatrix} \text{psia}$$

Notes:
- The reservoir pressure initially increases near the well in block #1 because it is injecting and decreases near the horizontal well in blocks #8 and #9 because it is producing.
- At later times, even the pressure in block #1 drops below its initial pressure. This is because the productivity index of the vertical well is orders of magnitude smaller than the horizontal well.

5.6 Practical considerations

A reservoir model may contain many different wells of different types (vertical vs. horizontal, producer vs. injector) and constraints (rate vs. BHP). A well schedule for all wells should be provided to the simulator that indicates when a well should begin producing or injecting, and when, if ever, it should be shut in. The well constraints can also change during the simulation, both in their value and from a constant rate to BHP, or vice versa. These changes can either occur on a set time schedule or when some condition is met in the reservoir. For example, if a constant rate well reaches its limiting pressure (P_{lim}), it is switched to a constant BHP well. In waterflooding, during multiphase flow (Chapter 9), a producer well may be shut-in or switched to an injector if the *water-cut* of the well reaches some limit, e.g., 95%.

A constant BHP well can be a producer (if $P_i > P_{wf}$) or injector (if $P_i < P_{wf}$). A well intended to be a constant BHP producer could mathematically switch to an injector if P_i dropped below the well pressure. However, it is unlikely this would be possible or allowed in practice. In the simulator, the well should be shut-in until the pressure of the well drops below the grid pressure.

5.7 Pseudocode for single-phase flow with constant BHP wells

The pseudocode presented in Chapter 4 for single-phase multidimensional flow remains applicable. However, the *well arrays* subroutine should be amended to include constant BHP wells and productivity index must also be computed. The productivity index subroutine requires inputs of the well number and the block number(s) that the well resides and is perforated. All well, reservoir, fluid, and numerical properties are inputs. The direction (x, y, or z) of the well must be sent or identified in the subroutine. Productivity index for the well in block i can then be computed using Eqs. (5.17) and (5.18).

The *well arrays* subroutine from Chapter 4 is adapted by first adding a nested loop, inside the wells loop, for all grids that the well is perforated. During preprocessing, the grids perforated by each well should be determined and saved to an array. If the well is a constant rate well, the rate is distributed to the grid block using Eqn. (4.14). If the well is a constant BHP well, the **J** matrix and Q vectors are updated using the productivity index and predefined well pressure.

During postprocessing it will be desired to plot the well pressure for constant rate wells and production/injection rates for constant BHP wells. This can be done using the well model, Eq. (5.9). Productivity indexes are needed for both calculations, even though they are not included in the **J** matrix for constant-rate wells.

```
SUBROUTINE PRODUCTIVITY INDEX
INPUTS: well #, block #, well, reservoir, fluid, petrophysical,
numerical, output properties
OUTPUTS: productivity index

IDENTIFY direction (x, y, or z) of well and compute direction-
dependent variables
CALCULATE equivalent radius using eqn 5.17
CALCULATE phase productivity index using eqn 5.18




SUBROUTINE WELL ARRAYS
INPUTS: numerical, petrophysical, fluid, well properties, Q
OUTPUTS: Q, J

FOR k = 1 to # wells
    FOR j = 1 to # perforated blocks of well k
        DETERMINE i, the (jth) block that well k is perforated
        SET prod_index = PRODUCTIVITY INDEX(inputs)
        Save prod_index for postprocessing
        IF constant rate well
            CALCULATE Q(i) using eqn. 4.14
        ELSEIF constant BHP well
            SET Q(i) = J*Pwf(i)
            SET J(i,i) = J(i,i) + prod_index
        ENDIF
    ENDFOR
ENDFOR




SUBROUTINE POSTPROCESS
INPUTS: reservoir, fluid, petrophysical, well, numerical, solution
properties
OUTPUTS: well data, plots

CALCULATE average reservoir pressure at all times
PLOT BHP vs time of constant rate wells using eqn 5.9
PLOT rate of constant BHP wells using eqn 5.9
PLOT average reservoir pressure vs time
PLOT (contour, surface, etc.) block pressure field at desired times
VALIDATE against analytical solution, if applicable
```

5.8 Exercises

Exercise 5.1. Radial flow with constant BHP well. Repeat the radial flow problem in Example 5.1, but use a constant BHP well, $P = 1000$ psia.

Exercise 5.2. Productivity indices. Calculate the productivity indices of a horizontal well perforated in blocks 6 and 9 and a horizontal well perforated in blocks 1 and 2 for a 2D reservoir with reservoir/fluid properties described by Example 4.2. The wellbore radius and skin are 0.25 ft and 0.0, respectively, for both wells.

Exercise 5.3. 1D with vertical wells. Repeat Example 5.3 to compute the pressure for the first three timesteps and at steady state, but use a constant BHP well in block #1 ($P_{wf} = 1000$ psia) and a constant rate injector well ($q_{wf} = 2000$ ft^3/day) in block #4.

Exercise 5.4. 2D with horizontal wells. Repeat Example 5.4, but use the two horizontal, constant BHP wells described in Exercise 5.2 instead of the wells presented in Example 5.4.

Exercise 5.5. Near-well simulator. Develop a reservoir simulator (computer code) to solve for 1D radial flow around a cylindrical, vertical well. Validate your code against Example 5.1 and/or Exercise 5.1 and then against the analytical solutions for infinite acting and pseudo-steady-state flow for a constant rate well presented in Chapter 2. Use N = 100 grids. You may need to use smaller time steps to match the analytical solution. Compare plots of well pressure vs. time. You should be able to adapt your simulator developed in Chapters 3 and 4 in Cartesian flow by changing the subroutine for creating the arrays (**T**, **A**, **J**, and *Q*). For an additional challenge, extend the simulator to 2D and 3D, including gravity.

Exercise 5.6. 2D reservoir simulator. Adapt your 2D reservoir simulator developed in Chapter 4 (Exercise 4.6) to allow for a user-specified number of wells, including both constant rate and BHP, and direction (i.e., vertical and horizontal). Test your code against any of the example problems or exercises.

Exercise 5.7. 2D project. Use your code to solve for pressure versus space and time in the synthetic Thomas oilfield during primary production. Reservoir, fluid, well, petrophysical, and numerical properties are given in the input file at https://github.com/mbalhof/Reservoir-Simulation. The boundary conditions are no-flow on all boundaries. If the pressure of a constant rate well reaches the bubble point, change the well constraint to a constant BHP well with $P_{wf} = p_b$. Be sure to initialize the reservoir oleic phase pressure using the methods in Chapter 1 (see Exercise 1.8). You may assume that the connate water is immobile so flow is single phase and capillary pressure is negligible.

Continue the simulation until 100 days after the last well is converted to a constant BHP well. Plot the production rate (STB/day) and bottomhole pressure (psia) versus time for all wells. Plot the cumulative oil recovery (as a % of original oil in place) versus time. Also plot the pressure field (all blocks) in the reservoir at each time a well is converted to a constant BHP well.

References

Abou-Kassem, J.H., Farouq-Ali, S.M., Rafiq Islam, M., 2013. Petroleum Reservoir Simulations. Elsevier.

Arps, J.J., 1945. Analysis of decline curves. Transactions of the AIME 160, 228–247. https://doi.org/10.2118/945228-G.

Aziz, K., Settari, A., 1979. Petroleum Reservoir Simulation. Blizprint, Ltd., Alberta, Canada.

Duong, A.N., 2011. Rate-decline analysis for fracture-dominated shale reservoirs. SPE Reservoir Evaluation and Engineering 14, 377–387. https://doi.org/10.2118/137748-PA.

Ertekin, T., Abou-Kassem, J.H., King, G.R., 2001. Basic Applied Reservoir Simulation, vol. 7. Society of Petroleum Engineers, Richardson.

Havlena, D., Odeh, A.S., 1963. The material balance as an equation of a straight line. Journal of Petroleum Technology 15 (8), 896–900. SPE-559-PA.

Ilk, Dilhan, R., Alan, J., Perego, A.D., Blasingame, T.A., September 2008. Exponential vs. Hyperbolic decline in tight gas sands: understanding the origin and implications for reserve estimates using Arps' decline curves. In: Paper Presented at the SPE Annual Technical Conference and Exhibition. https://doi.org/10.2118/116731-MS. Denver, Colorado, USA.

Jackson, G.T., Balhoff, M.T., Huh, C., Delshad, M., July 2011. CFD-based representation of non-Newtonian polymer injectivity for a horizontal well with coupled formation-wellbore hydraulics. Journal of Petroleum Science and Engineering 78 (Issue 1), 86–95.

Peaceman, D.W., 1978. Interpretation of well-block pressures in numerical reservoir simulation. Society of Petroleum Engineers Journal 265.

Peaceman, D.W., 1983. Interpretation of well-block pressures in numerical reservoir simulation with nonsquare grid blocks and anisotropic permeability. Society of Petroleum Engineers Journal 23, 531–543. https://doi.org/10.2118/10528-PA.

Valko, P.P., January 2009. Assigning value to stimulation in the Barnett Shale: a simultaneous analysis of 7000 plus production hystories and well completion records. In: Paper Presented at the SPE Hydraulic Fracturing Technology Conference, The Woodlands, Texas. https://doi.org/10.2118/119369-MS.

Van Poolen, H.K., Breitenback, E.A., Thurnau, D.H., 1986. Treatment of individual wells and grids in reservoir modeling. SPE Journal 8, 341–346.

Walsh, M.P., Lake, L.W., 2003. A Generalized Approach to Primary Hydrocarbon Recovery. Elsevier, ISBN 0-444-50683-7.

Chapter 6

Nonlinearities in single-phase flow through subsurface porous media

6.1 Introduction

Many reservoir and fluid properties are functions of pressure, so they also vary with time. Examples of such properties in single-phase flow include density, viscosity, total compressibility, permeability, and porosity. As a result, the transmissibility, T, accumulation, A, compressibility, c_t, productivity index, J, and even the source terms, Q and G, can be dependent on the unknown pressure, P^{n+1}. The pressure/time dependence of these properties makes the system of equations nonlinear and, therefore, more challenging to solve.

In this chapter, a few examples of nonlinear problems that can arise in single-phase flow are discussed, including the flow of compressible gases, non-Newtonian fluids, and Forchheimer flow. These nonlinearities, along with many more, can appear in multiphase problems. We introduce a few approaches for solving nonlinear problems, including explicit updates, Picard iteration, and the Newton—Raphson method.

6.2 Examples of nonlinearities in single-phase flow problems

Multiphase flow is especially nonlinear and will be discussed in Chapter 9. However, there are many single-phase flow problems that have pressure and/or time-dependent variables, which make the system of algebraic equations nonlinear. A few of these pressure/time-dependent variables include viscosity, compressibility, porosity, and permeability. In some cases, the dependency is negligible and the variables can be assumed constants.

An Introduction to Multiphase, Multicomponent Reservoir Simulation.
https://doi.org/10.1016/B978-0-323-99235-0.00009-9

Viscosity is pressure dependent for all fluids, but especially for gases. Non-Newtonian fluids, such as polymers used in enhanced oil recovery (EOR) or hydraulic fracturing, have velocity/pressure gradient-dependent viscosity. Fluid compressibility, particularly for gases, is pressure dependent. The compressibility includes the z-factor for real gases which is pressure dependent. The pressure dependence of porosity is incorporated in the rock compressibility which is often assumed constant. However, geochemical reactions and geomechanics can result in changes in porosity with time. Any process that changes porosity will also cause changes in permeability. Permeability is also affected by the Klinkenberg effect for gases at low pressures. In high-velocity, non-Darcy flow (e.g., near wells and fractures), velocity is a nonlinear function of pressure gradient and Darcy's law is not applicable. The nonlinearities are usually modeled with an effective, velocity-dependent permeability. Fluid density is also a function of pressure, but its effect is incorporated in the fluid compressibility. All fluid properties are temperature dependent, which is important in thermal problems, but in this text, only isothermal simulation is discussed.

6.2.1 Gas flow

6.2.1.1 Compressibility

Single-phase flow of a gaseous phase is more numerically challenging than slightly compressible liquids (i.e., aqueous and oleic) because the compressibility of gases is a strong function of pressure. At the high pressures and temperatures of interest at reservoir depth, a real gas law must be employed. Recall from Chapter 1 (Eq. 1.10), the following pressure-dependent gas phase compressibility,

$$c_g = \frac{1}{p} - \frac{1}{z}\frac{\partial z}{\partial p}\bigg|_T, \tag{1.10}$$

where the z-factor is pressure dependent for a real gas and various correlations have been developed. Fig. 1.5 showed the dependency of z and c_g on pressure. A compressibility matrix, c_t, is used with the accumulation terms of the balance equations.

6.2.1.2 Viscosity

The viscosity of a gas is also a function of pressure. A correlation was developed by Lee et al. (1966),

$$\mu_g = K_1 \exp\left(X\rho_g^Y\right), \tag{6.1}$$

where the gas density, ρ_g, is in g/cm^3 and K_1, X, and Y are functions of the molecular weight (M_g) and temperature (T),

$$K_1 = \frac{\left(0.00094 + 2 \times 10^{-6} M_g\right) T^{1.5}}{\left(209 + 19 M_g + T\right)},$$

$$X = 3.5 + \frac{986}{T} + 0.01 M_g; \quad Y = 2.4 - 0.2X,$$

where T is the temperature (Rankine). The interblock transmissibility (**T**) and productivity index (**J**), if applicable, are affected by viscosity.

6.2.1.3 Permeability and the Klinkenberg effect

For gases, slippage at the pore walls can affect the permeability. The Klinkenberg permeability is a function of pressure and the gas molecular weight,

$$k_g = k\left(1 + b/_p\right), \tag{6.2}$$

where k_g is the apparent permeability to gas, k is the single-phase permeability to a liquid phase, and b (psia) is a constant for a particular gas in a given rock type. Jones and Owens (1980) proposed correlations for b based on experiments. Eq. (6.2) shows that at high pressures, such as those at reservoir conditions, the Klinkenberg effect is negligible. However, if experiments are conducted at low (near atmospheric) pressure, the Klinkenberg effect could be significant. The transmissibilities (**T**) and well productivity (**J**) terms would then be pressure dependent.

6.2.2 Non-Newtonian flow

Many subsurface fluids are non-Newtonian, exhibiting a shear-dependent viscosity. Polymers used in enhanced oil recovery (Seright, 1995; Willhite and Green, 1998; Lake et al., 2014), hydraulic fracturing fluids (Howard and Fast, 1957; Gidley et al., 1989), some gels used during matrix acidizing (Paccaloni and Tambini, 1993), drilling muds (Jensen et al., 2004), and even some heavy oils (Smith, 1992) are often characterized as non-Newtonian. The bulk viscosity versus shear rate can be measured in a standard rheometer and fit to a common constitutive equation, such as power-law model,

$$\mu = m\dot{\gamma}^{n-1}, \tag{6.3}$$

where m and n represent the consistency (cp-s^{n-1}) and shear-thinning index (dimensionless), respectively, and $\dot{\gamma}$ is the shear rate (1/s). The shear-thinning index is less than one for shear-thinning fluids and $n = 1$ for a Newtonian fluid. A log–log plot of viscosity versus shear rate yields a straight line for power-law fluids, and the fitted constants can be determined from least-squares regression to the rheometer data. Another common constitutive rheological model used to describe shear-thinning fluids is the Carreau model which extends the power-law model to include a Newtonian plateau at both low and high shear rates,

$$\mu = \mu_\infty + (\mu_0 - \mu_\infty)\left[1 + \left(\lambda\dot{\gamma}\right)^2\right]^{(n-1)/2}, \tag{6.4}$$

where μ_0 represents the plateau viscosity at low shear rates, μ_∞ represents the plateau viscosity at high shear rates, and λ, n are other fitting constants. Fig. 6.1 qualitatively shows the viscosity versus shear rate of a Carreau fluid on a log–log plot.

The in situ viscosity of non-Newtonian fluids that requires calculation of an apparent/effective shear rate in the porous medium, $\dot{\gamma}_{\text{eff}}$, is taken to be,

$$\dot{\gamma}_{\text{eff}} = C\left(\frac{3n+1}{4n}\right)^{n/n-1}\frac{4u}{\sqrt{8kk_{rw}\phi S_w}}, \tag{6.5}$$

where C is a constant of the porous medium, k_{rw} and S_w are the relative permeability and saturation, respectively, of the non-Newtonian aqueous phase (both 1.0 if only one phase is present), and u is the Darcy velocity. The value of C is usually taken between 2.0 and 6.0, and Koh et al. (2018) determined a value of 4.0. However, the in situ rheology is often different than the bulk rheology (measured in a rheometer), especially at high shear rates. The in situ rheology follows the bulk rheology well at low shear rates but often diverges and increases at high shear rates. The apparent *shear-thickening* behavior is generally attributed to viscoelasticity as the polymer is squeezed through small pore constrictions (Xie and Balhoff, 2021). Delshad et al. (2008) proposed an empirical model (based on the Carreau model) for polymer flow in porous media with shear-thickening at high shear rates.

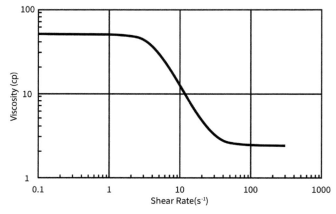

FIGURE 6.1 Viscosity versus shear rate for a polymer measured in a rheometer and in situ in a core.

Block viscosities and, therefore, interblock transmissibilities (**T**) are functions of the block pressures, *P*, although indirectly. Here we propose one iterative procedure for 1D flow to accurately compute transmissibilities at given vector of grid pressures, *P*.

a. Guess the block viscosities, μ, e.g., the values in the previous timestep and calculate interblock viscosity via upwinding from the grid of higher pressure (or potential with gravitational effects),

$$\mu_{i-\frac{1}{2}} = \begin{cases} \mu_{i-1} & \text{if } P_{i-1} > P_i \\ \mu_i & \text{if } P_i > P_{i-1} \end{cases}.$$

b. Calculate interblock transmissibilities, *T*, and well productivity indexes, *J*, using the interblock and block viscosities, respectively.
c. Compute interblock velocities, $u_{i-1/2}$, using Darcy's law.
d. Reverse upwind to determine block velocities, u_i,

$$u_i = \begin{cases} |u_{i-\frac{1}{2}}| & \text{if } P_{i-1} > P_i \quad \text{and } P_i > P_{i+1} \\ \\ |u_{i+\frac{1}{2}}| & \text{if } P_{i+1} > P_i \quad \text{and } P_i > P_{i-1} \\ \\ 0.5|u_{i-\frac{1}{2}}| + 0.5|u_{i+\frac{1}{2}}| & \text{otherwise} \end{cases}$$

e. Compute block shear rates (Eq. 6.5) and updated block viscosities (Eq. 6.4).

Steps a—e can then be iterated on until convergence. Alternatively, steps b—e can be skipped altogether, if sufficiently small timesteps are employed.

6.2.3 Forchheimer flow

Darcy's law adequately describes the slow (creeping) flow of Newtonian fluids in porous media and is usually applicable in engineering applications for $N_{Re} < 1$, where N_{Re} is the dimensonless Reynold's number,

$$N_{Re} = \frac{\rho v d_p}{\mu}, \tag{6.6}$$

and d_p is some length scale (e.g., the average grain size diameter) and $v \; (= u/\phi)$ is the intersticial or frontal velocity. It is generally acceptable to use Darcy's law for modeling flow in most subsurface applications because velocities are relatively low (e.g., 1 ft/day). However, higher velocities are often observed in fractures where the permeability is high and near wellbores where flow accelerates radially; a more complicated constitutive model is necessary to describe flow in these cases.

Forchheimer's equation (1901) is an extension to Darcy's law that is intended to capture nonlinearities that occur due to inertia in the laminar flow regime,

$$-\frac{\Delta P}{L} = \frac{\mu}{k}u + \rho\beta u^2, \tag{6.7a}$$

$$-\nabla P = \mu \mathbf{k}^{-1}\vec{u} + \beta\rho\left|\vec{u}\right|\vec{u}, \tag{6.7b}$$

where Eqs. (6.7a) and (6.7b) are 1D and multidimensional form, respectively. The quadratic term is small compared to the linear term at low velocities and Forchheimer's equation reduces to Darcy's law. The constant, β (ft^{-1}), is referred to as the non-Darcy coefficient and, like permeability, is an empirical value that is specific to the porous medium. It is often found experimentally through data reduction. A few authors have attempted to develop correlations relating the non-Darcy coefficient to medium properties, such as permeability, porosity, and tortuosity (Geertsma, 1974; Liu et al., 1995). While often assumed a scalar, the non-Darcy coefficient is likely a tensor for anisotropic media (Wang et al., 1999; Balhoff and Wheeler, 2009) since it is dependent on the medium morphology.

Eq. (6.7) can be rearranged to obtain a velocity-dependent permeability, k_{app},

$$k_{app} = \left[\frac{1}{k} + \beta\left(\frac{\rho u}{\mu}\right)\right]^{-1}, \tag{6.8}$$

and then substituted for permeability in Darcy's law. Eq. (6.8) can be used to obtain transmissibilities given the vector of pressures, but an iterative process, similar to that proposed for non-Newtonian flow is required. For example, apparent permeability is guessed, transmissibilities are computed using the apparent permeability, the interblock (and then block) velocities computed, and then apparent permeability is updated using Eq. (6.8). The process is repeated until convergence.

Although other models for non-Darcy flow have been developed, Eq. (6.7) is widely used in reservoir simulation. Forchheimer's equation has been found to fit some experimental data very well by Forchheimer (1901, 1930) and others (Ahmed, 1967; Sunada, 1965; Kim, 1985; Brownell et al., 1947; Fancher and Lewis, 1933; Blake, 1922; Lindquist, 1933; Mobasheri and Todd, 1963). However, the equation has been shown to be unacceptable for matching other experimental data (Forchheimer, 1930; Barree and Conway, 2004, 2005) and even Forchheimer (1930) added additional terms for some data sets.

6.3 Numerical methods for nonlinear problems

In all of the examples presented here (gas, non-Newtonian, and non-Darcy flow), and more, the arrays $\mathbf{T}$, $\mathbf{A}$, and/or $\mathbf{J}$, and even Q and G, are functions of the unknown block pressures, making them nonlinear; a system of nonlinear equations describes flow in the reservoir simulator,

$$\left(\mathbf{T}^* + \mathbf{J}^* + \mathbf{A}^*\mathbf{c}_t^*\right)\overrightarrow{P}^{n+1} = \mathbf{A}^*\mathbf{c}_t^*\overrightarrow{P}^n + \overrightarrow{Q}^* + \overrightarrow{G}^*,\tag{6.9}$$

where the asterisk superscript on the arrays indicates they are time dependent because they are functions of the solution. Here we only consider the implicit form of the matrix equations, but depending on how the arrays are computed, there may be a certain amount of explicitness to the problem. There are several techniques for solving the nonlinear problem and the best method for solution usually depends on the problem and degree of nonlinearity.

6.3.1 Explicit update of fluid and reservoir properties

If the nonlinearities are relatively weak, an explicit update of properties (and therefore arrays) is acceptable and stable and, therefore, preferred due to its simplicity. In this approach, the pressure solution in the previous timestep (P^n) is used to evaluate reservoir/fluid properties and then these are used to compute the updated arrays in Eq. (6.9). Since the arrays are computed at the end of the previous timestep, no iteration is required and the matrix equations are linear. New pressures, P^{n+1}, are computed by solving the linear system and then used to calculate the new array values for the next timestep. Mathematically, Eq. (6.9) can be written with the asterisk superscript set to n, which indicates they are updated at the end of the previous timestep. The method is only conditionally stable, despite the fact that this appears to be an implicit scheme, because of the explicit calculation of the arrays. Nonetheless, this method, explicit updates, is usually acceptable and the preferred method for single-phase problems and some multiphase problems.

The procedure/algorithm is very similar to that proposed in Chapters 3—5 with one important change. For linear problems where reservoir and fluid properties did not change with time, the arrays $\mathbf{T}, \mathbf{A}, \mathbf{c}_t, \mathbf{J}, Q, G$ only needed to be computed once, prior to the time-stepping while loop. Here, the arrays must be recomputed each timestep (i.e., inside the while loop in a computer code) and this adds computation time. If the change in reservoir/fluid properties or pressure from time n to $n + 1$ exceeds a predetermined tolerance, the timestep can be reduced.

6.3.2 Picard iteration

A more advanced approach than using explicit updates involves iterative updates of the reservoir and fluid properties within the timestep, referred to here as *Picard iteration*. In this technique (1) a guess for the pressure solution is made, $P^{(v)}$ (e.g., the solution in the previous timestep $P^{(v)} = P^n$), (2) the fluid and reservoir properties are calculated using the guessed pressures and the arrays can be evaluated, (3) the resulting linear system of equations is solved to obtain a new pressure field, $P^{(v+1)}$, and (4) steps 2—3 are repeated until the solution converges, that is, until pressure does not change significantly or the error in Eq. (6.9) is small. The converged solution is P^{n+1}, and then the simulation proceeds to the next timestep. The iterative updates in Picard iteration theoretically allow for the use of longer timesteps than explicit evaluation, but the added benefit of longer timesteps comes at the cost of more computational work per timestep. Furthermore, there is still no guarantee of convergence or stability in Picard iteration, although convergence usually does occur with the (usually) weak nonlinearities of single-phase flow. Example 6.1 shows the iterative solution of the flow of nonlinear gas using Picard iteration.

Example 6.1. Nonlinear gas flow (Picard method).
A 1D, single-phase coreflood flow experiment is conducted with a gas. The core has properties: $\phi = 0.2$, $k = 50$ mD, $L = 1.0$ ft, $a = 0.02$ ft^2. The initial pressure is atmospheric, $P = 14.7$ psia. A constant pressure, $P = 200$ psia, is imposed at $x = 0$ and the pressure is maintained at atmospheric, 14.7 psia, at $x = L$. The fluid viscosity is approximately constant, $\mu_g = 0.01$ cp, and the gas acts ideally at these low pressures. For the purposes of this problem, the formation volume factor, B_g, can be assumed uniform.[1]

Klinkenberg effects can be neglected. Determine the pressure field after three iterations of the first timestep using the Picard method. Use four uniform blocks, a timestep of $\Delta t = 1.0E{-}5$ days, and an initial guess of $P = 14.7$ psia.

Solution
The transmissibility is uniform with space and constant with time,

$$T = \frac{ka}{\mu_g \Delta x} = \frac{50 \cdot 0.02}{0.01 \cdot 0.25} = 400 \frac{mD - ft}{cp} = 2.532 \frac{ft^3}{psi - day}$$

There are constant pressure boundary conditions at both ends, so $J_1 = J_4 = 2T$ and $Q_1 = J_1 P_{B1}$, $Q_4 = J_4 P_{B4}$.

$$\mathbf{T} = \begin{bmatrix} 2.532 & -2.532 & & \\ -2.532 & 5.064 & -2.532 & \\ & -2.532 & 5.064 & -2.532 \\ & & -2.532 & 2.532 \end{bmatrix} \frac{ft^3}{psi - day}$$

$$\mathbf{J} = \begin{bmatrix} 5.064 & & & \\ & & & \\ & & & \\ & & & 5.064 \end{bmatrix} \frac{ft^3}{psi - day} \quad \vec{Q} = \begin{bmatrix} 1012.8 \\ \\ \\ 74.4 \end{bmatrix} \frac{ft^3}{day}$$

In this problem, **T**, **J**, and Q are not functions of pressure and do not change with time (in other problems, they may be functions of pressure). However, the accumulation terms are pressure dependent because gas compressibility is pressure dependent; for an ideal gas $c_g = 1/p$.

Example 6.1. Nonlinear gas flow (Picard method).—cont'd

$$A_i c_{t.i} = \frac{a \Delta x \phi c_{g.i}}{\Delta t} = \frac{0.02 \cdot 0.25 \cdot 0.2}{1E - 5} \frac{1}{p_i} = 100 p_i^{-1} \frac{ft^3}{psi - day}$$

The system of equations is therefore nonlinear (pressure dependent).

Picard's method can be used by evaluating Ac_t at the pressure computed in the previous iteration, solving the linear system of algebraic equations, and continuation until convergence. A residual vector, F, can be computed which is the difference in the right and left hand side of Eq. (6.9). Then the error is defined by some norm of F, e.g., sqrt($F^T F$).

Iteration #1, guess $P = [14.7\ 14.7\ 14.7\ 14.7]$ psia,

$$\mathbf{Ac_t} = \begin{bmatrix} 6.8027 & & & \\ & 6.8027 & & \\ & & 6.8027 & \\ & & & 6.8027 \end{bmatrix} \frac{ft^3}{psi - day}$$

$$\vec{F} = -\left(\mathbf{T}^{n+1} + \mathbf{J}^{n+1} + \mathbf{A}^{n+1} \mathbf{c}_t^{n+1}\right) \vec{P}^{guess} + \left(\mathbf{A}^{n+1} \mathbf{c}_t^{n+1}\right) \vec{P}^n + \vec{Q}^{n+1}$$

$$= \begin{bmatrix} 938.3592 \\ \\ \\ \end{bmatrix} \frac{ft^3}{day}$$

error = 938.3592 ft³/day

$$(\mathbf{T} + \mathbf{J} + \mathbf{Ac}_t) \vec{P}^{v+1} = \mathbf{Ac}_t \vec{P}^n + \vec{Q}^* + \vec{G}^*$$

$$\vec{P} = \begin{bmatrix} 82.5415 \\ 29.8940 \\ 18.0683 \\ 15.2923 \end{bmatrix} psia.$$

Continued

Example 6.1. Nonlinear gas flow (Picard method).—cont'd

Iteration #2, guess $P = [82.5\ 29.9\ 18.1\ 15.3]$ psia

$$\mathbf{Ac_t} = \begin{bmatrix} 1.2115 & & & \\ & 3.3452 & & \\ & & 5.5345 & \\ & & & 6.5392 \end{bmatrix} \frac{ft^3}{psi - day}$$

$$\vec{F} = \begin{bmatrix} 379.3159 \\ 52.5344 \\ 4.2716 \\ 0.1561 \end{bmatrix} \frac{ft^3}{day} \quad error = 382.9604 \frac{ft^3}{day}$$

Solving the system of equations gives,

$$\vec{P} = \begin{bmatrix} 132.2421 \\ 52.9678 \\ 24.2509 \\ 16.4108 \end{bmatrix} psia$$

Iteration #3, guess $P = [132.2\ 53.0\ 24.3\ 16.4]$ psia

$$\mathbf{Ac_t} = \begin{bmatrix} 0.7562 & & & \\ & 1.8879 & & \\ & & 4.1236 & \\ & & & 6.0935 \end{bmatrix} \frac{ft^3}{psi - day}$$

$$\vec{F} = \begin{bmatrix} 53.5196 \\ 55.7642 \\ 13.4762 \\ 0.7625 \end{bmatrix} \frac{ft^3}{day} \quad error = 78.4614 \frac{ft^3}{day}$$

Solving the system of equations gives,

$$\vec{P} = \begin{bmatrix} 142.8749 \\ 66.9047 \\ 29.8598 \\ 17.5039 \end{bmatrix} psia$$

Example 6.1. Nonlinear gas flow (Picard method).—cont'd

Continuing with the iterations until the error is small (here 21 iterations) gives,

$$\overrightarrow{P}^1 = [146.51 \quad 75.06 \quad 35.36 \quad 18.75] \, \text{psia}$$

We can then continue to the second timestep using P^1 as an initial guess. Note that the number of iterations required for convergence is somewhat high which can be computationally expensive. Newton's method is expected to converge in less iterations.

Summary of results for Example 6.1. Only a few iterations are shown per timestep.

Time (days)	Iteration	Error	Block #1 (psia)	Block #2 (psia)	Block #3 (psia)	Block #4 (psia)
0			14.7	14.7	14.7	14.7
1.0E−05	1	9.38E+02	82.54	29.89	18.07	15.29
	2	3.83E+02	132.24	52.97	24.25	16.41
	21	7.07E−07	146.51	75.06	35.36	18.75
2.0E−05	1	1.36E+02	165.7	102.26	53.15	23.9
	2	2.03E+01	167.13	106.32	57.58	25.35
	12	6.72E−07	167.38	107.08	58.59	25.75
3.0E−05	1	5.80E+01	172.77	119.6	71.03	30.86
	2	5.12E+00	173.03	120.39	72.14	31.42
	9	1.54E−07	173.05	120.45	72.24	31.48
∞			200	200	200	200

---

1 Although often an acceptable approximation for slightly compressible fluids, this is probably a poor approximation for a compressible gas.

6.3.3 Newton's method

6.3.3.1 1D Newton's method

The most widely used method for solving nonlinear equations is Newton's, or the Newton—Raphson, method. Before introducing the multidimensional form

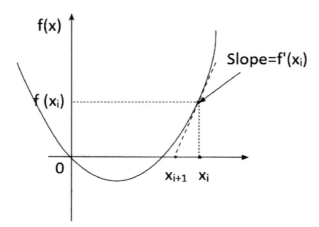

FIGURE 6.2   1D Newton's method for finding the root of a nonlinear equation.

of Newton's method, a review of the 1D (one nonlinear equation and one unknown) form is presented. Consider a 1D nonlinear equation, $f(x) = 0$, which is depicted in Fig. 6.2. The objective is to find the value of x that makes the function zero.

Newton's method is derived from a Taylor series (Eq. 3.1). Setting the function to zero and neglecting all higher order terms (second derivative and greater),

$$0 \approx f(x_i) + f'(x_i)(x_{i+1} - x_i) + \dots, \tag{6.10}$$

where the function $f(x_{i+1})$ was set to zero. Solving for x_{i+1} gives,

$$x_{i+1} = x_i - \frac{f(x_i)}{f'(x_i)}, \tag{6.11}$$

where x_i is the guess for the root and x_{i+1} is the updated guess in the iterative process. Newton's method works as follows: (1) guess a value of the root, x_i, (2) compute the function, $f(x)$, and the derivative, $f'(x)$ at the point x_i, (3) compute the new guess of the root, x_{i+1}, using Eq. (6.11), and (4) repeat steps 1–3 until converging on the root. One iteration of Newton's method is shown in Fig. 6.2.

There are several different criteria for convergence that can be used. For example, $abs(x_{i+1}-x_i) <$ tol or $abs((x_{i+1}-x_i)/x_i) <$ tol, where *tol* is a predefined small tolerance (e.g., 1E−6). An alternative is to require $abs(f) <$ tol. The latter is generally considered the better option because the ultimate goal is to make f close to zero. The former criteria can sometimes result in a local minimum where f is not close to zero. However, either (or both) convergence criteria are used depending on the situation. Example 6.2 demonstrates Newton's method to find the root of a nonlinear equation.

Example 6.2. Newton–Raphson for Rachford–Rice.
Solve the following nonlinear equation for v using the Newton–Raphson method and an initial guess $v = 0.6$

$$f(v) = \frac{0.65}{1 + 0.55 \cdot v} + \frac{0.30}{1 - 0.36 \cdot v} + \frac{0.05}{1 - 0.72 \cdot v} - 1 = 0$$

The equation is the Rachford–Rice equation for a flash calculation of a two-phase, three-component mixture which is discussd in Chapter 10. The solution, v, is the mole fraction of the mixture that is in the gaseous phase.

Solution
Plotting the function

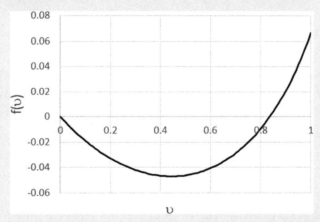

The plot shows there is a root is slightly above $v = 0.8$. There is also a nonphysical root at $v = 0$ and may be other nonphysical roots at $v < 0$ or $v > 1$.
The function has the analytical derivative,

$$f'(v) = \frac{-0.3575}{(1 + 0.55v)^2} + \frac{0.108}{(1 - 0.36v)^2} + \frac{0.036}{(1 - 0.72v)^2}$$

Newton's method uses the iterative formula, Eq. (6.11).
First iteration, $v = 0.6$,

$$f(0.6) = \frac{0.65}{1 + 0.55 \cdot 0.6} + \frac{0.30}{1 - 0.36 \cdot 0.6} + \frac{0.05}{1 - 0.72 \cdot 0.6} - 1 = -0.0406$$

$$f'(0.6) = \frac{-0.3575}{(1 + 0.55 \cdot 0.6)^2} + \frac{0.108}{(1 - 0.36 \cdot 0.6)^2} + \frac{0.036}{(1 - 0.72 \cdot 0.6)^2} = 0.08519$$

$$v_{i+1} = v_i - \frac{f(v_i)}{f'(v_i)} = 0.6 - \frac{-0.0406}{0.085} = 1.0764$$

The updated guess is not physical because it would mean that 107.6% by mole of the fluid is gas. But the solution has not converged.

Continued

Example 6.2. Newton–Raphson for Rachford–Rice.—cont'd

Second iteration, $v = 1.076$,

$$f(1.076) = \frac{0.65}{1 + 0.55 \cdot 1.076} + \frac{0.30}{1 - 0.36 \cdot 1.076} + \frac{0.05}{1 - 0.72 \cdot 1.076} - 1 = 0.12029$$

$$f'(1.076) = \frac{-0.3575}{(1 + 0.55 \cdot 1.076)^2} + \frac{0.108}{(1 - 0.36 \cdot 1.076)^2} + \frac{0.036}{(1 - 0.72 \cdot 1.076)^2}$$
$$= 0.08579$$

$$v_{i+1} = v_i - \frac{f(v_i)}{f'(v_i)} = 1.0764 - \frac{0.12029}{0.8579} = 0.937$$

With additional iterations, the solution eventually converges to $v = 0.8359$, so the fluid is 83.59 mol % gaseous and 16.41% liquid.

v	$f(v)$	df/dx	v new
0.6000	−0.0406	0.0852	1.0765
1.0765	0.12044	0.8587	0.9363
0.9363	0.03502	0.429	0.8547
0.8547	0.00547	0.3032	0.8366
0.8366	0.0002	0.2812	0.8359
0.8359	3E−07	0.2803	0.8359
0.8359	6.8E−13	0.2803	0.8359

Note that if a different guess were used, e.g., $v = 0.5$, the solution would have converged to a nonphysical value of $v = 1.7916$. This shows that while Newton's method may converge quickly, it may also converge to an unintended root or not converge at all.

6.3.3.2 Multidimensional Newton's method

The procedure in multidimensions is an extension of the 1D Newton's method and can be derived from a multidimensional Taylor series. First, the problem is posed as a residual vector, F, which is the null vector (zero) when evaluated at the solution pressure, P^{n+1},

$$\vec{F}\left(\vec{P}^{n+1}\right) = -\left(\mathbf{T}^{n+1} + \mathbf{J}^{n+1} + \mathbf{A}^{n+1}\mathbf{c}_t^{n+1}\right)\vec{P}^{n+1} + \left(\mathbf{A}^{n+1}\mathbf{c}_t^{n+1}\right)\vec{P}^{n} + \vec{Q}^{n+1}$$
$$+ \vec{G}^{n+1} = 0. \tag{6.12}$$

The goal is to determine the vector of pressure values, P^{n+1}, which make each of the elements of the vector, F^{n+1}, equal to, or near, zero. Multidimensional Newton's method works as follows:

(1) Guess the pressure field in the reservoir at time $n + 1$, P^{n+1}. The pressure at the previous timestep, P^n, can be used as a guess.

(2) Calculate pressure-dependent fluid and reservoir properties at P^{n+1} and use them to compute the arrays.

(3) Compute the residual vector, F, using Eq. (6.12) and Jacobian matrix of partial derivatives, $\Im$.

(4) Solve the linear system of equations $\Im * \delta P = -F$ to obtain the change in pressure vector, δP.

(5) Update the pressure in the reservoir, $P^{n+1} = P^n + \delta P$ and repeat steps 1–5 until the solution converges. Then proceed to the next timestep.

Step 3 involves computing the $N \times 1$ vector, F, and $N \times N$ Jacobian matrix, $\Im$. The residual vector, F, can be computed using Eq. (6.12). The Jacobian is the matrix of partial derivatives, evaluated at the current guess of pressures,

$$\Im_{ij} = \frac{\partial F_i}{\partial P_j} \approx \frac{F_i(P_j + \varepsilon) - F_i(P_j)}{\varepsilon}, \tag{6.13}$$

where F_i is the residual value Eq. (6.12) of the ith block and P_j is the pressure of the jth block. Thus $\Im_{ij}$ is the scalar value in the ith row and jth column of the Jacobian matrix and represents how much the residual, F_i, changes with a perturbation in pressure, P_j. The Jacobian matrix is a square, $N \times N$, matrix.

$$\Im = \begin{pmatrix} \partial F_1/\partial P_1 & \partial F_1/\partial P_2 & \partial F_1/\partial P_3 & \cdots & \partial F_1/\partial P_N \\ \partial F_2/\partial P_1 & \partial F_2/\partial P_2 & \partial F_2/\partial P_3 & \cdots & \partial F_2/\partial P_N \\ \partial F_3/\partial P_1 & \partial F_3/\partial P_2 & \partial F_3/\partial P_3 & \cdots & \partial F_3/\partial P_N \\ & & & \ddots & \\ \partial F_N/\partial P_1 & \partial F_N/\partial P_2 & \partial F_N/\partial P_3 & \cdots & \partial F_N/\partial P_N \end{pmatrix}$$

$$\tag{6.14}$$

$$= \begin{pmatrix} \partial F_1/\partial P_1 & \partial F_1/\partial P_2 & 0 & \cdots & 0 \\ \partial F_2/\partial P_1 & \partial F_2/\partial P_2 & \partial F_2/\partial P_3 & \cdots & 0 \\ 0 & \partial F_3/\partial P_2 & \partial F_3/\partial P_3 & \cdots & 0 \\ & & & \ddots & \\ 0 & 0 & 0 & \cdots & \partial F_N/\partial P_N \end{pmatrix}$$

As shown in Eq. (6.14), the Jacobian matrix mimics the structure of the transmissibility matrix, **T**, and is tridiagonal, pentadiagonal, and heptadiagonal in 1D, 2D, and 3D, respectively. This is because only perturbations in the pressure of blocks adjacent to or equal to block i affect F_i. The right-hand-side of Eq. (6.14) is tridiagonal (for 1D) with zero-entries where the partial derivatives are zero.

The partial derivatives can be computed analytically or numerically. It is preferred to use analytical derivatives, if amenable, because it reduces the computation time and is more accurate (less roundoff and truncation error). As a result, most reservoir simulators use analytical derivatives when possible. However, the derivation of such derivatives can be tedious. If necessary, numerical finite difference approximations can be used (Eq. 6.13) where ε is a small number in comparison to P_j, but not too small as to be affected by roundoff error. Eq. (6.13) is a forward difference approximation; equivalently a backward difference approximation could be employed. Despite the improved accuracy, centered differences are not generally used because it would require an additional computation of the residual vector, F.

As was the case in 1D Newton's method, convergence is usually determined when one (or both) of the following criteria are met: (1) the change in all block pressures is small enough between iterations that one can conclude the pressure has converged or (2) some vector norm of F (e.g., L_2, L_∞) is reduced below a predetermined tolerance. The latter is usually a better measure of convergence although the former may be easier and faster to compute. Upon convergence, Newton's method does so *quadratically* (meaning the error is squared in each iteration) near the root making Newton's method fast and the generally preferred method. There is no guarantee that Newton's method will converge, but if a reasonable guess value is provided, the problem is not extremely nonlinear, and moderate timesteps are used, Newton's method usually converges in these types of problems. Advanced, globally convergent Newton methods have been developed. The guess value for pressure is usually taken as the pressure in the previous timestep, but more aggressive methods can be employed which attempt to extrapolate the solution from several previous timesteps. Example 6.3 demonstrates Newton's method for the flow of gas.

Example 6.3. Nonlinear gas flow (Newton's method).
Repeat Example 6.1, but use Newton's method.

Solution
Newton's method can be used to find the value of pressure that makes the residual vector small,

$$\vec{F}\left(\vec{P}^{n+1}\right) = -\left(\mathbf{T}^{n+1} + \mathbf{J}^{n+1} + \mathbf{A}^{n+1}\mathbf{c}_t^{n+1}\right)\vec{P}^{n+1} + \mathbf{A}^{n+1}\mathbf{c}_t^{n+1}\vec{P}^{n} + \vec{Q}^{n+1} = 0,$$

where **T**, **J**, and Q are constant and were computed in Example 6.2. Newton's method requires the formation of a Jacobian of partial derivatives which can be computed numerically or analytically. The Jacobian in this problem is relatively simple because the **T**, **J**, and Q arrays are not pressure dependent and partial derivatives of the $\mathbf{Ac}_t$ matrix can be computed analytically.

Example 6.3. Nonlinear gas flow (Newton's method).—cont'd

$$\frac{\partial c_g}{\partial p} = \frac{\partial}{\partial p}\left(\frac{1}{p}\right) = -\frac{1}{p^2}$$

Therefore,

$$\frac{\partial}{\partial p_i^{n+1}}\left(A_i^{n+1} c_{t,i}^{n+1} p_i^{n+1}\right)$$

$$= A_i^{n+1} c_{t,i}^{n+1} + p_i^{n+1}\frac{\partial\left(A_i^{n+1} c_{t,i}^{n+1}\right)}{\partial p_i^{n+1}} = A_i^{n+1} c_{t,i}^{n+1} - A_i^{n+1} c_{t,i}^{n+1} = 0$$

$$\frac{\partial}{\partial p_i^{n+1}}\left(A_i^{n+1} c_{t,i}^{n+1} p_i^{n}\right) = p_i^{n}\frac{\partial\left(A_i^{n+1} c_{t,i}^{n+1}\right)}{\partial p_i^{n+1}} = \frac{p_i^{n}}{p_i^{n+1}}A_i^{n+1} c_{t,i}^{n+1}$$

which gives the Jacobian for this problem,

$$\mathfrak{J} = -(\mathbf{T}+\mathbf{J}) + \mathbf{A}^v\mathbf{c}_t^v\text{diag}\left(\overrightarrow{P}^n\right)[\text{diag}(P^v)]^{-1},$$

where v is the iteration number. The steps are to: (a) compute $\mathbf{Ac}_t$ and then F at the current guessed pressure, (b) compute the Jacobian of partial derivatives, (c) solve the system of equations to get a new pressure, and (d) compute the error and repeat steps until the error is below the predefined tolerance.

Iteration #1, guess $P = [14.7\ 14.7\ 14.7\ 14.7]$ psia

$$\mathbf{Ac}_t = \begin{bmatrix} 6.8027 & & & \\ & 6.8027 & & \\ & & 6.8027 & \\ & & & 6.8027 \end{bmatrix}\frac{\text{ft}^3}{\text{psi}-\text{day}}$$

$$\overrightarrow{F} = \begin{bmatrix} 938.3592 \\ \\ \\ \end{bmatrix}\frac{\text{ft}^3}{\text{day}} \quad \text{error} = 938.3592\frac{\text{ft}^3}{\text{day}}$$

$$\overrightarrow{P} = \begin{bmatrix} 82.5415 \\ 29.8940 \\ 18.0683 \\ 15.2923 \end{bmatrix}\text{psia}$$

Continued

Example 6.3. Nonlinear gas flow (Newton's method).—cont'd

Iteration #2, guess $P = [82.5\ 29.9\ 18.1\ 15.3]$ psia

$$\mathbf{Ac_t} = \begin{bmatrix} 1.2115 & & & \\ & 3.3452 & & \\ & & 5.5345 & \\ & & & 6.5392 \end{bmatrix} \frac{ft^3}{psi - day}$$

$$\vec{F} = \begin{bmatrix} 379.3159 \\ 52.5344 \\ 4.2716 \\ 0.1561 \end{bmatrix} \frac{ft^3}{day} \quad error = 382.9604 \frac{ft^3}{day}$$

$$\vec{P} = \begin{bmatrix} 142.1448 \\ 63.9744 \\ 28.0180 \\ 17.1183 \end{bmatrix} psia$$

Iteration #3, guess $P = [142.1\ 64.0\ 28.1\ 17.1]$ psia

$$\mathbf{Ac_t} = \begin{bmatrix} 0.7035 & & & \\ & 1.5631 & & \\ & & 3.5691 & \\ & & & 5.8417 \end{bmatrix} \frac{ft^3}{psi - day}$$

$$\vec{F} = \begin{bmatrix} 5.3924 \\ 29.8642 \\ 15.9097 \\ 1.2244 \end{bmatrix} \frac{ft^3}{day} \quad error = 34.2865 \frac{ft^3}{day}$$

$$\vec{P} = \begin{bmatrix} 146.3437 \\ 74.5619 \\ 34.7015 \\ 18.5571 \end{bmatrix} psia$$

Example 6.3. Nonlinear gas flow (Newton's method).—cont'd

Time (days)	Iteration	Error	Block #1 (psia)	Block #2 (psia)	Block #3 (psia)	Block #4 (psia)
0			14.7	14.7	14.7	14.7
1.0E−05	1	9.38E+02	82.54	29.89	18.07	15.29
	2	3.83E+02	142.14	63.97	28.02	17.12
	6	7.52E−07	146.51	75.06	35.36	18.75
2.0E−05	1	1.36E+02	165.7	102.26	53.15	23.9
	2	2.03E+01	167.34	106.96	58.4	25.67
	5	1.73E−09	167.38	107.08	58.59	25.75
3.0E−05	1	5.80E+01	172.77	119.6	71.03	30.86
	2	5.12E+00	173.05	120.45	72.23	31.48
	5	1.98E−13	173.05	120.45	72.24	31.48
∞			200	200	200	200

Where the error was calculated using the pressure at the previous iteration. Newton's method converged in significantly fewer iterations than Picard.

6.4 Pseudocode for Newton's method

A simple, albeit not necessarily the most computationally efficient, subroutine is presented here for creation of the Jacobian. We first calculate the residual vector, F, at the guessed pressure, P, using Eq. (6.12). We then loop through the $icol = 1$ to N grids and perturb the pressure in grid icol by ε, and recompute the arrays and then the residual vector at the perturbed pressure. The partial derivatives are computed for all rows ($irow = 1$ to N) in column $icol$ using Eq. (6.13).

The main code follows the Newton algorithm described by steps 1−5 in Section 6.3.3.2. There are nested while loops; the outer loop for time-stepping and the inner loop for convergence of Newton's method. In the inner loop, the new arrays and residual vector are computed, the Jacobian is formed by calling the subroutine, the system of linear equations is solved, pressures are updated, and finally error is computed.

```
MAIN CODE
CALL PREPROCESS
INIT tolerance
INIT error (>tolerance)
SET t = 0
WHILE t < t_final
     WHILE error > tolerance
          CALL WELL ARRAYS
          CALL GRID ARRAYS
          IF explicit update method
               CALCULATE P using eqn 6.9
               SET error = 0
          ELSEIF picard method
               SET P_old = P
               CALCULATE P using eqn 6.9
               CALCULATE error e.g. norm(P-P_old)
          ELSEIF newton method
               CALCULATE F using eqn 6.12
               CALL JACOBIAN
               SET P_old = P
               SET δP = LinearSolve(ℑ,-F)
               SET P = P_old + δP
               CALCULATE error, norm(F) or norm(δP)
          ENDIF
     ENDWHILE
     INCREMENT t by Δt
     UPDATE pressure-dependent properties in all grids at new P
ENDWHILE
CALL POSTPROCESS


SUBROUTINE JACOBIAN
INPUTS: F, reservoir, fluid, petrophysical properties
OUTPUTS: ℑ

FOR ICOL = 1 to # grids
     SET P_temp = P(ICOL)
     SET P(ICOL) = P_temp + ε
     UPDATE pressure-dependent properties at new P
     CALL WELL ARRAYS
     CALL GRID ARRAYS
     CALCULATE F2 at new P using eqn 6.12
     FOR IROW = 1 to # grids
          ℑ(IROW,ICOL) = [F2(IROW)-F1(IROW)]/ε
     ENDFOR
     SET P(ICOL) = P_temp
ENDFOR
```

6.5 Exercises

Exercise 6.1. Nonlinear gas flow and the Picard method. Repeat Example 6.1 (gas flow) using the Picard method, but now assume that the viscosity is also pressure dependent and given by Eq. (6.1). The reservoir temperature is 150 °F and molecular weight is 19 lb_m/lbmole. The initial pressure (p_{init} = 2000 psia) and boundary conditions (p_1 = 3000 psia and p_2 = 2000 psia) on the core are also changed. The higher pressure means that the gas does not act ideally, but the z-factor is approximately constant (z = 0.8). Perform two iterations in the first timestep.

Exercise 6.2. Nonlinear gas flow and the Newton-Raphson method. Repeat Exercise 6.1 but use the Newton–Raphson method. You may compute partial derivatives numerically.

Exercise 6.3. Non-Newtonian viscosity. Calculate the viscosity, shear rate, and velocity of a reservoir grid block with pressure of 3000 psia if its upwinded neighbor block has a pressure of 3400 psia. Both blocks have a permeability and porosity of 200 mD and 25%, respectively. The distance between block centers is 80 ft. The viscosity is described by a Carreau model with n = 0.5, μ_0 = 100 cp, μ_∞ = 1 cp, and λ = 1.0 1/s. Use C = 4 in the shear rate equation and an initial guess of 100 cp for viscosity. Perform three iterations.

Exercise 6.4. Computer code for Non-Newtonian viscosity. Write a code (subroutine or function) to calculate the viscosity of a Carreau fluid given the shear rate and rheological properties, n, μ_0, μ_∞, λ. Test your code by making a log–log plot of viscosity (cp) versus shear rate (1/s) with n = 0.5, μ_0 = 100 cp, μ_∞ = 1 cp, and λ = 1.0 1/s.

Exercise 6.5. Computer code for iterating on block properties in non-Newtonian flow. Write a code (subroutine or function) to determine the velocity, shear rate, and viscosity of a grid block given the block pressures. Test your code against Exercise 6.3.

Exercise 6.6. Computer code for Picard method. Develop (or adapt a previous) a reservoir simulator to solve a nonlinear problem either by explicit updates or Picard iteration (user-defined option). Validate your code against Example 6.1 and/or Exercise 6.1.

Exercise 6.7. Computer code for Newton's method. Adapt the reservoir simulator developed in Exercise 6.6 to include an option for the Newton–Raphson method. Validate your code against Example 6.3 and/or Exercise 6.2.

Exercise 6.8. Reservoir simulator for non-Newtonian flow. Use the reservoir simulator developed in Exercise 6.6 (Picard method) and/or 6.7 (Newton method) along with the subroutines developed in Exercises 6.4 and 6.5 to solve for single-phase, non-Newtonian flow in a 1D ($L = 4000$ ft) reservoir with initial pressure of 3000 psia. There is a constant rate injector well at $x = 0$ (1000 scf/day of polymer) and a constant rate producer well at $x = L$ (1000 scf/day of polymer). The reservoir has uniform permeability ($k = 200$ md), porosity ($\phi = 0.25$), formation volume factor ($B_w = 1$), and total compressibility ($c_t = 1 \times 10^{-6}$ psi^{-1}). The rheology of the polymer is described by a Carreau model with parameter $n = 0.5$, $\mu_0 = 100$ cp, $\mu_\infty = 1$ cp, and $\lambda = 1.0$ 1/s. Assume the reservoir is saturated with aqueous polymer ($\rho_w^{sc} = 62.4$ lb$_m$/ft^3). The cross-sectional area of the reservoir is 10,000 ft^2.

Use $N = 50$ grid blocks and $\Delta t = 1$ day in the simulation. Plot the block pressure, velocity, shear rate, and apparent viscosity versus distance at 2, 10, and 30 days.

References

Ahmed, N., 1967. Physical Properties of Porous Medium Affecting Laminar and Turbulent Flow of Water. PhD dissertation. Colorado State University, Fort Collins, Colorado.

Balhoff, M.T., Wheeler, M.F., 2009. A predictive pore-scale model for non-Darcy flow in porous media. SPE Journal 14 (4), 579–587.

Barree, R.D., Conway, M.W., 2004. Beyond Beta Factors: A Complete Model for Darcy, Forchheimer, and Trans-Forchheimer Flow in Porous Media. Paper presented at the. SPE Annual Technical Conference and Exhibition, Houston, 26-29 September. SPE 89325.

Barree, R.D., Conway, M.W., 2005. Reply to discussion of beyond beta factors: a complete model for Darcy, Forchheimer, and trans-Forchheimer flow in porous media. Journal of Petroleum Technology 57 (8), 73–74.

Blake, F.C., 1922. The resistance of packing to fluid flow. Transactions of the American Institute of Chemical Engineers 14, 415–421.

Brownell, L.E., Dombrowski, H.S., Dickey, C.A., 1947. Pressure drop through porous media. Chemical Engineering Progress 43, 537–548.

Delshad, M., Kim, D.H., Magbagbeola, O.A., Huh, C., Pope, G.A., Farhad, T., 2008. Mechanistic interpretation and utilization of viscoelastic behavior of polymer solutions for improved polymer-flood efficiency. In: SPE Symposium on Improved Oil Recovery.

Fancher, G.H., Lewis, J.A., 1933. Flow of simple fluids through porous materials. Industrial and Engineering Chemistry Research 25 (10), 1139–1147.

Forchheimer, P., 1901. Wasserbewegung durch boden. Zeit. Ver. Deutsch. Ing 45, 1781–1788.

Forchheimer, P., 1930. Teubner Verlagsgesellschaft. Leipzig and Berlin. pp. 139–163.

Geertsma, J., 1974. Estimating the coefficient of inertial resistance in fluid flow through porous media. Society of Petroleum Engineers Journal 445–450.

Gidley, J.L., Holditch, S.A., Nierode, D.E., et al., 1989. Fracturing Fluids and Additives in Recent Advances in Hydraulic Fracturing. SPE Monograph Series, Richardson, Texas, 12 (Chapter 7): 131.

Howard, C.C., Fast, C.R., 1957. Optimum fluid characteristics for fracture extension. API Drilling and Production Practice 24, 261.

Jensen, B., Paulsen, J.E., Saasen, A., Statoil, A.S.A., Prebensen, O.I., Balzer, H., M-I/Swaco Norge, A.S., 2004. Application of water based drilling fluid. In: Total Fluid Management. IADC/SPE Drilling Conference, 2—4 March, Dallas, Texas.

Jones, F.O., Owens, W.W., 1980. A laboratory study of low-permeability gas sands. Journal of Petroleum Technology 32 (9), 1631—1640. SPE-7551-PA.

Kim, B.Y.K., 1985. The Resistance to Flow in Simple and Complex Porous Media Whose Matrices Are Composed of Spheres. MSc thesis. University of Hawaii at Manoa, Honolulu, Hawaii.

Koh, H., Lee, V.B., Pope, G.A., 2018. Experimental investigation of the effect of polymers on residual oil saturation. SPE Journal 23, 1—17. https://doi.org/10.2118/179683-PA.

Lake, L.W., Johns, R.T., Rossen, W.R., Pope, G., 2014. Fundamentals of Enhanced Oil Recovery. Society of Petroleum Engineers.

Lee, A.L., Gonzalez, M.H., Eakin, B.E., 1966. The viscosity of natural gases. Journal of Petroleum Technology 18 (08), 997—1000.

Lindquist, E., 1933. On the flow of water through porous soil. In: Proceedings of 1er Congres des Grands Barrages. Commission internationale des grands barrages, Stockholm, Sweden, pp. 81—101.

Liu, X., Civan, F., Evans, R.D., 1995. Correlation of the non-Darcy flow coefficient. Journal of Canadian Petroleum Technology 34 (10) (November).

Mobasheri, F., Todd, D.K., 1963. Investigation of the hydraulics of flow near recharge wells. In: Water Resources Center Contribution No. 72. University of California Berkeley, Berkeley, California.

Paccaloni, G., Tambini, M., 1993. Advances in matrix stimulation technology. Journal of Petroleum Technology 45 (3), 256—263. SPE-20623-PA.

Seright, R.S., 1995. Improved Techniques for Fluid Diversion in Oil Recovery, Second Annual Report. Contract No. DE-AC22-92BC14880. US DOE, Washington, DC.

Smith, G.E., 1992. Waterflooding Heavy Oils. Society of Petroleum Engineers.

Sunada, D.K., 1965. Laminar and Turbulent Flow of Water through Homogeneous Porous Media. PhD dissertation. University of California at Berkeley, Berkeley, California.

Wang, X., Thauvin, F., Mohanty, K.K., 1999. Non-Darcy flow through anisotropic porous media. Chemical Engineering Science 54, 1859—1869.

Willhite, G.P., Green, D.W., 1998. Enhanced Oil Recovery, vol. 6. Society of Petroleum Engineers.

Xie, C., Balhoff, M.T., 2021. Lattice Boltzmann modeling of the apparent viscosity of thinning—elastic fluids in porous media. Transport in Porous Media 137 (1), 63—86.

Chapter 7

Component transport in porous media

7.1 Introduction

In Chapter 2, mass balance equations were derived for flow in porous media for a single or multiple fluid phases. In this chapter, the relevant partial differential equations (PDEs) that describe solute transport in porous media of one or more components are presented. The equations are the basis for single-phase flow of components such as tracers as well as multiphase flow of water, oil, and gas components (*Black Oil* model), or a fully compositional model which is discussed in Chapter 10. We finish the chapter with a few analytical solutions to these PDEs.

7.2 Transport mechanisms

7.2.1 Advection

Advective (also referred to as *convective*) transport refers to the movement of a component as a result of being "carried" by the bulk fluid. The convective transport is proportional to the velocity of the bulk fluid. Advective flux ($\vec{N}_{\kappa,\alpha}^{ad}$ [=] mass/area-time) of component κ in phase α is the velocity of the bulk fluid multiplied the concentration of the component,

$$\vec{N}_{\kappa,\alpha}^{ad} = M_\kappa \rho_{M,\alpha} x_{\kappa,\alpha} \vec{u}_\alpha = \rho_\alpha \omega_{\kappa,\alpha} \vec{u}_\alpha = C_{\kappa,\alpha} \vec{u}_\alpha, \tag{7.1}$$

where M_κ is the molecular weight of component κ, $\rho_{M,\alpha}$ and ρ_a are the molar and mass densities, respectively, of phase α, $x_{\kappa,\alpha}$ and $\omega_{\kappa,\alpha}$ are the molar and mass fractions, respectively, of component κ in phase α, $C_{\kappa,\alpha} = \rho_\alpha \omega_{\kappa,\alpha}$ is the mass concentration of component κ in phase α, and u_α is the Darcy velocity of phase α. The component is advected in the direction of positive potential of the phase, Φ_α.

An Introduction to Multiphase, Multicomponent Reservoir Simulation.
https://doi.org/10.1016/B978-0-323-99235-0.00001-4

7.2.2 Hydrodynamic dispersion

Hydrodynamic dispersion is the transport mechanism that accounts for the combined mechanisms of molecular diffusion and mechanical dispersion in the porous medium. It is the phenomenon that causes a component, such as a tracer, in a porous medium to "spread out" as it is transported from an injector to a producer. Molecular *Diffusion* and mechanical *Dispersion* are different transport mechanisms as discussed below, but both can be mathematically described by Fick's law,

$$\overrightarrow{N}^{d}_{\kappa,\alpha} = -M_{\kappa}\phi S_{\alpha}\mathbf{D}_{\kappa,\alpha}\nabla\left(\rho_{M,\alpha}x_{\kappa,\alpha}\right) = -\phi S_{\alpha}\mathbf{D}_{\kappa,\alpha}\nabla\left(\rho_{\alpha}\omega_{\kappa,\alpha}\right) = -\phi S_{\alpha}\mathbf{D}_{\kappa,\alpha}\nabla C_{\kappa,\alpha},$$

$$(7.2)$$

where ϕ is the porosity, S_{α} is the saturation of phase α, and $\mathbf{D}_{\kappa,\alpha}$ is the molecular diffusion or dispersion coefficient/tensor of component κ in phase α. According to Eq. (7.2), the flux of a component is proportional to concentration gradient. Therefore, the component transports via diffusion/dispersion from high concentration to low concentration and may be in the same or opposite direction of advective transport. Diffusion coefficients are scalar properties but dispersion coefficients can be second-order tensors. Fick's law is analogous to, and has the same form as, Fourier's law for heat flux and Darcy's law for momentum flux.

7.2.2.1 Diffusive transport

Molecular diffusion is the spreading of a component due to the random motion of molecules and results in the transport of components from a region of higher concentration to one of lower concentration. It occurs in all directions and does not depend on the direction and magnitude of flow. The molecular diffusion coefficient ($D_{m,\kappa,\alpha}$) of a dilute component, κ, in a phase, α, is given theoretically by the Stokes—Einstein equation,

$$D_{m,\kappa,\alpha} = \frac{k_{B}T}{6\pi\mu_{\alpha}r_{\kappa}},$$

$$(7.3)$$

where, k_{B} is the Boltzmann constant (1.38×10^{-23} J/K), T is the absolute temperature, μ_{α} is the phase viscosity, and r_{κ} is the radius of the particle/component. Importantly, the diffusion coefficient is inversely proportional to phase viscosity. Typical values for the diffusion coefficient of a component in a gaseous phase is 10^{-6} to 10^{-5} m^2/s ($1-10$ ft^2/day) and in liquids are 10^{-10} to 10^{-9} m^2/s ($10^{-4}-10^{-3}$ ft^2/day). Diffusion in liquids is much slower than in gases because of the high molecular interaction and density of the solvent. The diffusion coefficient can be measured experimentally and values are published for many component-fluid systems. For example, the diffusion coefficient of carbon dioxide (CO_2) in water is 1.92 cm^2/s (1.58×10^{-3} ft^2/day) at 25°C (Cussler, 1997).

In porous media, an effective diffusion coefficient ($D_{m,\text{eff}}$) is defined to model molecular diffusion when affected by *tortuosity*, the dimensionless length that a particle or component travels ($\tau = \Delta l/L$) due to the locally heterogeneous pore structure (Fig. 7.1).

The tortuous path results in a longer effective length for a molecule to diffuse. The effective diffusion coefficient, $D_{m,\text{eff}}$, in a porous medium is given by Eq. (7.4),

$$\frac{D_{m,\text{eff}}}{D_{m,\kappa,\alpha}} = D_r = f(\tau, \phi), \tag{7.4}$$

where D_r is the *restricted diffusion coefficient* and is the ratio of the effective diffusion coefficient in the porous medium and the molecular diffusion coefficient in bulk solution. It is a function of the porosity and tortuosity, τ. Various models have been developed to describe the tortuosity/porosity dependence (Van Brakel, 1974), such as $D_r = \phi/\tau^2$. Tortuosity has also been related to the porosity of the porous medium and can be estimated by an empirical model, e.g.,

$$\tau^2 = (F\phi)^n = \left(A\phi^{1-m}\right)^n, \tag{7.5}$$

where F is the *formation resistivity factor* and A, n, and m are empirical constants.

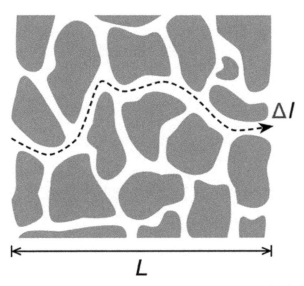

FIGURE 7.1 Schematic of a 2D porous medium and the tortuous path (*dashed line*) of a component.

7.2.2.2 Mechanical dispersion

Mechanical dispersion is caused by the variation in velocities that occur in porous media. Unlike molecular diffusion, it does depend on the direction and magnitude of flow. The two principal directions of mechanical dispersion are: (a) longitudinal, or parallel to mean flow, and (b) transverse, or perpendicular to mean flow (Sahimi, 2012). The longitudinal dispersion coefficient (D_L) is usually greater than the transverse dispersion coefficient (D_T) and both increase with the dimensionless Peclet number,

$$N_{Pe,d} = \frac{u d_p}{D_{m,\kappa,\alpha}} \tag{7.6}$$

where d_p is a characteristic length scale (e.g., grain size diameter). In Eq. (7.6) the diffusive Peclet number ($N_{Pe,d}$) is defined using the molecular diffusion coefficient but later a dispersive Peclet number (N_{Pe}) is defined with the dispersion coefficient and the length of the domain. Theoretical studies (e.g., Saffman, 1959; Koch and Brady, 1985; Mehmani & Balhoff, 2015) identified the underlying physical mechanisms of dispersion and developed equations to describe these regimes. However, simpler relationships are typically employed in reservoir simulation.

7.2.2.3 Combined effects and hydrodynamic dispersion

As previously mentioned, mechanical dispersion and molecular diffusion are different transport mechanisms although their coefficients have the same units (e.g., ft^2/day) and can be treated mathematically in a similar way (Fick's law, Eq. 7.2). The effects of the two transport mechanisms are additive and can be combined to give a hydrodynamic dispersion coefficient, D. In the absence of advection, hydrodynamic dispersion is simplified to only molecular diffusion ($D = D_{m,\kappa,\alpha}$). However, when advective transport is present (as in most subsurface applications), dispersive flux is generally much greater than diffusive flux and $D \gg D_{m,\text{eff}}$. Fig. 7.2 shows the relationship between dimensionless dispersion coefficient and Peclet number ($N_{Pe,d}$) on a log–log scale.

At low $N_{Pe,d}$ (advective transport is small), the dispersion coefficient approximately equals the restricted diffusion coefficient, D_r. As $N_{Pe,d}$ increases (higher advective transport), mechanical dispersion coefficients increase and are much greater, orders of magnitude, than diffusion coefficients for large $N_{Pe,d}$. In typical subsurface applications the $N_{Pe,d}$ can be 10^3 or larger.

Empirical equations for hydrodynamic dispersion have been developed. For example, in 1D,

$$D_{\kappa,\alpha} = D_r + a \cdot N_{Pe,d}^b. \tag{7.7a}$$

where a and b are empirical constants, unique to the porous medium, and are different for longitudinal and transverse dispersion. The exponent b is usually between 1 and 2 (Freeze, 1979) for longitudinal dispersion and often assumed close to unity. For relatively high Peclet number, the second term (mechanical dispersion) dominates over the first (molecular diffusion). A second-order

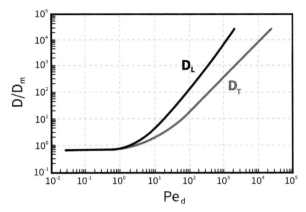

FIGURE 7.2   Qualitative relationships of dimensionless dispersion (D/D_m) versus Pe_d in a porous medium.

tensor form is often applied in reservoir simulation that includes both longitudinal and transverse dispersion,

$$\mathbf{D}_{\kappa,\alpha} = \frac{D_{m,\kappa,\alpha}}{\tau}\delta_{ij} + \frac{\alpha_{T,\alpha}}{\varphi S_\alpha}\left|\overrightarrow{u}_\alpha\right|\delta_{ij} + \frac{\left(\alpha_{L,\alpha} - \alpha_{T,\alpha}\right)}{\varphi S_\alpha}\frac{\overrightarrow{u}_{\alpha,i}\,\overrightarrow{u}_{\alpha,j}}{\left|\overrightarrow{u}_\alpha\right|}, \tag{7.7b}$$

where $u_{\alpha,i}$ and $u_{\alpha,j}$ are the components of Darcy velocity of phase α in directions i and j, and δ_{ij} is the Kronecker delta function. The coefficients $\alpha_{L,\alpha}$ and $\alpha_{T,\alpha}$ are phase longitudinal and transverse dispersivities, respectively, and have been shown to be dependent on the heterogeneity of the medium as a function of the length scale of measurement (John et al., 2010; Lake et al., 2014) which can be 10^{-1} to 10^3 ft. The magnitude of vector flux in Eq. (7.7b) for each phase is computed as:

$$\left|\overrightarrow{u}_\alpha\right| = \sqrt{\left(u_{x,\alpha}\right)^2 + \left(u_{y,\alpha}\right)^2 + \left(u_{z,\alpha}\right)^2}. \tag{7.8}$$

Eq. (7.7b) shows that hydrodynamic dispersion is a second-order, 3×3 tensor. It can be substituted into Eq. (7.2) to describe dispersive transport. Although dispersive flux is much greater than diffusive flux alone, advective flux dominates transport in most subsurface applications. Example 7.1 demonstrates the calculation of the hydrodynamic dispersion tensor in a porous medium, and Example 7.2 calculates advective and dispersive fluxes. Throughout most of the remainder of this text, the terms *dispersion* and *dispersion coefficient* refer to hydrodynamic dispersion, that is the combined effects of mechanical dispersion and molecular diffusion.

Example 7.1. Dispersion tensor.

Calculate the dispersion tensor for a component with molecular diffusion coefficient of $D_m = 0.001$ ft^2/day in an aqueous phase that is flowing in a porous medium at a Darcy velocity of $u_x = 0.5$ ft/day, $u_y = 0.4$ ft/day, and $u_z = 0.2$ ft/day. The porosity of the porous medium is 0.20, the tortuosity is 1.5, and the longitudinal and transverse dispersivities are 1.0 and 0.1 ft, respectively. The saturation of the aqueous phase is 50%.

Solution

The dispersion tensor is given by Eq. (7.7b), where the phase velocity is given by,

$$\left| \vec{u}_\alpha \right| = \sqrt{\left(u_{x,\alpha}\right)^2 + \left(u_{y,\alpha}\right)^2 + \left(u_{z,\alpha}\right)^2} = \sqrt{0.5^2 + 0.4^2 + 0.2^2} = 0.6708 \frac{\text{ft}}{\text{day}}$$

Therefore the 9 components of the dispersion tensor are

$$D_{xx,w} = \frac{D_m}{\tau} + \frac{\alpha_{T,w}}{\phi S_w}\left|\vec{u}_w\right| + \frac{\left(\alpha_{L,w} - \alpha_{T,w}\right)}{\phi S_w}\frac{u_{w,x} u_{w,x}}{\left|\vec{u}_w\right|} = \frac{0.001\frac{\text{ft}^2}{\text{day}}}{1.5}$$

$$+ \frac{0.1\text{ft}}{0.2 \cdot 0.5} \cdot 0.6708 \frac{\text{ft}}{\text{day}} + \frac{(1.0 - 0.1)\text{ft}}{0.2 \cdot 0.5}\frac{0.5\frac{\text{ft}}{\text{day}} \cdot 0.5\frac{\text{ft}}{\text{day}}}{0.6708\frac{\text{ft}}{\text{day}}} = 4.0257 \frac{\text{ft}^2}{\text{day}}$$

$$D_{xy,w} = \frac{\left(\alpha_{L,w} - \alpha_{T,w}\right)}{\phi S_w}\frac{u_{w,x} u_{w,y}}{\left|\vec{u}_w\right|} = \frac{(1.0 - 0.1)\text{ft}}{0.2 \cdot 0.5}\frac{0.5\frac{\text{ft}}{\text{day}} \cdot 0.4\frac{\text{ft}}{\text{day}}}{0.6708\frac{\text{ft}}{\text{day}}} = 2.6834 \frac{\text{ft}^2}{\text{day}}$$

$$D_{xz,w} = \frac{\left(\alpha_{L,w} - \alpha_{T,w}\right)}{\phi S_w}\frac{u_{w,x} u_{w,z}}{\left|\vec{u}_w\right|} = \frac{(1.0 - 0.1)\text{ft}}{0.2 \cdot 0.5}\frac{0.5\frac{\text{ft}}{\text{day}} \cdot 0.2\frac{\text{ft}}{\text{day}}}{0.6708\frac{\text{ft}}{\text{day}}} = 1.3417 \frac{\text{ft}^2}{\text{day}}$$

$$D_{yx,w} = \frac{\left(\alpha_{L,w} - \alpha_{T,w}\right)}{\phi S_w}\frac{u_{w,y} u_{w,x}}{\left|\vec{u}_w\right|} = \frac{(1.0 - 0.1)\text{ft}}{0.2 \cdot 0.5}\frac{0.4\frac{\text{ft}}{\text{day}} \cdot 0.5\frac{\text{ft}}{\text{day}}}{0.6708\frac{\text{ft}}{\text{day}}} = 2.6834 \frac{\text{ft}^2}{\text{day}}$$

$$D_{yy,w} = \frac{D_m}{\tau} + \frac{\alpha_{T,w}}{\phi S_w}\left|\vec{u}_w\right| + \frac{\left(\alpha_{L,w} - \alpha_{T,w}\right)}{\phi S_w}\frac{u_{w,y} u_{w,y}}{\left|\vec{u}_w\right|} = \frac{0.001\frac{\text{ft}^2}{\text{day}}}{1.5}$$

$$+ \frac{0.1\text{ft}}{0.2 \cdot 0.5} \cdot 0.6708 \frac{\text{ft}}{\text{day}} + \frac{(1.0 - 0.1)\text{ft}}{0.2 \cdot 0.5}\frac{0.4\frac{\text{ft}}{\text{day}} \cdot 0.4\frac{\text{ft}}{\text{day}}}{0.6708\frac{\text{ft}}{\text{day}}} = 2.8182 \frac{\text{ft}^2}{\text{day}}$$

Example 7.1. Dispersion tensor.—cont'd

$$D_{yz,w} = \frac{(\alpha_{L,w} - \alpha_{T,w})}{\phi S_w} \frac{u_{w,y} u_{w,z}}{|\vec{u}_w|} = \frac{(1.0 - 0.1)\text{ft}}{0.2 \cdot 0.5} \frac{0.4 \frac{\text{ft}}{\text{day}} \cdot 0.2 \frac{\text{ft}}{\text{day}}}{0.6708 \frac{\text{ft}}{\text{day}}} = 1.0733 \frac{\text{ft}^2}{\text{day}}$$

$$D_{zx,w} = \frac{(\alpha_{L,w} - \alpha_{T,w})}{\phi S_w} \frac{u_{w,x} u_{w,z}}{|\vec{u}_w|} = \frac{(1.0 - 0.1)\text{ft}}{0.2 \cdot 0.5} \frac{0.2 \frac{\text{ft}}{\text{day}} \cdot 0.5 \frac{\text{ft}}{\text{day}}}{0.6708 \frac{\text{ft}}{\text{day}}} = 1.3417 \frac{\text{ft}^2}{\text{day}}$$

$$D_{zy,w} = \frac{(\alpha_{L,w} - \alpha_{T,w})}{\phi S_w} \frac{u_{w,z} u_{w,y}}{|\vec{u}_w|} = \frac{(1.0 - 0.1)\text{ft}}{0.2 \cdot 0.5} \frac{0.2 \frac{\text{ft}}{\text{day}} \cdot 0.4 \frac{\text{ft}}{\text{day}}}{0.6708 \frac{\text{ft}}{\text{day}}} = 1.0733 \frac{\text{ft}^2}{\text{day}}$$

$$D_{zz,w} = \frac{D_m}{\tau} + \frac{\alpha_{T,w}}{\phi S_w} |\vec{u}_w| + \frac{(\alpha_{L,w} - \alpha_{T,w})}{\phi S_w} \frac{u_{w,z} u_{w,z}}{|\vec{u}_w|}$$

$$= \frac{0.001 \frac{\text{ft}^2}{\text{day}}}{1.5} + \frac{0.1\text{ft}}{0.2 \cdot 0.5} \cdot 0.6708 \frac{\text{ft}}{\text{day}} + \frac{(1.0 - 0.1)\text{ft}}{0.2 \cdot 0.5} \frac{0.2 \frac{\text{ft}}{\text{day}} \cdot 0.2 \frac{\text{ft}}{\text{day}}}{0.6708 \frac{\text{ft}}{\text{day}}}$$

$$= 1.2081 \frac{\text{ft}^2}{\text{day}}$$

Therefore,

$$\mathbf{D}_w = \begin{pmatrix} 4.0257 & 2.6834 & 1.3417 \\ 2.6834 & 2.8182 & 1.0733 \\ 1.3417 & 1.0733 & 1.2081 \end{pmatrix} \frac{\text{ft}^2}{\text{day}}$$

Example 7.2. Fluxes.

Calculate the advective and dispersive fluxes for a component in the aqueous phase (density of 65 lb_m/ft^3) with mass fraction of 0.01. The gradients of mass fractions are $\nabla\omega = [0.0015, -0.0007, 0.001]$ lb_m/lb_m-ft. Use the Dispersion tensor computed in Example 7.1.

<u>Solution</u>

$$\vec{N}_{\kappa,\alpha}^{ad} = \rho_\alpha \omega_{\kappa,\alpha} \vec{u}_\alpha = 65 \frac{\text{lb}_m}{\text{ft}^3} \cdot 0.01 \frac{\text{lb}_m}{\text{lb}_m} \cdot \begin{bmatrix} 0.5 \\ 0.4 \\ 0.2 \end{bmatrix} \frac{\text{ft}}{\text{day}} = \begin{bmatrix} 0.325 \\ 0.26 \\ 0.13 \end{bmatrix} \frac{\text{lb}_m}{\text{ft}^2 - \text{day}}$$

Continued

Example 7.2. Fluxes.—cont'd

$$\vec{N}_{\kappa,\alpha}^d = -\phi S_\alpha \rho_\alpha \mathbf{D}_{\kappa,\alpha} \cdot \nabla \omega_{\kappa,\alpha} = -0.2 \cdot 0.5 \cdot 65 \frac{\text{lb}_m}{\text{ft}^3} \cdot \begin{pmatrix} 4.0257 & 2.6834 & 1.3417 \\ 2.6834 & 2.8182 & 1.0733 \\ 1.3417 & 1.0733 & 1.2081 \end{pmatrix}$$

$$\frac{\text{ft}^2}{\text{day}} \cdot \begin{bmatrix} 0.0015 \\ -0.0007 \\ 0.001 \end{bmatrix} \frac{\text{lb}_m}{\text{lb}_m - \text{ft}} = \begin{bmatrix} -0.0358 \\ -0.0203 \\ -0.0161 \end{bmatrix} \frac{\text{lb}_m}{\text{ft}^2 - \text{day}}$$

The dispersive fluxes are smaller and in the opposite direction as the advective fluxes.

7.2.3 Reactive transport and other source terms

Chemical components can be generated or consumed via a chemical reaction. Although many subsurface components are inert (have small or no reactivity), geochemical reactions are common in many subsurface applications (Frenier and Ziauddin, 2014), such as matrix stimulation using acids and precipitation and dissolution in carbon capture, utilization, and storage (DePaolo and Cole, 2013). The reaction rate of component κ, $r_{\kappa,\alpha}$, is the mass of κ generated/consumed per unit volume per time in the phase. For example, for a first-order reaction,

$$r_{\kappa,\alpha} = -k\rho_\kappa \omega_{\kappa,\alpha}, \tag{7.9}$$

where k is the rate constant of the reaction. Most geochemical reactions are not first order and have more complicated rate expressions, but detailed discussion of geochemistry is outside the scope of this text.

Generation and consumption of components can also occur via a source or sink (i.e., they can be injected or produced from the reservoir from a well). The mass of component κ in phase α per unit volume per time is given by $r_{\kappa,\alpha} = C_{\kappa,\alpha}^* \widetilde{q}_\alpha$, where $\widetilde{q}_\alpha$ is the source term for the phase and $C_{\kappa,\alpha}^*$ is defined as,

$$C_{\kappa,\alpha}^* = \begin{cases} C_{\kappa,\alpha,inj} & \text{if injector well} \\ C_{\kappa,\alpha} & \text{if producer well} \end{cases},$$

where $C_{\kappa,\alpha,inj}$ is the concentration of κ that is injected into phase α, and $C_{\kappa,\alpha}$ is the concentration subsurface that is produced. $C_{\kappa,\alpha,inj}$ is an input defined at the injector well, but $C_{\kappa,\alpha}$ depends on the concentration at the location of the producer well. The injected concentration is specified but partitions between subsurface phases. Likewise, the subsurface concentration of each phase is defined, but production of the respective phases depends on relative mobilities. Well compositions and distribution among phases are discussed more in Chapter 10.

7.3 Component mass balance equations

Multiple components and fluid phases may be present, and we perform a mass balance of a component, κ, in a phase, α. The local equilibrium assumption is used in the derivation. The accumulation of mass of a component in a phase equals the net mass of the component that enters/leaves the control volume via flux plus the mass of the component that enters/leaves from a source or sink,

$$\underbrace{W_{\kappa,\alpha}\Delta V\big|_{t+\Delta t} - W_{\kappa,\alpha}\Delta V\big|_t}_{\text{accumulation } \kappa} = \underbrace{\overrightarrow{N}_{\kappa,\alpha,x}\Big|_x a\Delta t}_{\text{mass in of } \kappa} - \underbrace{\overrightarrow{N}_{\kappa,\alpha,x}\Big|_{x+\Delta x} a\Delta t}_{\text{mass out of } \kappa} + \underbrace{r_{\kappa,\alpha}\Delta V \Delta t}_{\text{mass generation } \kappa},$$

(7.10)

where $W_{\kappa,\alpha}$ is the bulk concentration of κ (mass κ/bulk volume) in phase α, $\overrightarrow{N}_{\kappa,\alpha}$ is the mass flux (sum of advective and dispersive) of the component (mass κ/area-time) in phase α, and $r_{\kappa,\alpha}$ is a source of the component (mass κ/volume-time) in phase α. Each term in the equation has units of mass, e.g., kg or lb$_\mathrm{m}$. If Eq. (7.10) is divided by the control volume ($\Delta V = a\,\Delta x$) and the time interval (Δt), Eq. (7.11) is obtained for 1D transport,

$$\frac{W_{\kappa,\alpha}\big|_{t+\Delta t} - W_{\kappa,\alpha}\big|_t}{\Delta t} = \frac{\overrightarrow{N}_{\kappa,\alpha,x}\big|_x - \overrightarrow{N}_{\kappa,\alpha,x}\big|_{x+\Delta x}}{\Delta x} + r_{\kappa,\alpha}.$$

(7.11)

The control volume can be chosen to be infinitesimally small over an infinitesimally small time step; the limit as both Δx and Δt go to zero results in partial derivatives,

$$\lim_{\Delta x \to 0} \frac{\overrightarrow{N}_{\kappa,\alpha,x}\big|_x - \overrightarrow{N}_{\kappa,\alpha,x}\big|_{x+\Delta x}}{\Delta x} = -\frac{\partial \overrightarrow{N}_{\kappa,\alpha,x}}{\partial x}$$

$$\lim_{\Delta t \to 0} \frac{W_\kappa\big|_{t+\Delta t} - W_\kappa\big|_t}{\Delta t} = \frac{\partial W_\kappa}{\partial t}.$$

Substituting the partial derivatives into (Eq. 7.11), the one-dimensional component mass balance equation in Cartesian coordinates is obtained,

$$\frac{\partial W_{\kappa,\alpha}}{\partial t} = -\frac{\partial \overrightarrow{N}_{\kappa,\alpha}}{\partial x} + r_{\kappa,\alpha},$$

(7.12)

where the bulk concentration ($W_{\kappa,\alpha}$) is defined as,

$$W_{\kappa,\alpha} = \phi S_\alpha \rho_\alpha \omega_{\kappa,\alpha},$$

(7.13)

and the net flux of component in the phase is the sum of the advective and dispersive fluxes,

$$\overrightarrow{N}_{\kappa,\alpha} = \overrightarrow{N}^{ad}_{\kappa,\alpha} + \overrightarrow{N}^{d}_{\kappa,\alpha} = \rho_\alpha \omega_{\kappa,\alpha} u_\alpha - D_{\kappa,\alpha} S_\alpha \phi \frac{\partial}{\partial x}\left(\rho_\alpha \omega_{\kappa,\alpha}\right).$$

(7.14)

The velocity and dispersion coefficient are scalars in Eq. (7.14) because the medium is 1D but are vectors and tensors, respectively, in multidimensional transport. Substituting Eqs. (7.13a) and (7.13b) into Eq. (7.12) and assuming component generation/consumption comes from a well but not reaction or adsorption,

$$\frac{\partial}{\partial t}\left(\phi S_\alpha \rho_\alpha \omega_{\kappa,\alpha}\right) = -\frac{\partial}{\partial x}\left[\rho_\alpha \omega_{\kappa,\alpha} u_\alpha - S_\alpha \phi D_{\kappa,\alpha}\frac{\partial}{\partial x}\left(\rho_\alpha \omega_{\kappa,\alpha}\right)\right] + C^*_{\kappa,\alpha}\tilde{q}_\alpha, \quad (7.15a)$$

where Eq. (7.15a) can be generalized for any coordinate system and dimensions,

$$\frac{\partial}{\partial t}\left(\phi S_\alpha \rho_\alpha \omega_{\kappa,\alpha}\right) = -\nabla\cdot\left[\rho_\alpha \omega_{\kappa,\alpha}\overrightarrow{u}_\alpha - S_\alpha \phi \mathbf{D}_{\kappa,\alpha}\cdot\nabla\left(\rho_\alpha \omega_{\kappa,\alpha}\right)\right] + C^*_{\kappa,\alpha}\tilde{q}_\alpha. \quad (7.15b)$$

7.3.1 Single-phase flow

Eq. (7.15) can be simplified in the case of single-phase flow without reaction,

$$\frac{\partial}{\partial t}\left(\phi C_\kappa\right) = -\nabla\cdot\left[C_\kappa \overrightarrow{u} - \phi \mathbf{D}_\kappa\cdot\nabla C_\kappa\right] + C^*_\kappa\tilde{q}, \quad (7.16)$$

where C_κ is the concentration (mass/volume) of the component in the single phase. It is assumed the phase saturation is unity in Eq. (7.16), but a constant saturation can be included if there are immobile phases present. The equation is known as the *advection−dispersion equation* or ADE. For the special case of one dimension, no sources or sinks, constant porosity, velocity, and dispersion coefficients, the equation can be simplified,

$$\phi\frac{\partial C_\kappa}{\partial t} = D_\kappa\phi\frac{\partial^2 C_\kappa}{\partial x^2} - u\frac{\partial C_\kappa}{\partial x} \quad (7.17a)$$

$$\phi\frac{\partial C_\kappa}{\partial t} = \frac{D_\kappa\phi}{r}\frac{\partial}{\partial r}\left(r\frac{\partial C_\kappa}{\partial r}\right) - \frac{u}{r}\frac{\partial(rC_\kappa)}{\partial r} \quad (7.17b)$$

where Eqs. (7.17a) and (7.17b) are the 1D ADE in Cartesian and cylindrical coordinates, respectively. Analytical solutions exist for some simple boundary conditions and are shown in Section 7.5.

7.3.2 Overall compositional equations

Eq. (7.15) is the general mass balance equation for a component κ in a phase α. The component may be present in multiple phases. For example, many hydrocarbon components will be present in both the oleic and gaseous phase. An overall component balance can be found by summation of Eq. (7.15) over all N_p phases,

$$\frac{\partial}{\partial t}\left(\phi\sum_{\alpha=1}^{N_p} S_\alpha \rho_\alpha \omega_{\kappa,\alpha}\right) = -\nabla\cdot\sum_{\alpha=1}^{N_p}\left[\rho_\alpha \omega_{\kappa,\alpha}\overrightarrow{u}_\alpha - S_\alpha \phi \mathbf{D}_{\kappa,\alpha}\cdot\nabla\left(\rho_\alpha \omega_{\kappa,\alpha}\right)\right] + \sum_{\alpha=1}^{N_p} C^*_{\kappa,\alpha}\tilde{q}_\alpha.$$

$$(7.18)$$

Eq. (7.18) is used in multiphase compositional reservoir simulation as discussed in Chapter 10. The equation can also be written in terms of mole fractions, $x_{\kappa,\alpha}$, instead of mass fractions. The equation, as shown, assumes no adsorption onto the solid phase, but can easily be added in the accumulation term (see Eq. 1.2).

A special case of a multicomponent, multiphase system is the *black oil* model. As discussed in Chapter 1, a black oil model uses the approximation that all components can be lumped into three pseudocomponents (water, oil, and gas) and three phases (aqueous, oleic, and gaseous). It is further assumed that only water is present in the aqueous phase, only gas is present in the gaseous phase, and both oil and gas are present in the oleic phase. Eq. (7.18), without chemical reactions, can be written for each of the three pseudocomponents,

$$\frac{\partial}{\partial t}(\phi S_w \rho_w) + \nabla \cdot \rho_w \overrightarrow{u}_w = \rho_w \widetilde{q}_w$$

$$\frac{\partial}{\partial t}(\phi S_o \rho_o) + \nabla \cdot \rho_o \overrightarrow{u}_o = \rho_o \widetilde{q}_o$$

$$\frac{\partial}{\partial t}\left[\phi\left(S_g \rho_g \omega_{gas,g} + S_o \rho_o \omega_{gas,o}\right)\right] + \nabla \cdot \left(\rho_g \omega_{gas,g} \overrightarrow{u}_g + \rho_o \omega_{gas,o} \overrightarrow{u}_o\right)$$

$$= \rho_g \omega_{gas,g} \widetilde{q}_g + \rho_o \omega_{gas,o} \widetilde{q}_o, \tag{7.19}$$

where the mass fraction of gas in the gaseous phase is unity ($\omega_{gas,g} = 1$) and the amount of gas dissolved in the oleic phase is described by the solution−gas ratio, R_s. Dispersion is ignored because it is assumed transport is advection dominated and concentration gradients are relatively small. Also, recognizing that,

$$\rho_w = \frac{\rho_{wat}^{sc}}{B_w}; \ \rho_o \omega_{oil,o} = \frac{\rho_{oil}^{sc}}{B_o}; \ \rho_g \omega_{gas,g} = \frac{\rho_{gas}^{sc}}{B_g}; \ \rho_g \omega_{gas,o} = \frac{\rho_{gas}^{sc}}{B_o} R_s,$$

where B_α is the formation volume factor of phase α, we obtain the black oil equations,

$$\frac{\partial}{\partial t}\left(\frac{\varphi S_w}{B_w}\right) + \nabla \cdot \left(\frac{\overrightarrow{u}_w}{B_w}\right) = \widetilde{q}_w^{sc}$$

$$\frac{\partial}{\partial t}\left(\frac{\varphi S_o}{B_o}\right) + \nabla \cdot \left(\frac{\overrightarrow{u}_o}{B_o}\right) = \widetilde{q}_o^{sc}$$

$$\frac{\partial}{\partial t}\left[\varphi\left(\frac{S_g}{B_g} + \frac{S_o}{B_o} R_s\right)\right] + \nabla \cdot \left(\frac{\overrightarrow{u}_g}{B_g} + \frac{\overrightarrow{u}_o}{B_o} R_s\right) = \widetilde{q}_g^{sc} + R_s \widetilde{q}_o^{sc}. \tag{7.20}$$

The phase velocities in Eq. (7.20) are described by the multiphase form of Darcy's law, and auxiliary equations are used for capillary pressure and relative permeability. Saturations of all phases sum to unity. A semianalytical solution is given in Section 7.5.2 for the case of oil and water, under certain assumptions, and numerical solutions are described in Chapter 9.

7.4 Analytical solutions

7.4.1 1D Cartesian ADE in a semi-infinite domain

The ADE has an analytical solution in a few limiting cases, many of which are reviewed by van Genuchten et al. (2013). Consider constant velocity (steady-state flow) with constant dispersion coefficient, D, and no sinks/sources (wells or reaction) in 1D (Fig. 7.3). Dimensionless variables, x_D (dimensionless distance), t_D (dimensionless time), C_D (dimensionless concentration), and N_{Pe} (Peclet number) are introduced to generalize the PDE Eq. (7.17a),

$$x_D = x/L; \quad t_D = u_x t/L\phi; \quad C_D = \frac{C - C_{init}}{C_{inj} - C_{init}}; \quad N_{Pe} = u_x L/D\phi.$$

Importantly, the Peclet number is defined here in terms of the dispersion coefficient, D, instead of the molecular diffusion coefficient. The dimensionless ADE can then be written as,

$$\frac{\partial C_D}{\partial t_D} = \frac{1}{N_{Pe}} \frac{\partial^2 C_D}{\partial x_D^2} - \frac{\partial C_D}{\partial x_D}. \tag{7.21}$$

7.4.1.1 Constant concentration at $x_D = 0$

The boundary conditions here include a constant initial concentration ($C_D = 0$) and a semi-infinite domain, so the concentration far from the inlet (the outer

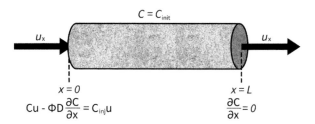

$$C = C_{init}$$

$$x = 0 \qquad\qquad x = L$$

$$Cu - \Phi D \frac{\partial C}{\partial x} = C_{inj} u \qquad\qquad \frac{\partial C}{\partial x} = 0$$

FIGURE 7.3 1D transport in a semi-infinite porous medium.

boundary condition) is constant and equal to the initial condition. The inner boundary condition assumes the concentration at the inlet is a constant, C_{inj} ($C_D = 1$). The dimensionless initial and boundary conditions are therefore,

$$C_D(x_D, 0) = 0$$
$$C_D(0, t_D) = 1; \quad C_D(\infty, t_D) = 0. \tag{7.22}$$

The solution to the PDE using Laplace transforms under these boundary conditions is given by Eq. (7.23) (Marle, 1981),

$$C_D(x_D, t_D) = \frac{1}{2} erfc\left[\sqrt{\frac{N_{Pe}}{4t_D}}(x_D - t_D)\right] + \frac{1}{2}e^{x_D N_{Pe}} erfc\left(\sqrt{\frac{N_{Pe}}{4t_D}}(x_D + t_D)\right)$$

$$\tag{7.23}$$

Eq. (7.23) is valid for all N_{Pe} and the boundary conditions given in Eq. (7.22). Lake et al. (2014) showed that for high N_{pe}, the second term in Eq. (7.20) can be ignored. For advection-dominated flow without diffusion/dispersion ($N_{Pe} = \infty$), the solution is simply,

$$C_D(x_D, t_D, N_{Pe} = \infty) = \begin{cases} 1 & \text{if} \quad x_D \leq t_D \\ 0 & \text{if} \quad x_D > t_D \end{cases}. \tag{7.24}$$

The equation shows that there is a sharp front at $x_D = t_D$ where the concentration changes from $C_D = 1$ to $C_D = 0$. For diffusion-dominated flow without advection ($N_{Pe} = 0$),

$$C_D(x_D, t_D, N_{Pe} \rightarrow 0) = erfc(\xi_D) = erfc\left(\sqrt{\frac{N_{Pe}}{4t_D}}x_D\right), \tag{7.25}$$

where $\xi_D = x/(4Dt)^{1/2}$ is the similarity variable. Several points should be made about Eq. (7.25). First, the solution is identical to the solution given in Chapter 2 for the 1D diffusivity equation in a semi-infinite domain because the PDE and boundary conditions are the same. Second, diffusion/dispersion-dominated flow rarely occurs in subsurface applications; as mentioned previously, the N_{Pe} is generally assumed high (10^3 or greater). Finally, if flow is diffusion/dispersion-dominated, a constant-concentration boundary condition is difficult to maintain in subsurface applications.

7.4.1.2 Mixed boundary condition at $x_D = 0$

The solution given in Eq. (7.23) assumes a constant-concentration boundary condition at $x_D = 0$. A useful application of the solution is for verification of a reservoir simulator with a constant-concentration injection well at $x_D = 0$. However, when the N_{pe} is relatively small ($< \sim 10$), the concentration at the well is not constant because component is quickly dispersed away when injected. The appropriate boundary condition at $x_D = 0$ requires the use of a third-order or flux-type boundary condition (Yagi and Miyauchi 1953; Aris and Amundson 1957; Brenner 1962; van Genuchten and Parker 1984),

$$C_D(x_D, 0) = 0$$

$$C_D(0, t_D) - \frac{1}{N_{Pe}} \frac{\partial C_D}{\partial x_D} = 1 \qquad (7.26)$$

$$\frac{\partial C_D}{\partial x_D}(L, t_D) = 0$$

The boundary condition at $x_D = 0$ reduces to those in Eq. (7.22) for large N_{pe}. The domain is also assumed finite, L, instead of semi-infinite. The solution to the PDE with these boundary conditions is provided by Brenner (1962).

$$C_D(x_D, t_D) = \frac{1}{2} erfc \left[\sqrt{\frac{N_{pe}}{4t_D}} (x_D - t_D) \right] + \sqrt{\frac{N_{pe} t_D}{\pi}} exp \left[\frac{-N_{pe}}{4t_D}(x_D - t_D)^2 \right]$$

$$- \frac{1 + N_{pe}(x_D + t_D)}{2} exp(N_{pe}x_D) erfc \left[\sqrt{\frac{N_{pe}}{4t_D}}(x_D + t_D) \right]$$

$$+ 2\sqrt{\frac{N_{pe} t_D}{\pi}} \left[1 + \frac{N_{pe}(2 - x_D + t_D)}{4} \right] exp \left[N_{pe} - \frac{N_{pe}(2 - x_D + t_D)^2}{4t_D} \right]$$

$$- \frac{1}{2} N_{pe} \left[2(2 - x_D + t_D) + t_D + \frac{N_{pe}(2 - x_D + t_D)^2}{2} exp(N_{pe}) \right]$$

$$erfc \left[\sqrt{\frac{N_{pe}}{4t_D}}(2 - x_D + t_D) \right] \qquad (7.27)$$

Eq. (7.27) is more cumbersome than Eq. (7.23) but is more accurate for the application of a constant-concentration injection well, especially for low N_{Pe}. In fact, the equation reduces to Eq. (7.23) for high N_{Pe}. Fig. 7.4 compares the solutions at various Peclet numbers. There is a distinct difference at $N_{Pe} = 10$ but the difference is negligible at $N_{Pe} = 1000$. It must be pointed out that the behavior of the curves for $N_{Pe} = 10$ at $x_D = 1$ is different since Eq. (7.23) assumes a semi-infinite domain. In contrast, Eq. (7.27) is defined for a finite domain and a boundary condition of zero concentration derivative at the exit. Although this difference could seem small, it could help to correctly model experimental results since solute is collected at the outlet of the system.

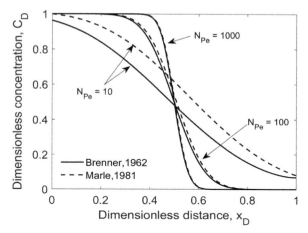

FIGURE 7.4 Comparison of dimensionless concentration versus distance at $t_D = 0.5$ of Marle (1981), Eq. (7.23), and Brenner (1962), Eq. (7.27), solutions for $N_{Pe} = 10, 100$ and 1000.

The differences between the two soluions are observed in Fig. 7.4. particularly when dispersive transport is relatively large compared to advective transport.

7.4.2 Semianalytical solution to two-phase flow

Analytical solutions are rare for multiphase flow, but one exception is the Buckley–Leverett solution. Although the solution is only valid under idealized assumptions, it provides valuable insight into flow physics and also as a benchmark for numerical simulations. The assumptions required for the Buckley–Leverett solution are as follows: (1) two phases (undersaturated oleic and aqueous), (2) incompressible fluids, (3) 1D flow, (4) constant and homogeneous fluid and reservoir properties, (5) no capillary pressure, (6) reservoir initially at uniform water saturation and water is injected at the inner boundary ($x = 0$) and produced at the outer boundary ($x = L$), (7) no additional sources or sinks (wells). Gravity is also ignored in the final solution as presented here.

7.4.2.1 Fractional flow

As the name suggests, *fractional flow* is the fraction of the total flow that pertains to a particular phase,

$$f_\alpha = \frac{u_\alpha}{u} = \frac{u_\alpha}{\displaystyle\sum_{\alpha=1}^{N_p} u_\alpha}, \qquad (7.28)$$

where u is the total velocity and the phase Darcy velocities are given by the multiphase form of Darcy's law (Eq. 1.21). For two-phase flow of an oleic and aqueous phase, the fractional flow of the aqueous phase, f_w, is given by,

$$f_w = \frac{1 + \frac{kk_{ro}}{u\mu_o}\left(\frac{\partial p_c}{\partial x} - \Delta\rho g \sin\theta\right)}{1 + \frac{1}{M}}. \tag{7.29}$$

where M is the ratio of the mobility of the aqueous phase to the oleic phase and θ is the dip angle,

$$M = \frac{\lambda_w}{\lambda_o} = \frac{kk_{rw}/\mu_w}{kk_{ro}/\mu_o} = \frac{k_{rw}\mu_o}{k_{ro}\mu_w}, \tag{7.30}$$

and λ_α is the mobility of the phase. Fractional flow is a function of saturation because relative permeabilities are functions of saturation. In the absence of gravity or capillary pressure, Eq. (7.30) reduces to $f_w = M/(M+1)$. Fig. 7.5A shows the effect of endpoint mobility ratio (i.e., Eq. (7.30) evaluated at endpoint relative permeabilities), M_o, on fractional flow in the absence of gravity, and Fig. 7.5B shows the effect of gravity number, $N_g^0 = \frac{kk_{ro}^0\Delta\rho g}{u\mu_o}$, a ratio of gravitational to viscous effects.

The fractional flow of all phases must sum to unity, so for an aqueous and oleic phase, $f_o = 1 - f_w$. In the presence of gravity, fractional flow of one phase can be greater than one or less than zero (Fig. 7.5B). The derivative of fractional flow with saturation ($f_w' = df_w/dS_w$) is shown in Fig. 7.6 for no gravity or capillary pressure. The derivative has a maximum because the fractional flow curve (Fig. 7.5) has an inflection point. An important consequence of the shape of the curve for the analytical solution for two-phase flow is that two water saturations often correspond to the same fractional flow derivative.

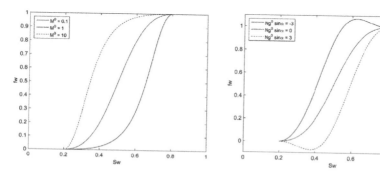

FIGURE 7.5 Fractional flow of aqueous phase versus saturation for (A) three different endpoint mobility ratios (endpoint relative permeabilities used in Eq. 7.30) and no gravity or capillary pressure and (B) three different gravity ratios ($N_g^0 = kk_{ro}^0\Delta\rho g/u\mu_o$) at an endpoint mobility ratio of 1.0 and no capillary pressure.

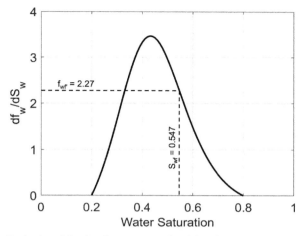

FIGURE 7.6 Derivative of fractional versus saturation. Viscosity of water and oil are 0.5 and 10 cp, respectively. Corey parameters are $S_{wr} = S_{or} = 0.2$, $n_w = n_o = 2$, $k_{rw}^0 = 0.1$, $k_{ro}^0 = 1.0$. The water saturation at the front, $S_{wf} = 0.547$.

7.4.2.2 Buckley–Leverett solution

The mass balance Eq. (7.20) for the water component can be simplified under the assumptions of the Buckley–Leverett problem,

$$\phi \frac{\partial S_w}{\partial t} = \frac{\partial u_w}{\partial x}. \tag{7.31}$$

Substituting the definition of fractional flow of water, Eq. (7.29), into Eq. (7.31),

$$\phi \frac{\partial S_w}{\partial t} = u \frac{\partial f_w}{\partial x} = u \frac{\partial f_w}{\partial S_w} \frac{\partial S_w}{\partial x}, \tag{7.32}$$

where the chain rule of differentiation was used on the fractional flow derivative to obtain the expression on the right. Eq. (7.32) is a 1D PDE that describes saturation as a function of space and time, $S_w = S_w(x, t)$. The equation is analogous to the advection Eq. (7.21) (without dispersion/diffusion), $N_{Pe} = \infty$, for single-component transport, although the coefficient in the advection equation is a constant, while f_w' is a function of saturation (Fig. 7.5). Nonetheless, similarities (and important differences) in the PDEs and solution are apparent.

Water saturation is a function of space and time, therefore the total differential of saturation is given by,

$$dS_w = \frac{\partial S_w}{\partial t} dt + \frac{\partial S_w}{\partial x} dx. \tag{7.33}$$

For a velocity front of constant saturation, $dS_w = 0$, and then from Eqs. (7.33) and (7.32) it follows,

$$\left(\frac{dx}{dt}\right)_{S_w} = -\frac{\left(\frac{\partial S_w}{\partial t}\right)_x}{\left(\frac{\partial S_w}{\partial x}\right)_t} = \frac{u}{\phi}\frac{\partial f_w}{\partial S_w}. \tag{7.34}$$

Eq. (7.34) is the Buckley–Leverett equation. Physically, this is a front of constant water saturation. The equation can be integrated in dimensionless form,

$$x_D = f'_w t_D, \tag{7.35}$$

where it was assumed that $x_{S_w} = 0$ at $t_{S_w} = 0$ in the integration, and the dimensionless variables were defined previously in the chapter.

Eq. (7.35) can be used to plot the water saturation (S_w) versus dimensionless distance, x_D, at specific time, t_D. The procedure works as follows: (1) choose, t_D, the dimensionless time (pore volumes injected) to plot the saturation profile; (2) choose a saturation, S_w, between the initial water saturation, S_{wi} (which may be greater than S_{wr}), and maximum oil saturation, $1 - S_{or}$; (3) compute the derivative of fractional flow, f'_w, at S_w; (4) compute the distance, x_D, that corresponds to that water saturation using Eq. (7.35); (5) repeat steps 2–4 for all water saturations between $S_{wi} < S_w < 1 - S_{or}$ and plot S_w versus x_D at the chosen dimensionless time.

7.4.2.3 Shock fronts

With the exception of straight-line relative permeability curves, the fractional flow curve has an inflection point. As a result, a shock front exists and the solution procedure described above must be modified. Fig. 7.7A shows a typical solution to Eq. (7.35). Three water saturations (triple valued function) correspond to the same distance, which is nonphysical because it would mean that some slow downstream waves would be overtaken by faster upstream waves. In the physical solution, a shock front exists at some front position (x_{Df}) with front saturation, S_{wf}, as shown in Fig. 7.7B.

The front position and saturation can be determined by performing a mass balance at the shock where S_w decreases from S_{wf} to S_{wi} (shock front) and therefore the fractional flow reduces from f_{wf} to f_{wi}. The balance can be used to derive an equation for the velocity at the shock front which is equal to the fractional flow derivative at the shock saturation,

$$f_{wf} = f'_w\big|_{S_{wf}}(S_{wf} - S_{wi}) + f_{wi}. \tag{7.36}$$

Eq. (7.36) is the *breakthrough tangent*, a straight line that passes through the x-axis at $S_w = S_{wi}$ and has a slope equal to the derivative of the fractional

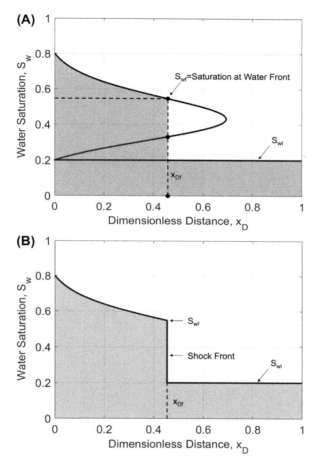

FIGURE 7.7 Water saturation (S_w) versus dimensionless distance (x_D) at $t_D = 0.2$ for (A) mathematical solution and (B) physical solution with shock front. Viscosity of water and oil are 0.5 and 10 cp, respectively. Corey parameters are $S_{wr} = S_{or} = 0.2$, $n_w = n_o = 2$, $k_{rw}^0 = 0.1$, $k_{ro}^0 = 1.0$. The water saturation at the front, $S_{wf} = 0.547$.

flow curve at $S_w = S_{wf}$. The breakthrough tangent (Fig. 7.8) can be drawn by hand on graph paper or computationally using a root solving method to find the value of S_w for which the tangent line and fractional curve are equal.

The tangent point that corresponds to S_{wf} (Fig. 7.8A) can be used to determine x_{Df}, the position of the front. The saturation is equal to the initial water saturation, S_{wi}, at positions greater than x_{Df} as shown in Fig. 7.7B and for $x_D < x_{Df}$, the aforementioned procedure for Eq. (7.35) can be used. Fig. 7.9 shows the water saturation versus distance at $t_D = 0.1$, 0.2, 0.4, and 2.0 pore volumes injected.

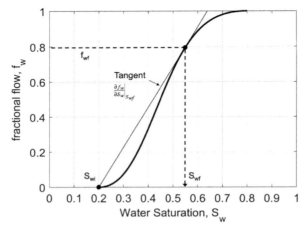

FIGURE 7.8 Fractional flow curve with breakthrough tangent. The initial water saturation is the residual water saturation. Viscosity of water and oil are 0.5 and 10 cp, respectively. Corey parameters are $S_{wr} = S_{or} = 0.2$, $n_w = n_o = 2$, $k_{rw}^0 = 0.1$, $k_{ro}^0 = 1.0$. The water saturation at the front, $S_{wf} = 0.547$.

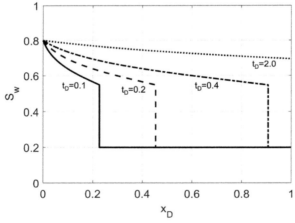

FIGURE 7.9 Water saturation versus dimensionless distance at $t_D = 0.1$, 0.2, 0.4, and 2.0. Viscosity of water and oil are 0.5 and 10 cp, respectively. Corey parameters are $S_{wr} = S_{or} = 0.2$, $n_w = n_o = 2$, $k_{rw}^0 = 0.1$, $k_{ro}^0 = 1.0$. The water saturation at the front, $S_{wf} = 0.547$.

7.4.2.4 Breakthrough time and oil recovery

Oil and water are produced at the producer well which is at $x_D = 1$. At early times, prior to the shock front reaching the producer, the oil rate, $q_o = u_o a = (1 - f_{wi})ua$, is constant. During this time, no injected water is produced; the only water that is produced is the initial water above the residual water saturation.

The dimensionless breakthrough time, $t_{D,bt}$, is the dimensionless time at which the water saturation at $x_D = 1$ equals the front saturation, S_{wf}, and can be found from Eq. (7.35) and using the derivative of the fractional flow curve at the front saturation, f'_{wf},

$$t_{D,bt} = \frac{1}{f'_{wf}}. \tag{7.37}$$

After the breakthrough time, the fractional flow (and therefore production rate) of oil at the producer decreases with time. The dimensionless cumulative oil produced (N_{pD}) can be found by integrating the oil rate with time which leads to,

$$N_{pd} = \frac{N_p}{V_p} = \begin{cases} f_{oi}t_D & \text{if } t_D \leq t_{D,bt} \\ \dfrac{(1 - f_{wp})}{f'_{wp}} + S_{wp} - S_{wi} & \text{if } t_D > t_{D,bt} \end{cases}, \tag{7.38}$$

where N_p is the dimensional cumulative oil produced, V_p is the pore volume, $f'_{wp} = 1/t_D$ and is the derivative of the fractional flow at the producer well, and a derivative plot (Fig. 7.6) can be used to find the corresponding saturation, S_{wp}. The dimensionless time, or pore volumes injected, was defined in Section 7.4.1. The saturation can then be used to determine the fractional flow at the producer, f_{wp}. Fig. 7.10 shows a typical recovery plot which has constant slope, f_{oi}, prior to breakthrough and after breakthrough asymptotically approaches the maximum recovery. The figure also shows the water saturation at the producer well which suddenly increases, from the initial water saturation to the front saturation, at the breakthrough time. After breakthrough the water saturation at the producer gradually increases.

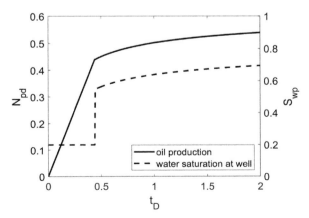

FIGURE 7.10 Dimensionless oil recovery and water saturation at the producer well ($x_D = 1$) verus dimensionless time obtained using Eq. (7.38). Viscosity of water and oil are 0.5 and 10 cp, respectively. Corey parameters are $S_{wr} = S_{or} = 0.2$, $n_w = n_o = 2$, $k^0_{rw} = 0.1$, $k^0_{ro} = 1.0$. The water saturation at the front, $S_{wf} = 0.547$, $f_{wf} = 0.79$, $f_{wf}' = 2.2675$, and $t_{D,bt} = 0.44$.

7.4.2.5 Capillary pressure and smearing of the shock front

For cases in which the capillary pressure is not negligible, but all other assumptions of the Buckley−Leverett apply, the mass balance equation for water reduces to,

$$\phi \frac{\partial S_w}{\partial t} + u \frac{\partial f_w^*}{\partial S_w} \frac{\partial S_w}{\partial x} + \frac{\partial}{\partial x}\left[D_1 \frac{\partial S_w}{\partial x}\right] = 0 \ , \tag{7.39}$$

where $f_w^* = M/(M+1)$ is the fractional flow without capillary pressure and the capillary diffusion coefficient is defined as,

$$D_1(S_w) = f_w^* \frac{kk_{ro}}{\mu_o} \frac{dp_c}{dS_w}. \tag{7.40}$$

Eq. (7.39) is analogous to the advection−dispersion equation; the second term is advective and the third dispersive. However, the coefficients f_w' and D_1 are both nonlinear functions of the primary variable, S_w, rather than constants as in the ADE. The dispersive term has a similar effect as it did in the ADE, that is, smearing of the front.

7.4.2.6 Pseudocode for semianalytical Buckley−Leverett solution

Developing a code for the Buckley−Leverett problem follows the procedure described in Section 7.4.2. Preprocessing involves defining the fluid viscosities and Corey parameters for the relative permeability curves. Dimensionless times (pore volumes injected) to make plots of saturation fields should be defined. During preprocessing, plots of relative permeability and fractional flow versus water saturation can be made. Pseudocode to calculate relative permeability was presented in Chapter 1. A new subroutine to compute fractional flow Eq. (7.29) of water and its derivative at a given water saturation must be developed. The derivative can be computed analytically or numerically.

A subroutine must be developed and called to determine the front saturation, S_{wf}, the corresponding fractional flow, f_{wf}, and derivative of fractional flow, f'_{wf}, at the front saturation. Determination of S_{wf} involves finding the point of tangency on the fractional flow curve using a line starting at S_{wi}. The pseudocode presented here is a simple brute force method, although more elegant strategies are possible. It involves starting at $S_w > S_{wi}$, calculating f_w using the fractional subroutine and the tangent line, Eq. (7.36), and then computing an error between the two fractional flow values. S_w is incrementally increased, and S_{wf} corresponds to the minimum error.

The saturation profile (S_w vs. x_D) can be determined by computing x_D at all saturations between $S_w = 1 - S_{or}$ and S_{wf} using Eq. (7.35). Then $S_w = S_{wi}$ at all positions, $x_D > x_{Df}$. Saturation profiles can be computed at all pre-determined t_D. Dimensionless oil recovery, N_{pD}, can be computed using Eq. (7.38). Plots can be made of S_w versus x_D, and N_{pd}, f_{wp}, S_{wp} versus t_D. Dimensionless recoveries can be converted to dimensional recoveries if desired.

```
MAIN CODE BUCKLEY_LEVERETT
CALL PREPROCESSING
CALL FRONT SATURATION
CALL SATURATION PROFILE
CALL DIMENSIONLESS OIL RECOVERY
PLOT Sw vs xD at various tD and NpD vs tD
```

```
SUBROUTINE FRACTIONAL FLOW
INPUTS: Sw, petrophysical, fluid properties
OUTPUTS: fw, fw'

CALL RELATIVE PERMEABILITY
CALCULATE fw using eqn 7.29
CALL RELATIVE PERMEABILITY at Sw + δSw
CALCULATE fw2 at Sw + δSw
SET fw' = (fw2-fw)/δSw
```

```
SUBROUTINE FRONT SATURATION
INPUTS: petrophysical and fluid properties
OUTPUTS: Swf, fwf, fwf', tD,bt

SET min_error = ∞
SET ε as a small increment, e.g. 0.001
SET Swf = Swi + ε
WHILE Swf < 1-Sor
      CALCULATE fwf using eqn 7.36
      CALL FRACTIONAL FLOW at Swf to get fwf,2 and fwf,2'
      CALCULATE error = abs(fwf − fwf,2)
      IF error < min_error
          min_error = error
          SET Swf,min = Swf, fwf,min = fwf,2 , fwf,min' = fwf,2'
      END
      SET Swf = Swf + ϵ
ENDWHILE
SET Swf = Swf,min fwf = fwf,min and fwf' = fwf,min'
SET tD,bt = 1/fwf'
```

```
SUBROUTINE SATURATION PROFILE
INPUTS: t_D, S_wf, S_wi, petrophysical and fluid properties
OUTPUTS: S_w, x_D (both vectors)

INIT  S_w = 1 to S_wf with N segments
FOR i = 1 to N
        CALL FRACTIONAL FLOW
            SET x_D = f'_w * t_D
ENDFOR
FOR j = N+1 to M segments
            SET S_w(j) = S_wi
            SET x_D(j) = x_D + 1/(M − N + 1)
ENDFOR
```

```
SUBROUTINE DIMENSIONLESS OIL RECOVERY
INPUTS: t_D, Swi, petrophysical and fluid properties
OUTPUTS: N_pD

CALL FRACTIONAL FLOW to get f_wi

FOR i = 1 to length(t_D)
        IF t_D < t_{D,bt}
            SET N_pD = (1 − f_wi) * t_D
        ELSE
            SET f'_wp = 1/t_D
            DETERMINE S_wp such that f'_wp = f'_w from FRACTIONAL FLOW
            SET f_wp = FRACTIONAL FLOW(S_wp)
            SET N_pD(i) = (1 − f_wp)/f'_wp + S_wp − S_wi
        ENDIF
ENDFOR
```

7.5 Exercises

Exercise 7.1. Balance equations. Derive the component balance for a phase (Eq. 7.15) for a 3D system in Cartesian coordinates.

Exercice 7.2. Solution to ADE for semi-infinite domain. Develop a computer code to compute the dimensionless concentration using the solution by Marle (1981). Use your code to make plots of C_D versus t_D at $x_D = 1$ at $N_{Pe} = 10$, 100, and 1000.

Exercice 7.3. Solution to ADE with third-oder boundary conditions. Develop a computer code to compute the dimensionless concentration using the solution by Brenner (1962). Use your code to make plots of C_D versus t_D at $x_D = 1$ at $N_{Pe} = 10$, 100, and 1000. Compare your results to the solution by Marle (1981) used in Exercise 7.2.

Exercise 7.4. Computer code for fractional flow. Develop a subroutine/function to compute the fractional flow and the derivative of fractional flow for a two-phase (aqueous-oleic) system at a given water saturation. Your code should call the subroutine for relative permeability you created in the Chapter 1 exercises. Make plots of f_w and f_w' versus S_w for a case with no capillary pressure and gravity, $\mu_w = 0.5$ cp, $\mu_o = 5$ cp, and Corey parameters $n_w = 2$, $n_o = 2$, $k_{rw}^0 = 0.2$, $k_{ro}^0 = 1.0$, $S_{wr} = 0.1$, $S_{or} = 0.2$.

Exercise 7.5. Determine front saturation. Use the fractional flow plots created in Exercise 7.4 to determine the front saturation, S_{wf}, front fractional flow, f_{wf}, and derivative, f_{wf}', by drawing tangent lines. Perform your calculations using $S_{wi} = S_{wr} = 0.1$ and also $S_{wi} = 0.2$. Then develop a subroutine/function to determine the front saturation and validate against your solution obtained by hand.

Exercise 7.6. Computer code to solve Buckley-Leverett. Develop a computer code to solve the two-phase Buckley–Leverett problem semi-analytically. Test your code using the parameters in Exercise 7.4. Use an initial water saturation equal to the residual water saturation and then repeat for $S_{wi} = 0.2$. In particular your code should,

a. Determine the front saturation, S_{wf}, and dimensionless breakthrough time, $t_{D,bt}$.
b. Make plots of S_w versus x_D at $t_D = 0.1$, 0.3, 0.5, and 1.0.
c. Make a plot of N_{pD} versus t_D up to $t_D = 3.0$.
d. Calculate the cumulative oil recovered in RB if 1000 STB/day ($B_o = 1.0$ RB/STB) of water are injected into a reservoir of pore volume 1×10^6 RB for 2500 days.

References

Aris, R., Amundson, N., 1957. Some remarks on longitudinal mixing or diffusion in fixed beds. AIChE Journal 3 (2), 280–282. https://doi.org/10.1002/aic.690030226.

Brenner, H., 1962. The diffusion model of longitudinal mixing in beds of finite length. Numerical values. Chemical Engineering Science 17 (4), 229–243.

Cussler, E.L., 1997. Diffusion: Mass Transfer in Fluid Systems, second ed. Cambridge University Press, New York.

DePaolo, D.J., Cole, D.R., 2013. Geochemistry of geologic carbin sequestration: an overview. Reviews in Mineraology and Geochemistry 77 (1), 1–14.

Freeze, R.A., 1979. Ground~ ater. Prentice-Hall.

Frenier, W., Ziauddin, M., 2014. Chemistry for Enhancing the Production of Oil and Gas. Society of Petroleum Engineers, ISBN 978-1-61399-317-0.

John, A.K., Lake, L.W., Bryant, S.L., Jennings, J.W., 2010. Investigation of mixing in field scale miscible displacement using particle tracking simulations of tracer floods with flow reversal. SPE Journal 15 (3), 598–609.

Koch, D.L., Brady, J.F., 1985. Dispersion in fixed beds. Journal of Fluid Mechanics 154, 399–427.

Lake, L.W., Johns, R., Rossen, B., Pope, G., 2014. Fundamentals of Enhanced Oil Recovery. Society of Petroleum Engineers, ISBN 978-1-61399-328-6.

Marle, C., 1981. Multiphase Flow in Porous Media. Éditions technip.

Mehmani, Y., Balhoff, M.T., 2015. Mesoscale and hybrid models of fluid flow and solute transport. Reviews in Mineralogy and Geochemistry 80 (1), 433–459.

Saffman, P.G., 1959. A theory of dispersion in a porous medium. Journal of Fluid Mechanics 6, 321–349.

Sahimi, M., 2012. Dispersion in porous media, continuous-time random walks, and percolation. Physical Review E 85 (1), 016316.

Van Brakel, J., 1974. Analysis of diffusion in macroporous media in terms of a porosity, a tortuosity and a constrictivity factor. International Journal of Heat and Mass Transfer 17 (9), 1093–1103.

van Genuchten, M.T., et al., 2013. Exact analytical solutions for contaminant transport in rivers. Journal of Hydrology and Hydromechanics 61 (3), 250.

van Genuchten, M.T., Parker, J.C., 1984. Boundary conditions for displacement experiments though short laboratory soil columns. Soil Science Society of America Journal 48, 703–708. https://doi.org/10.2136/sssaj1984.03615995004800040002x.

Yagi, S., Miyauchi, T., 1953. On the residence time curves of the continuous reactors. Chemical Engineering 17 (10), 382–386. https://doi.org/10.1252/kakoronbunshu1953.17.382.

Chapter 8

Numerical solution to single-phase component transport

8.1 Introduction

The relevant PDEs for single-phase flow and component transport were derived in Chapters 2 and 7, respectively,

$$\varphi c_t \frac{\partial p}{\partial t} = \frac{1}{\rho_\alpha} \nabla \cdot \left(\rho_\alpha \frac{k_{r\alpha}^0 \mathbf{k}}{\mu_\alpha} \nabla \Phi_\alpha \right) + \tilde{q}_\alpha \tag{2.11}$$

$$\frac{\partial}{\partial t}(\varphi C_\kappa) = -\nabla \cdot \left[C_\kappa \vec{u}_\alpha - \varphi \mathbf{D}_{\kappa,\alpha} \cdot \nabla(C_\kappa) \right] + C_\kappa^* \tilde{q}_\alpha, \tag{7.16}$$

where Eq. (2.11) is the diffusivity/pressure equation for the single flowing phase, α, and Eq. (7.16) is the balance equation for each component, κ. In this chapter, numerical methods and limitations are discussed for solving coupled flow and transport in a single phase. Most of the chapter is devoted to 1D problems but the approach can be easily extended to 2D and 3D.

8.2 Finite difference solution to the ADE in 1D for a single component

The ADE (Eq. 7.16) for 1D, single-phase transport of a component can be expanded using a chain rule on the time derivative. In 1D,

$$\varphi \frac{\partial C_\kappa}{\partial t} + C_\kappa \varphi c_f \frac{\partial p}{\partial t} = \frac{\partial}{\partial x} \left(D_\kappa \varphi \frac{\partial C_\kappa}{\partial x} \right) - \frac{\partial (C_\kappa u)}{\partial x} + C_\kappa^* \tilde{q}_\alpha, \tag{8.1}$$

An Introduction to Multiphase, Multicomponent Reservoir Simulation.
https://doi.org/10.1016/B978-0-323-99235-0.00006-3

where the definition of formation compressibility, $\varphi c_f = d\varphi/dp$ was substituted into the equation. The PDE can be discretized using finite differences as described in Chapter 3,

$$\varphi_i \frac{C_i^{n+1} - C_i^n}{\Delta t} = \frac{1}{\Delta x_i}\left(D\varphi \frac{\partial C}{\partial x}\bigg|_{i+\frac{1}{2}} - D\varphi \frac{\partial C}{\partial x}\bigg|_{i-\frac{1}{2}}\right) - \frac{1}{\Delta x_i}\left(uC|_{i+\frac{1}{2}} - uC|_{i-\frac{1}{2}}\right)$$
$$+ \frac{Q_i^{rc}}{V_i}C_i^* - \frac{\varphi_i c_f\left(P_i^{n+1} - P_i^n\right)}{\Delta t}C_i, \tag{8.2}$$

where C_i^* is the injected concentration (C_{inj}) into block i if the well is an injector and the block concentration (C_i) if the well is a producer; Q_i^{rc} is the rate of the well at reservoir conditions. In Eq. (8.2), the subscript "i" refers to the grid block and the subscript "κ" (denoting the component) has been dropped for brevity. The interblock velocity of the flowing phase is given by Darcy's law,

$$u_{\alpha,i\pm\frac{1}{2}} = -\frac{k k_{r\alpha}^0}{\mu_\alpha}\bigg|_{i\pm\frac{1}{2}}\left[\frac{(P_{i\pm1} - P_i)}{\Delta x_{i\pm\frac{1}{2}}} - \rho_\alpha g\frac{(D_{i\pm1} - D_i)}{\Delta x_{i\pm\frac{1}{2}}}\right], \tag{8.3}$$

where the pressure field can be computed using the methods discussed in Chapters 3−6. The final term in Eq. (8.2), that includes formation compressibility, is usually small and negligible and therefore the compressibility term is neglected throughout this chapter. Eqs. (8.2) and (8.3) imply that the velocity (or pressure) field must be known to solve the transport equation.

Discretizing the concentration gradients in Eq. (8.2) and rearranging gives,

$$C_i^{n+1} - C_i^n = \frac{\Delta t}{\Delta x_i}\left[\frac{D_{i+\frac{1}{2}}}{\Delta x_{i+1/2}}(C_{i+1} - C_i) + \frac{D_{i-\frac{1}{2}}}{\Delta x_{i-1/2}}(C_{i-1} - C_i)\right]$$
$$- \frac{\Delta t}{\Delta x_i \varphi_i}(uC|_{i+\frac{1}{2}} - uC|_{i-\frac{1}{2}}) + \frac{Q_i^{rc}\Delta t}{V_i \varphi_i}C_i^*. \tag{8.4a}$$

Introducing dimensionless variables, the algebraic block equations can then be written,

$$C_i^{n+1} - C_i^n = \eta_{i+\frac{1}{2}}(C_{i+1} - C_i) + \eta_{i-\frac{1}{2}}(C_{i-1} - C_i) \pm \zeta_{i+\frac{1}{2}}C_{i+\frac{1}{2}} \pm \zeta_{i-\frac{1}{2}}C_{i-\frac{1}{2}}$$
$$+ \tau_i C_i^*,$$

(8.4b)

where the sign on ζ is positive if flow is into block i and negative if flow is out of the block. The dimensionless variables are defined as,

$$\eta_{i\pm\frac{1}{2}} = \frac{\Delta t D_{i\pm\frac{1}{2}}}{\Delta x_i \Delta x_{i\pm\frac{1}{2}}} = \frac{D\Delta t}{\Delta x^2},$$

(8.5a)

$$\pi_{i\pm\frac{1}{2}} = \frac{|u_{i\pm\frac{1}{2}}|\Delta x_i}{D_{i\pm\frac{1}{2}}\varphi_i} = \frac{|u|\Delta x}{D\varphi},$$

(8.5b)

$$\zeta_{i\pm\frac{1}{2}} = \pi_{i\pm\frac{1}{2}}\eta_{i\pm\frac{1}{2}} = \frac{|u_{i\pm\frac{1}{2}}|\Delta t}{\Delta x_{i\pm\frac{1}{2}}\varphi_i} = \frac{|u|\Delta t}{\Delta x\varphi},$$

(8.5c)

$$\tau_i = \frac{Q_i^{rc}\Delta t}{V_i\varphi},$$

(8.5d)

where the furthest right-hand side of the variable definitions are for homogenous properties.

Each of the dimensionless variables in Eq. (8.4b) has physical meaning; η is a dimensionless dispersion coefficient (sometimes called the *Fourier number*) and is analogous to the dimensionless diffusivity constant introduced for single-phase flow in Chapter 3; π is a local/block Peclet number and represents the relative advection to diffusion; ζ is the Courant number which is important for numerical stability and discussed more later; and τ is the dimensionless time or pore volumes injected/produced of fluid from a well in the grid block. Wells are discussed in Section 8.3.1.

Superscripts (n or $n + 1$) on concentration are intentionally not included on the right-hand side of Eq. (8.4b) because the equation can be solved explicitly or implicitly. Eq. (8.4) can be applied to each grid block which results in a system of equations. Solution methods are discussed in Section 8.5, but first the interblock concentrations must be defined (Section 8.3) and boundary conditions imposed (Section 8.4).

8.3 Discretization of advective terms

The discretized mass balance for a component (Eq. 8.4) included interblock concentrations. However, concentrations are generally defined at block centers; the interblock values can be defined in terms of block properties in several different ways. Two common approaches, cell-centered and single-point upwinding, are described here.

8.3.1 Cell-centered

A *cell-centered* approach for discretizing advective terms effectively involves averaging block concentrations in adjacent cells,

$$C_{i-\frac{1}{2}} = \frac{C_{i-1} + C_i}{2} \text{ and } C_{i+\frac{1}{2}} = \frac{C_i + C_{i+1}}{2}, \tag{8.6}$$

where it is assumed that the blocks have the same volume (otherwise Eq. (8.6) should be weighted by bulk volumes). Substitution into the advective derivatives gives,

$$\frac{\partial(uC)}{\partial x} = \frac{uC|_{i+\frac{1}{2}} - uC|_{i-\frac{1}{2}}}{\Delta x_i} = \frac{1}{\Delta x_i} \left[u_{i+\frac{1}{2}} \frac{C_{i+1} + C_i}{2} - u_{i-\frac{1}{2}} \frac{C_i + C_{i-1}}{2} \right]$$

$$= \frac{u_{i+\frac{1}{2}} C_{i+1} + (u_{i+\frac{1}{2}} - u_{i-\frac{1}{2}}) C_i - u_{i-\frac{1}{2}} C_{i-1}}{2\Delta x_i}.$$

The finite difference equation then becomes,

$$C_i^{n+1} - C_i^n = \eta_{i+\frac{1}{2}}(C_{i+1} - C_i) + \eta_{i-\frac{1}{2}}(C_{i-1} - C_i) \pm \zeta_{i+\frac{1}{2}} C_{i+1}$$

$$- (\pm \zeta_{i+\frac{1}{2}} \pm \zeta_{i-\frac{1}{2}}) C_i \pm \zeta_{i-\frac{1}{2}} C_{i-1} + \tau_i C_i^*, \tag{8.7}$$

The cell-centered approach has the advantage of being second-order accurate in truncation error. However, the solution often results in undesirable oscillations. An example of these oscillations is shown later in this chapter.

8.3.2 Upwinding

An alternative approach to cell-centered differencing of advective terms is to use upwinding for interblock concentrations. *Upwinding* is a numerical technique that evaluates interblock properties (in this case concentration) in the block upwind of flow; that is in the block which flow comes *from*. Fluid always flows from higher to lower potential (or, in the absence of gravity, from higher to lower pressure). Therefore, the upwinded concentration is defined as,

$$C_{i-\frac{1}{2}} = \begin{cases} C_{i-1} & \text{if} \quad \Phi_{i-1} > \Phi_i \\ C_i & \text{if} \quad \Phi_i > \Phi_{i-1} \end{cases} \quad \text{and} \quad C_{i+\frac{1}{2}} = \begin{cases} C_{i+1} & \text{if} \quad \Phi_{i+1} > \Phi_i \\ C_i & \text{if} \quad \Phi_i > \Phi_{i+1} \end{cases},$$

$$(8.8)$$

where Φ_i is the potential in block i. Substitution into the discretized mass balance gives,

$$C_i^{n+1} - C_i^n = \eta_{i+\frac{1}{2}}(C_{i+1} - C_i) + \eta_{i-\frac{1}{2}}(C_{i-1} - C_i) - \zeta_{i+\frac{1}{2}}C_i + \zeta_{i-\frac{1}{2}}C_{i-1}$$

$$+ \tau_i C_i^* - c_f\left(P_i^{n+1} - P_i^n\right)C_i \qquad (8.9)$$

if flow is left to right ($\Phi_{i-1} > \Phi_i > \Phi_{i+1}$), but flow could be right to left or be flowing into or out of block i from both directions (for example, if a well is in the block).

Upwinding makes physical sense; the fluid brings the component along with it as it flows from upstream to downstream. A numerical advantage of single-point upwinding is that oscillations do not occur in the solution, as they can in the centered difference approach. However, this *single-point* upwinding approach is only *first-order accurate* in the truncation error. More advanced, accurate upwinding methods (Todd et al., 1972; Saad et al., 1990; Jianchun et al., 1995) can be employed to improve accuracy by using higher-order methods with flux limiters.

8.3.3 Matrices

The algebraic equations can be written for every block in the reservoir model and the coefficients summarized into matrices,

$$
\boldsymbol{\eta} = - \begin{bmatrix}
\eta_{3/2} & -\eta_{3/2} & & & & \\
-\eta_{3/2} & \eta_{3/2} + \eta_{5/2} & -\eta_{5/2} & & & \\
& -\eta_{5/2} & \eta_{5/2} + \eta_{7/2} & \eta_{7/2} & & \\
& & & \ddots & -\eta_{N-1/2} & \\
& & & & -\eta_{N-1/2} & \eta_{N-1/2}
\end{bmatrix} ;
$$

$$
\boldsymbol{\zeta}_{up} = \begin{bmatrix}
\zeta_{3/2} & & & & \\
-\zeta_{3/2} & \zeta_{5/2} & & & \\
& -\zeta_{5/2} & \zeta_{7/2} & & \\
& & & \ddots & \\
& & & -\zeta_{N-1/2} & 0
\end{bmatrix} ;
$$

$$
\boldsymbol{\zeta}_{cen} = \frac{1}{2} \begin{bmatrix}
\zeta_{3/2} & \zeta_{3/2} & & & \\
-\zeta_{3/2} & -\zeta_{3/2} + \zeta_{5/2} & \zeta_{5/2} & & \\
& -\zeta_{5/2} & -\zeta_{5/2} + \zeta_{7/2} & \zeta_{7/2} & \\
& & & \ddots & \\
& & & -\zeta_{N-1/2} & -\zeta_{N-1/2}
\end{bmatrix}
$$

where the $\boldsymbol{\eta}$ matrix is tridiagonal in 1D and symmetric. Flow is assumed left ($i = 1$) to right ($i = N$) throughout the reservoir in both $\boldsymbol{\zeta}$ matrices for illustration purposes but must be adjusted for different flow directions. The upwinded courant matrix, $\boldsymbol{\zeta}_{up}$, will always be two entries per row for 1D flow but the column may vary if the flow direction changes from grid block to grid block. The centered courant matrix, $\boldsymbol{\zeta}_{cen}$, is tridiagonal but the off-diagonal terms have opposite signs. The main diagonal is the sum of the off-diagonals.

8.4 Wells and boundary conditions

8.4.1 Wells

Wells are treated as a source/sink term in the mass balance equations. If block i contains a well, the change in concentration of that block due to the well is

$\tau_i C_i^*$, where τ_i is the dimensionless pore volumes of the block injected/produced, as previously defined in Eq. (8.5d), and C_i^* defined as,

$$C_i^* = \begin{cases} C_{inj} & \text{if} \quad \text{injector} \\ C_i & \text{if} \quad \text{producer} \end{cases},$$

where C_{inj} is the specified concentration injected into the well and C_i is the concentration of block i. As always, wells can be operated at constant rate or constant bottomhole pressure. If the well is operated at constant bottomhole pressure, the total well rate was defined in Chapter 5 as $Q_i = -J_i(P_i - P_{wf})$.

8.4.2 No flux boundary condition

The most common boundary condition for component transport in reservoir simulation is to assume no dispersive flux across the boundary, therefore $dC/dx = 0$. Moreover, the boundary is often completely sealed to flow ($u = 0$), so advective transport is also zero. If flow is advected through the boundary it can be modeled as a well. The boundary flux terms on the algebraic block balances are therefore zero. For example, at the $x = 0$ boundary, block $i = 1$ can be written,

$$C_1^{n+1} - C_1^n = \eta_{1/2}\cancel{(C_0 - C_1)} + \cancel{\varsigma_{1/2} C_{1/2}} + \eta_{3/2}(C_2 - C_1) \pm \varsigma_{3/2} C_{3/2} + \tau_1 C_1^*. \tag{8.10}$$

where the flux terms at the boundary vanish and $C_{3/2}$ can be defined with a centered difference or upwinding scheme as defined in Section 8.3. Example 8.1 demonstrates the formation of the algebraic block equations for no-flux boundary conditions, injector wells near the boundaries, and a producer well at the center of the reservoir. Example 8.2 demonstrates the formation of the matrices including the well/boundary terms in τ for the same reservoir.

Example 8.1 Block mass balances for component transport.
A component, κ, is transported in a 1D, homogeneous single-phase reservoir by both advection and dispersion. The reservoir is discretized into five uniform blocks. There is an injector well at both $x = 0$ and $x = L$ with rates Q_{inj} and injected concentrations, C_{inj}. There is a producer at $x = L/2$ with rate $2Q_{inj}$. Flow is steady state and has constant velocity throughout the reservoir. Write the algebraic mass balance equations for all five grid blocks for the following discretization schemes for advection. The boundaries are sealed (no advection) and $dC/dx = 0$ at the boundaries. You may ignore rock compressibility.

a. Upwinding
b. Centered

Solution
All properties are homogeneous and uniform which lead to the constant dimensionless properties,

$$\eta = \frac{D\Delta t}{\Delta x^2} = \frac{D\Delta t N^2}{L^2}; \zeta = \frac{|Q_{inj}|\Delta t N}{aL\varphi}$$

The general balance equation is

$$C_i^{n+1} - C_i^n = \eta_{i-\frac{1}{2}}(C_{i-1} - C_i) + \eta_{i+\frac{1}{2}}(C_{i+1} - C_i) \pm \zeta_{i-\frac{1}{2}}C_{i-\frac{1}{2}} \pm \zeta_{i+\frac{1}{2}}C_{i+\frac{1}{2}},$$

where the advection terms are positive if flow is into block and negative if is flow out of block. Flow goes from block #1 to #2 to #3 and from block #5 to #4 to #3 because there is a producer in block #3.

a. Upwinding
Block #1

$$C_1^{n+1} - C_1^n = \eta_{\frac{1}{2}}(C_0 - C_1) + \eta_{\frac{3}{2}}(C_2 - C_1) \pm \zeta_{\frac{1}{2}}C_{\frac{1}{2}} \pm \zeta_{\frac{3}{2}}C_{\frac{3}{2}}$$

There is no dispersive or advective flux through the boundary, but there is an injector well, and the pore volumes injected, $\tau = \zeta$.

$$C_1^{n+1} - C_1^n = \diagup\eta_{\frac{1}{2}}(C_0 - C_1) + \eta_{\frac{3}{2}}(C_2 - C_1) + \diagup\zeta C_0 - \zeta C_1 + \zeta C_{inj}$$

$$C_1^{n+1} = C_1^n + [-\eta - \zeta]C_1 + [\eta]C_2 + [\zeta]C_{inj}$$

Block #2

$$C_2^{n+1} - C_2^n = \eta_{\frac{3}{2}}(C_1 - C_2) + \eta_{\frac{5}{2}}(C_3 - C_2) \pm \zeta_{\frac{3}{2}}C_{\frac{3}{2}} \pm \zeta_{\frac{5}{2}}C_{\frac{5}{2}}$$

$$C_2^{n+1} - C_2^n = \eta_{\frac{3}{2}}(C_1 - C_2) + \eta_{\frac{5}{2}}(C_3 - C_2) + \zeta_{\frac{3}{2}}C_1 - \zeta_{\frac{5}{2}}C_2$$

$$C_2^{n+1} = C_2^n + [\eta + \zeta]C_1 + [-2\eta - \zeta]C_2 + [\eta]C_3$$

Block #3

$$C_3^{n+1} - C_3^n = \eta_{\frac{5}{2}}(C_2 - C_3) + \eta_{\frac{7}{2}}(C_4 - C_3) \pm \zeta_{\frac{3}{2}}C_{\frac{3}{2}} \pm \zeta_{\frac{5}{2}}C_{\frac{5}{2}}$$

Flow comes into the block #3 from #2 and #4. Flow leaves block #3 from a producer well with rate $2Q_{inj}$

$$C_3^{n+1} - C_3^n = \eta_{\frac{5}{2}}(C_2 - C_3) + \eta_{\frac{7}{2}}(C_4 - C_3) + \zeta_{\frac{3}{2}}C_2 + \zeta_{\frac{5}{2}}C_4 - \left(\frac{2Q_{inj}\Delta t}{a\Delta x\varphi}\right)C_3$$

$$C_3^{n+1} - C_3^n = [\eta + \zeta]C_2 + [-2\eta - 2\zeta]C_3 + [\eta + \zeta]C_4$$

Block #4

$$C_4^{n+1} - C_4^n = \eta_{\frac{7}{2}}(C_3 - C_4) + \eta_{\frac{9}{2}}(C_5 - C_4) \pm \zeta_{\frac{7}{2}}C_{\frac{7}{2}} \pm \zeta_{\frac{9}{2}}C_{\frac{9}{2}}$$

Flow goes from block 5 to 4 and block 4 to 3,

$$C_4^{n+1} - C_4^n = \eta_{7/2}(C_3 - C_4) + \eta_{9/2}(C_5 - C_4) - \zeta_{7/2}C_4 + \zeta_{9/2}C_5$$

$$C_4^{n+1} = C_4^n + [\eta]C_3 + [-2\eta - \zeta]C_4 + [\eta + \zeta]C_5$$

Block #5

$$C_5^{n+1} - C_5^n = \eta_{9/2}(C_4 - C_5) + \eta_{11/2}(C_6 - C_5) \pm \zeta_{9/2}C_{9/2} \pm \zeta_{11/2}C_{11/2}$$

There is no dispersive or advective flux through the boundary, but there is an injector well

$$C_5^{n+1} - C_5^n = \eta_{9/2}(C_4 - C_5) - \zeta_{9/2}C_5 + \zeta C_{inj}$$

$$C_5^{n+1} - C_5^n = [\eta]C_4 + [-\eta - \zeta]C_5 + \zeta C_{inj}$$

a. Centered

Block #1

$$C_1^{n+1} - C_1^n = \eta_{1/2}(C_0 - C_1) + \eta_{3/2}(C_2 - C_1) \pm \zeta_{1/2}C_{1/2} \pm \zeta_{3/2}C_{3/2}$$

$$C_1^{n+1} - C_1^n = \eta_{1/2}(C_0 - C_1) + \eta_{3/2}(C_2 - C_1) + \zeta C_o - \zeta(0.5C_1 + 0.5C_2) + \zeta C_{inj}$$

$$C_1^{n+1} = C_1^n + [-\eta - 0.5\zeta]C_1 + [\eta - 0.5\zeta]C_2 + [\zeta]C_{inj}$$

Block #2

$$C_2^{n+1} - C_2^n = \eta_{3/2}(C_1 - C_2) + \eta_{5/2}(C_3 - C_2) \pm \zeta_{3/2}C_{3/2} \pm \zeta_{5/2}C_{5/2}$$

$$C_2^{n+1} - C_2^n = \eta_{3/2}(C_1 - C_2) + \eta_{5/2}(C_3 - C_2) + 0.5\zeta_{3/2}(C_1 + C_2)$$
$$- 0.5\zeta_{5/2}(C_2 + C_3)$$

$$C_2^{n+1} = C_2^n + [\eta + 0.5\zeta]C_1 + [-2\eta]C_2 + [\eta - 0.5\zeta]C_3$$

Block #3

$$C_3^{n+1} - C_3^n = \eta_{5/2}(C_2 - C_3) + \eta_{7/2}(C_4 - C_3) \pm \zeta_{3/2}C_{3/2} \pm \zeta_{5/2}C_{5/2}$$

Flow comes into the block #3 from #2 and #4. Flow leaves block #3 from a producer well with rate $2*Q_{inj}$

$$C_3^{n+1} - C_3^n = \eta_{5/2}(C_2 - C_3) + \eta_{7/2}(C_4 - C_3) + 0.5\zeta_{3/2}(C_2 + C_3)$$
$$+ \zeta_{5/2}(C_3 + C_4) - \left(\frac{2Q_{inj}\Delta t}{A\Delta x\varphi}\right)C_3$$

$$C_3^{n+1} = C_3^n + [\eta + 0.5\zeta]C_2 + [(2\eta + \zeta) - (2\zeta)]C_3 + [\eta + 0.5\zeta]C_4$$

Block #4

$$C_4^{n+1} - C_4^n = \eta_{7/2}(C_3 - C_4) + \eta_{9/2}(C_5 - C_4) \pm \zeta_{7/2} C_{7/2} \pm \zeta_{9/2} C_{9/2}$$

Flow goes from block #5 to #4 and block #4 to #3,

$$C_4^{n+1} - C_4^n = \eta_{7/2}(C_3 - C_4) + \eta_{9/2}(C_5 - C_4) - 0.5\zeta_{7/2}(C_3 + C_4)$$
$$+ 0.5\zeta_{9/2}(C_4 + C_5)$$

$$C_4^{n+1} - C_4^n = [\eta - 0.5\zeta]C_3 + [-2\eta]C_4 + [\eta + 0.5\zeta]C_5$$

Block #5

$$C_5^{n+1} - C_5^n = \eta_{9/2}(C_4 - C_5) + \eta_{11/2}(C_6 - C_5) \pm \zeta_{9/2} C_{9/2} \pm \zeta_{11/2} C_{11/2}$$

$$C_5^{n+1} - C_5^n = \eta_{9/2}(C_4 - C_5) - 0.5\zeta_{9/2}(C_4 + C_5) + \zeta C_{inj}$$

$$C_5^{n+1} - C_5^n = [\eta - 0.5\zeta]C_4 + [-\eta - 0.5\zeta] C_5 + \zeta C_{inj}$$

Example 8.2 Matrices and vectors for transport.
Create the relevant matrices and vectors required for the upwinded and centered schemes for the 5-block problem described by Example 8.1.

Solution
The matrices can be evaluated by collecting the algebraic equations found in Example 8.1,

$$\eta = - \begin{bmatrix} \eta & -\eta & & & \\ -\eta & 2\eta & -\eta & & \\ & -\eta & 2\eta & -\eta & \\ & & -\eta & 2\eta & -\eta \\ & & & -\eta & \eta \end{bmatrix}, \zeta_{up} = \begin{bmatrix} \zeta & & & & \\ -\zeta & \zeta & & & \\ & -\zeta & 0 & -\zeta & \\ & & \zeta & -\zeta & \\ & & & & \zeta \end{bmatrix}$$

$$\zeta_{cen} = \frac{1}{2} \begin{bmatrix} \zeta & \zeta & & & \\ -\zeta & 0 & \zeta & & \\ & -\zeta & -2\zeta & -\zeta & \\ & & \zeta & 0 & -\zeta \\ & & & \zeta & \zeta \end{bmatrix};$$

$$\tau_{\text{inj}} = \begin{bmatrix} \zeta & & \\ & & \\ & & \zeta \end{bmatrix} \quad \tau_p = \begin{bmatrix} 2\zeta & \\ & \\ & \end{bmatrix}.$$

8.4.3 Constant concentration (Dirichlet)

It is difficult to maintain a constant concentration at the boundary, particularly for low to moderate Peclet numbers, because solute is immediately dispersed away from the boundary. However, many analytical solutions do assume a Dirichlet boundary condition ($C = C_{\text{inj}}$) at the injection site (e.g., Eq. 7.23, Marle, 1981). Fig. 8.1 depicts either Dirichlet or a mixed boundary condition at $x = 0$.

Most subsurface applications are advection dominated, so a Dirichlet boundary condition, although strictly not appropriate, may be an acceptable approximation.

The boundary condition can be imposed by substitution for the ghost cell as was done in Chapter 3, e.g., $C_0 = 2C_B - C_1$ at the $x = 0$ boundary. The algebraic equations can then be written for a centered difference Eq. (8.11a) or upwinding Eq. (8.11b) for block $i=1$ with a Dirichlet boundary condition as,

$$C_1^{n+1} - C_1^n = -\eta_{3/2}C_1 + \eta_{3/2}C_2 - 0.5\zeta_{3/2}(C_1 + C_2) - \tau_{1,p}C_1 + \tau_{1,\text{inj}}C_{\text{inj}}, \tag{8.11a}$$

$$C_1^{n+1} - C_1^n = -\eta_{3/2}C_1 + \eta_{3/2}C_2 - \zeta_{3/2}C_1 - \tau_{1,p}C_1 + \tau_{1,\text{inj}}C_{\text{inj}}, \tag{8.11b}$$

where,

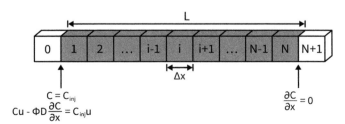

FIGURE 8.1 1D reservoir with either Dirichlet or third-order boundary conditions at inlet and zero concentration derivative at exit.

$$\tau_{1,p} = \begin{cases} 2\eta_B + \zeta_B & \text{if centered} \\ 2\eta_B + 2\zeta_B & \text{if upwinded} \end{cases} \tau_{1,inj} = \begin{cases} 2\eta_B & \text{if centered} \\ 2\eta_B + \zeta_B & \text{if upwinded} \end{cases},$$

and the subscript B refers to the value at the boundary. τ_p and τ_{inj} are diagonal matrices that include the well and boundary terms.

8.5 Solution methods

The component balances require knowledge of interblock velocities (and therefore block pressures) to compute local Peclet/Courant numbers as well as dispersion coefficients, if they are velocity dependent (Chapter 7). For steady-state flow problems, the pressure/velocity field does not change with time and does not need to be updated in each timestep. For unsteady-state problems, the diffusivity/pressure equation can be used to determine the pressure field which can be used with Eq. (8.3), Darcy's law, to compute the velocity field. The pressure equation for each block, i, is discretized implicitly, as described in Chapters 3–6. The coupling between flow and transport can be done using various numerical approaches.

8.5.1 Implicit pressure, explicit concentration (IMPEC)

A common approach to solving compositional problems is to first solve the pressure equation to obtain the pressure and velocity field, followed by solving

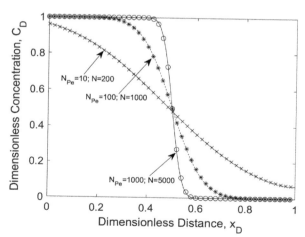

FIGURE 8.2 Comparison of numerical to analytical (Brenner, 1962) solution of dimensionless concentration versus dimensionless distance at $t_D = 0.5$ for $N_{Pe} =$ (A) 10, (B) 100, and (C) 1000 using $N = 200$, 1000, and 5000, respectively. An explicit, upwinded scheme was employed. In all cases, the maximum Courant number (0.024, 0.047, 0.09, respectively) to ensure stability was used. Only 50 evenly spaced grid points are shown.

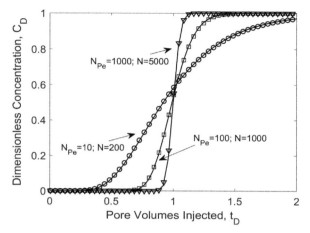

FIGURE 8.3 Dimensionless concentration (c_D) versus dimensionless time (pore volumes injected, t_D) at $x_D = 1$ for $N_{Pe} = $ (A) 10, (B) 100, and (C) 1000 using $N = 200$, 1000, and 5000, respectively. An explicit, upwinded scheme was employed. In all cases, the maximum Courant number (0.024, 0.047, 0.09, respectively) to ensure stability was used. Only 50 evenly spaced grid points are shown. Analytical solution taken from Brenner (1962).

the compositional equation to obtain component concentrations. The timestep is advanced and then the steps are repeated. The method is known as *Implicit Pressure, Explicit Concentration* or IMPEC. As the name implies, grid pressures are first computed implicitly and then component concentrations are computed explicitly in each timestep. Fluid, petrophysical, and other grid properties are updated, if necessary, before proceeding to the next time level.

The pressure field is solved implicitly in the usual way, using the linear system of Eq. (4.12). Arrays **T**, **J**, **A**, Q and G may change with time and, if so,

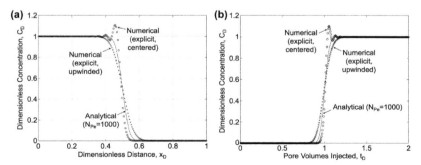

FIGURE 8.4 (A) Dimensionless concentration versus dimensionless distance for $t_D = 0.5$ and (B) dimensionless concentration versus dimensionless time (pore volumes injected, t_D) at $x_D = 1$, $N_{Pe} = 1000$, $N = 200$, and $\zeta = 0.35$. An explicit scheme was employed and centered differences result in oscillations, but the upwinded scheme does not have oscillations.

need to be recomputed at the beginning of each timestep. For example, fluid properties may depend on the component concentration (e.g., the viscosities of polymer solutions are strong functions of polymer concentration).

Once the pressure field is computed implicitly at a given time level ($n+1$), interblock velocities can be computed from Eq. (8.3), Darcy's law. The velocities can be used to compute interblock dispersion coefficients using D versus velocity correlations (Eq. 7.7). Block pressures, interblock velocities, and interblock dispersion coefficients can be substituted into the explicit form of the component balance (Eq. 8.4) to update concentrations *explicitly*. The equations can be written in matrix form as,

$$\vec{C}^{n+1} = \left[\mathbf{I} - \left(\mathbf{\eta} + \mathbf{\zeta} + \mathbf{\tau}_p\right)\right]\vec{C}^{n} + \mathbf{\tau}_{inj}\vec{C}_{inj},\qquad(8.12)$$

where $\mathbf{\eta}$, $\mathbf{\zeta}$, $\mathbf{\tau}_p$, and $\mathbf{\tau}_{inj}$ are the dimensionless $N{\times}N$ matrices of coefficients. The Courant matrix, $\mathbf{\zeta}$, can be either upwinded (preferred) or centered.

Fig. 8.2 shows the dimensionless concentration versus dimensionless distance at $t_D = 0.5$ for $N_{Pe} =$ (a) 10, (b) 100, and (c) 1000, using no flux boundary conditions. The numerical solution used $N = 200$, 1000, and 5000 grids, respectively, and an explicit, upwinded scheme was employed. Only 50 evenly spaced grid points are shown. The analytical solution (Eq. 7.27) is taken from Brenner (1962).

Excellent agreement is observed between the numerical and analytical solution. However, at high N_{Pe}, more grids are required to obtain accuracy (agreement with the analytical solution) and reduce numerical dispersion (discussed in Section 8.7).

Fig. 8.3 shows the dimensionless concentration (C_D) versus dimensionless time (pore volumes injected, t_D) at the exit, $x_D = 1$ for $N_{Pe} =$ (a) 10, (b) 100, and (c) 1000 using $N = 200$, 1000, and 5000, respectively and an explicit, upwinded scheme was employed.

Fig. 8.4 shows the dimensionless concentration versus (a) dimensionless distance for $t_D = 0.5$ and (b) dimensionless time (pore volumes injected, t_D) at $x_D = 1$, $N_{Pe} = 1000$, $N = 200$, and $\zeta = 0.35$ but using *centered differences* (with an explicit scheme) in addition to upwinding. Using the centered differences results in oscillations.

8.5.2 Implicit pressure, implicit concentration

In the implicit pressure, implicit concentration (IMPIC) method, the pressure field is computed implicitly and then the component concentration field is also solved implicitly,

$$\left[\mathbf{I} + \mathbf{\eta} + \mathbf{\zeta} + \mathbf{\tau}_p\right]\vec{C}^{n+1} = \vec{C}^{n} + \mathbf{\tau}_{inj}\vec{C}_{inj}.\qquad(8.13)$$

The IMPIC method is more stable (unconditionally stable if flow is steady) than IMPEC and thus allows for larger timesteps but is more computationally demanding. The computational effort is particularly high for multicomponent problems since the system of equations is of size $(N_c - 1)N$, where N_c is the total number of components and N is the number of grids. Example 8.3 provides an example of steady-state flow of both an explicit and implicit update of concentration for a tracer component transported in 1D. Upwinding and centered differences are compared for both schemes.

Example 8.3 Single-phase transport of a component (tracer) in 1D.
The 1D, homogeneous reservoir described by Example 8.1 has the following properties: $\varphi = 0.2$, $L = 5000$ ft (length), $a = 10,000$ ft^2. The injector wells placed near the boundaries each have a rate of $Q_{inj} = 1000$ scf/day ($B_w = 1$) and tracer is injected at a concentration, C_{inj}, of 1.0 lb$_m$/ft^3. The dispersion coefficient is $D = 125$ ft^2/day. Numerically solve for the concentration of tracer using 5 equal-sized grid blocks and a timestep of 500 days using the following methods:
(a) Explicit, upwinded
(b) Explicit, centered
(c) Implicit, upwinded
(d) Implicit, centered

Solution:
Calculate the velocity in the x-direction

$$u_x = \frac{Q_{sc}B_w}{a} = \frac{1000\,\frac{scf}{day}\cdot 1.0\,\frac{ft^3}{scf}}{10,000\text{ ft}^2} = 0.1\,\frac{ft}{day}$$

Compute the dimensionless parameters. Use $L = 5000/2 = 2500$ ft, the well-to-well distance.

$$N_{Pe} = \frac{u_x L}{D\varphi} = \frac{0.1\,\frac{ft}{day}\cdot\left(\frac{5000\text{ ft}}{2}\right)}{125\,\frac{ft^2}{day}\cdot 0.2} = 10; \quad \pi = \frac{u_x \Delta x}{D\varphi} = \frac{0.1\,\frac{ft}{day}\cdot 1000\text{ ft}}{125\,\frac{ft^2}{day}\cdot 0.2} = 4$$

$$\zeta = \frac{u_x \Delta t}{\Delta x \cdot \varphi} = \frac{0.1\,\frac{ft}{day}\cdot 500\text{ days}}{1000\text{ ft}\cdot 0.2} = 0.25 \quad \eta = \frac{D\Delta t}{\Delta x^2} = \frac{125\,\frac{ft^2}{day}\cdot 500\text{ days}}{(1000\text{ ft})^2}$$

$$= 0.0625\tau$$

$$\tau = \frac{Q\Delta t}{a\varphi L} = \frac{2000\,\frac{ft^3}{day}\,500\text{ days}}{10,000\text{ ft}^2\cdot 0.2\cdot 5000\text{ ft}} = 0.1 \text{ pore volumes injected}$$

The dimensionless matrices and vectors were created in Example 8.3 and the numerical values can be substituted.

(a) Explicit upwinded
    Check stability requirements (Section 8.6) and numerical dispersion (Section 8.7)

$$\zeta < \frac{\pi}{\pi+2}; \quad \zeta = 0.25 < \frac{4}{4+2} = 0.667 \quad \text{(stable)}$$

$$\frac{D_{num}}{D} = \frac{\pi}{2}(1 - \zeta) = \frac{4}{2}(1 - 0.25) = 1.5 \quad \text{(high numerical dispersion)}$$

Concentrations are explicitly updated using the following matrix equations

$$\vec{C}^{n+1} = (\mathbf{I} - \zeta_{\textbf{up}} - \eta - \tau_p)\vec{C}^n + \tau_{inj}\vec{C}_{inj}$$

(b) Explicit centered

Check stability requirements and numerical dispersion

$$\pi < 2\left(\frac{1 - \eta}{\eta}\right); \quad 4 < 2\left(\frac{1 - 0.0625}{0.0625}\right) = 30;$$

$$\eta < \frac{1}{2}; \quad 0.0625 < \frac{1}{2}; \text{(Stable)}$$

$$\frac{D_{num}}{D} = -\frac{\pi}{2}\zeta = -\frac{4}{2}0.25 = -0.5$$

Concentrations are explicitly updated using the following matrix equations

$$\vec{C}^{n+1} = (\mathbf{I} - \zeta_{\textbf{cen}} - \eta - \tau_p)\vec{C}^n + \tau_{inj}\vec{C}_{inj}$$

(c) Implicit upwinded

Check stability requirements and numerical dispersion.

Implicit methods are unconditionally stable

$$\frac{D_{num}}{D} = \frac{\pi}{2}(1 + \zeta) = \frac{4}{2}(1 + 0.25) = 2.5$$

Concentrations are implicitly updated using the following matrix equations

$$(\mathbf{I} + \zeta_{\textbf{up}} + \eta + \tau_p)\vec{C}^{n+1} = \vec{C}^n + \tau_{inj}C_{inj}$$

(d) Implicit centered

Check stability requirements and numerical dispersion.

Implicit methods are unconditionally stable

$$\frac{D_{num}}{D} = \frac{\pi}{2}\zeta = \frac{4}{2} \cdot 0.25 = 0.5$$

Concentrations are implicitly updated using the following matrix equations

$$(\mathbf{I} + \zeta_{\textbf{cen}} + \eta + \tau_p)\vec{C}^{n+1} = \vec{C}^n + \tau_{inj}\vec{C}_{inj}$$

$$\tau = \frac{Q\Delta t}{a\varphi L} = \frac{2000\,\frac{ft^3}{day}\cdot 500\text{ days}}{10,000\text{ ft}^2 \cdot 0.2 \cdot 5000\text{ ft}} = 0.1 \text{ pore volumes injected}$$

The results are summarized in Table E.1.

Mixed methods which are a hybrid of explicit and implicit formulations are also possible.

$$(1-\theta)\left[\mathbf{I} + \mathbf{\eta} + \mathbf{\zeta} + \tau_p\right]\vec{C}^{n+1} = \theta\left[\mathbf{I} - \left(\mathbf{\eta} + \mathbf{\zeta} + \tau_p\right)\right]\vec{C}^{n} + \tau_{inj}\vec{C}_{inj} \qquad (8.14)$$

where Eq. (8.14) reduces to the implicit method if $\theta = 0$ and the explicit method if $\theta = 1$; it is a mixed method for $0 < \theta < 1$. If $\theta = \frac{1}{2}$ and has accuracy $O(\Delta t^2)$.

TABLE E.1 Comparison of solutions (concentration, lb_m/ft^3) for Example 8.3 using the 4 numerical approaches.

PVI	Method	Block #1 500 ft	Block #2 1500 ft	Block #3 2500 ft	Block #4 3500 ft	Block #5 4500 ft
Initial		0.0	0.0	0.0	0.0	0.0
0.1	Explicit, upwinded	0.25	0.0	0.0	0.0	0.25
	Explicit, centered	0.25	0.0	0.0	0.0	0.25
	Implicit, upwinded	0.1926	0.044551	0.017135	0.044551	0.1926
	Implicit, centered	0.20872	0.034268	0.0093458	0.034268	0.20872
0.2	Explicit, upwinded	0.42188	0.078125	0.0	0.078125	0.42188
	Explicit, centered	0.45313	0.046875	0.0	0.046875	0.45313
	Implicit, upwinded	0.34258	0.11271	0.053895	0.11271	0.34258
	Implicit, centered	0.38144	0.092258	0.031958	0.092258	0.38144
0.3	Explicit, upwinded	0.54492	0.18066	0.048828	0.18066	0.54492
	Explicit, centered	0.61523	0.12598	0.017578	0.12598	0.61523
	Implicit, upwinded	0.46061	0.19151	0.10682	0.19151	0.46061
	Implicit, centered	0.52303	0.16538	0.068346	0.16538	0.52303

Continued

TABLE E.1 Comparison of solutions (concentration, lb_m/ft^3) for Example 8.3 using the 4 numerical approaches.—cont'd

PVI	Method	Block #1 500 ft	Block #2 1500 ft	Block #3 2500 ft	Block #4 3500 ft	Block #5 4500 ft
0.4	Explicit, upwinded	0.63593	0.28625	0.13123	0.28625	0.63593
	Explicit, centered	0.742	0.22449	0.058228	0.22449	0.742
	Implicit, upwinded	0.55442	0.27305	0.17076	0.27305	0.55442
	Implicit, centered	0.63798	0.24684	0.11703	0.24684	0.63798
∞		1.0	1.0	1.0	1.0	

Notes: In this problem, large grid sizes and timesteps were used which resulted in inaccuracies. The large grid sizes also contributed to very high numerical dispersion. Each of the four methods gives a different result but would all converge on the same solution with small grids and timesteps. In the problem, a realistic Darcy velocity (0.1 ft/day) was used but an unrealistically high dispersion coefficient (125 ft^2/day) was used to create a relatively low Peclet number (10). Most practical subsurface applications involve large Peclet numbers (>1000).

8.5.3 Fully implicit

The fully implicit solution requires solution of both flow and transport equations simultaneously. Balances for grid block, i, can be written as,

$$f_i\left(\overrightarrow{C}^{n+1}, \overrightarrow{P}^{n+1}\right) = T_{i-\frac{1}{2}}^{n+1}\left(P_{i-1}^{n+1} - P_i^{n+1}\right) + T_{i+\frac{1}{2}}^{n+1}\left(P_{i+1}^{n+1} - P_i^{n+1}\right)$$

$$-(Ac_t)_i^{n+1}\left(P_i^{n+1} - P_i^n\right) + Q_i + G_i = 0$$

$$g_i\left(\overrightarrow{C}^{n+1}, \overrightarrow{P}^{n+1}\right) = C_i^{n+1} - C_i^n - \eta_{i+\frac{1}{2}}^{n+1}\left(C_{i+1}^{n+1} - C_i^{n+1}\right) - \eta_{i-\frac{1}{2}}^{n+1}\left(C_{i-1}^{n+1} - C_i^{n+1}\right)$$

$$\pm \zeta_{i+\frac{1}{2}}^{n+1} C_{i+\frac{1}{2}}^{n+1} \pm \zeta_{i-\frac{1}{2}}^{n+1} C_{i-\frac{1}{2}}^{n+1} + \tau_i^{n+1} C_i^{*n+1} = 0$$

At first glance, the equations do not appear to be strongly coupled, and in many cases they are not (in which case IMPEC or IMPIC can be used with sufficiently small timesteps). However, the equations can be coupled and nonlinear in several ways. For example, the interblock velocity (and thus ζ) is a function of pressure via Darcy's law (Eq. 8.3). Dispersion coefficients (and thus η) may also be velocity dependent. Petrophysical and fluid properties, such as viscosity (and thus T, A), may be a function of component concentration.

The coupled, nonlinear system of equations must be solved using the techniques described in Chapter 6, e.g., Newton–Raphson method, $\overrightarrow{X}^{new} = \overrightarrow{X}^{old} - \overrightarrow{\mathfrak{J}}^{-1} \overrightarrow{F}$, where X is the vector of unknown pressures and concentrations, F is the vector of balance errors, and $\mathfrak{J}$ is the Jacobian of partial derivatives.

$$\mathfrak{J} = \begin{bmatrix} \dfrac{\partial f_1}{\partial X_1} & \dfrac{\partial f_1}{\partial X_2} & \dfrac{\partial f_1}{\partial X_N} \\[2mm] \dfrac{\partial g_1}{\partial X_1} & \dfrac{\partial g_2}{\partial X_2} & \dfrac{\partial g_2}{\partial X_N} \\[2mm] \dfrac{\partial g_N}{\partial X_1} & \dfrac{\partial g_N}{\partial X_2} & \dfrac{\partial g_N}{\partial X_N} \end{bmatrix}; \overrightarrow{X} = \begin{bmatrix} P_1 \\ C_1 \\ P_2 \\ C_2 \\ \\ P_N \\ C_N \end{bmatrix}; \overrightarrow{F} = \begin{bmatrix} f_1 \\ g_1 \\ f_2 \\ g_2 \\ \\ f_N \\ g_N \end{bmatrix} \qquad (8.15)$$

The Newton–Raphson method is implemented in the usual way. First, a guess of the pressure and concentration field is made (usually the values from the previous timestep). Then, using the guesses, the solution vector, F, and Jacobian, $\mathfrak{J}$, are computed. The solution is updated and the procedure is repeated until a small error in the norm of solution vector is found and the next timestep is advanced.

The fully implicit method is unconditionally stable. However, for multicomponent systems, both IMPIC and fully implicit methods become computationally challenging. For an N_c component system in the fully implicit method, a system of N_cN equations and unknowns must be solved multiple times (iterations) per time step. For this reason, IMPEC, with relatively small timesteps to maintain stability, is often the method of choice. However, the numerical scheme is always problem dependent.

8.6 Stability

Any numerical scheme with some explicitness is at risk of becoming unstable. The analysis here is restricted to steady-state flow, 1D reservoirs with homogeneous reservoir/fluid properties, and uniform grids. Stability criteria can be extended to more complex problems. The stability criterion for the 1D, homogenous ADE using an explicit scheme and centered difference for the second derivative and first-order, backward/forward difference (upwinded) for advective terms is,

$$2\eta + \zeta < 1 \quad \text{OR} \quad \zeta < \frac{\pi}{\pi + 2}. \tag{8.16a}$$

The first term of the first criterion ($2\eta < 1$) is similar to the stability condition for the 1D diffusivity equation (Chapter 3) and the second term ($\zeta < 1$) is the Courant–Friedrichs–Lewy (CFL) (1928) condition for advection-dominated problems; the second criterion can be shown to be equivalent to the first. If a second-order, centered difference is used for the advection terms (Hogarth et al., 1990),

$$\frac{\zeta^2}{2} < \eta < \frac{1}{2}. \tag{8.16b}$$

Fig. 8.5 shows the dimensionless concentration versus dimensionless distance at $t_D = 0.5$ for $N_{Pe} = 100$ and $N = 100$. The solution is unstable for the explicit, upwinded scheme but stable for the implicit, upwinded scheme because implicit methods are unconditionally stable. The Courant number ($\zeta = 0.34$) was only ~2% larger than the value required for stability of the explicit, upwinded scheme; larger Courant numbers result in much more exaggerated instabilities.

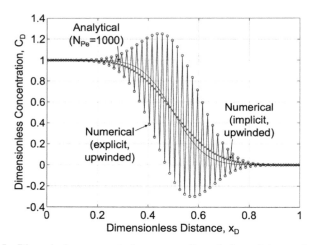

FIGURE 8.5 Dimensionless concentration versus dimensionless distance at $t_D = 0.5$ for $N_{Pe} = 100$ and $N = 100$. The Courant number (0.34) was 2% larger than the value required for stability of the explicit, upwinded scheme. The solution is unstable for the explicit, upwinded scheme but stable for the implicit, upwinded scheme. *Analytical solution taken from Brenner (1962), Eq.(7.27).*

8.7 Numerical dispersion

Advection-dominated flow (very large N_{Pe}) is characterized by a sharp front in the concentration profile (C_D vs. x_D). The sharp front is impossible to reproduce perfectly using numerical approaches and a finite number of grids. This numerical phenomenon is referred to as *numerical dispersion* and exaggerates any *physical dispersion* because it produces artificial spreading in the solution. Numerical dispersion manifests itself mathematically the same as physical dispersion. Numerical solutions with finite grids and numerical dispersion can be described exactly by an applicable analytical solution, but only when a nonphysical, finite dispersion coefficient is introduced. For steady-state flow without reaction and uniform grids, the value of the numerical dispersion coefficient, D_{num}, can be found using a Taylor series expansion (Ataie-Ashtiani et al., 1999),

$$\frac{D_{num}}{D} = \left(\frac{1}{2} - \beta\right)\pi + \omega\pi\zeta - \frac{\pi}{2}\zeta, \qquad (8.17)$$

where β indicates the spatial differencing scheme for advective terms (upwinding or centered) and ω is the temporal differencing scheme (explicit, implicit, or Crank−Nicolson),

$$\beta = \begin{cases} 0 & \text{upwind} \\ \frac{1}{2} & \text{centered} \end{cases}; \quad \omega = \begin{cases} 0 & \text{explicit} \\ \frac{1}{2} & \text{Crank} - \text{Nicolson.} \\ 1 & \text{implicit} \end{cases}$$

The effective dispersion coefficient is the sum of the physical and numerical dispersion. Therefore the effective Peclet number, $N_{Pe,\text{eff}}$, can be computed as,

$$N_{Pe,eff} = \left[\frac{1}{N_{Pe}}\left(1 + \frac{D_{num}}{D}\right)\right]^{-1}. \qquad (8.18)$$

Fig. 8.6 shows the dimensionless concentration versus dimensionless distance at $t_D = 0.5$. An explicit, upwinded scheme was employed and a Courant number, $\zeta = 0.1$, $N_{Pe} = 1000$, and $N = 500$ grids for the numerical simulations. Numerical dispersion causes a mismatch with the analytical solution at $N_{Pe} = 1000$ but gives an excellent match to the analytical solution using the effective Peclet number obtained from Eqs. (8.17) and (8.18) ($N_{Pe,eff} = 526.3$).

Consider the explicit, upwinded method; if the maximum Courant number for stability Eq. (8.16a) is used and substituted into Eq. (8.17), one can solve for the local Peclet (π) and number of grids (N) required to incur a specified amount of numerical dispersion,

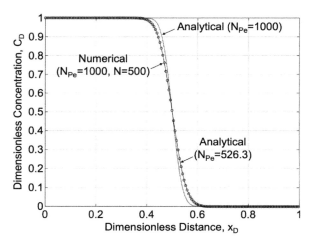

FIGURE 8.6 Dimensionless concentration versus dimensionless distance at $t_D = 0.5$. An explicit, upwinded scheme was employed and a Courant number, $\zeta = 0.1$, $N_{Pe} = 1000$, and $N = 500$ grids for the numerical simulations. Numerical dispersion causes a mismatch with the analytical solution at $N_{Pe} = 1000$ but gives an excellent match to an effective $N_{Pe} = 526.3$. *Analytical solution taken from Brenner (1962).*

$$\pi = \frac{N_{Pe}}{N} < \frac{2\left(D_{\text{num}}/D\right)}{1 - \left(D_{\text{num}}/D\right)}. \tag{8.19}$$

Eq. (8.19) shows that an infinite number of grids (or $\pi = 0$) is required to eliminate numerical dispersion. The number of grids in one direction required to achieve a numerical dispersion that is less than 10% of the physical dispersion is $N > \sim 5N_{Pe}$. Most reservoir applications are advection dominated and have high Peclet numbers ($>10^3$). This implies that at least 5000 grids are needed in each direction for D_{num}/D to be small. In three dimensions this would be 5000^3, over 100 billion, grids which is not computationally feasible. Table 8.1 summarizes the numerical dispersion and stability requirements for various methods for steady-state flow in 1D with uniform grids. Example 8.4 demonstrates the calculation of effective dispersion coefficients for different numerical schemes.

In many applications, we are forced to accept the numerical dispersion incurred. However, techniques for minimization of numerical dispersion have been developed. *Adaptive meshing* techniques (Brandt, 1977; Schmidt and Jacobs, 1988) are used and involve using finer gridding near the velocity front (and coarser grids far away from the front), thus making more efficient use of the limited number of grids. Adaptive meshing schemes require remeshing the

TABLE 8.1 Comparison of numerical schemes for the 1D, homogeneous advection-dispersion equation.

Method	Stability	D_{num}/D	Truncation error[a]	Oscillations
Explicit, upwind	$\zeta < \frac{\pi}{\pi+2}$	$\frac{\pi}{2}(1-\zeta)$	$O(\Delta x)$	No
Explicit, center	$\frac{\zeta^2}{2} < \eta < \frac{1}{2}$	$-\frac{\pi}{2}\zeta$	$O(\Delta x^2)$	Yes
Implicit, upwind	Unconditional	$\frac{\pi}{2}(1+\zeta)$	$O(\Delta x)$	No
Implicit, center	Unconditional	$\frac{\pi}{2}\zeta$	$O(\Delta x^2)$	Yes

[a]Truncation error for spatial terms refers to advection; dispersive term is $O(\Delta x^2)$ for all methods. The temporal truncation error is Δt for explicit and implicit methods and Δt^2 for Crank–Nicolson.

domain periodically as the front moves. Example 8.4. demonstrates the calculation of effective dispersion coefficients for different numerical schemes.

Example 8.4 Numerical dispersion.
Calculate the effective Peclet number for advection-dispersion in a 1D homogenous porous medium if the actual/physical Peclet number is 1000, $N = 500$, and $\zeta = 0.1$. Perform the calculations for the following cases,
    a. Explicit, upwinded
    b. Explicit, centered difference
    c. Implicit, upwinded
    d. Implicit, centered difference

Solution

$$\pi = \frac{N_{Pe}}{N} = \frac{1000}{500} = 2$$

    a. Explicit, upwinded

$$\frac{D_{num}}{D} = \frac{\pi}{2}(1-\zeta) = \frac{2}{2}(1-0.1) = 0.9$$

$$N_{Pe,eff} = \left[\frac{1}{N_{Pe}}\left(1 + \frac{D_{num}}{D}\right)\right]^{-1} = \left[\frac{1}{1000}(1+0.9)\right]^{-1} = 526.3$$

    b. Explicit, centered difference

$$\frac{D_{num}}{D} = -\frac{\pi}{2}\zeta = -\frac{2}{2}\cdot 0.1 = -0.1$$

$$N_{Pe,eff} = \left[\frac{1}{N_{Pe}}\left(1 + \frac{D_{num}}{D}\right)\right]^{-1} = \left[\frac{1}{1000}(1 - 0.1)\right]^{-1} = 1111.1$$

c. Implicit, upwinded

$$\frac{D_{num}}{D} = \frac{\pi}{2}(1 + \zeta) = \frac{2}{2}(1 + 0.1) = 1.1$$

$$N_{Pe,eff} = \left[\frac{1}{N_{Pe}}\left(1 + \frac{D_{num}}{D}\right)\right]^{-1} = \left[\frac{1}{1000}(1 + 1.1)\right]^{-1} = 476.2$$

d. Implicit, centered difference

$$\frac{D_{num}}{D} = \frac{\pi}{2}\zeta = \frac{2}{2}\cdot 0.1 = 0.1$$

$$N_{Pe,eff} = \left[\frac{1}{N_{Pe}}\left(1 + \frac{D_{num}}{D}\right)\right]^{-1} = \left[\frac{1}{1000}(1 + 0.1)\right]^{-1} = 909.1$$

8.8 Channeling and viscous fingering

Channeling occurs when a phase or component is transported more quickly through a high-permeability or low-resistance pathway. Channeling is caused by strong heterogeneities in the reservoir and is exacerbated if the displacing fluid has a lower viscosity than the displaced fluid. Flow through fractures is an example of channeling but it can occur in unfractured, heterogeneous media.

Saffman–Taylor instability (1958) or *viscous fingering* is the propagation of a less viscous fluid in a 2D or 3D porous medium caused by perturbations in the displacing front (Fig. 8.7). Viscous fingering occurs even in homogeneous porous media and is often studied in flow in between flat plates, i.e., Hele-Shaw cells (Homsy, 1987). Once a perturbation is initiated, a path of least resistance (lower viscosity) is formed which then propagates. Viscous fingering occurs when the mobility ratio (*M*) is greater than 1.0. Viscous fluids, such as water-based polymers as discussed in Chapter 6, are often injected in enhanced oil recovery to reduce the mobility ratio and thus mitigate viscous fingering. Polymers can also reduce channeling because the viscous fluid increases the resistance in the high-permeability zones. Polymer flooding is an

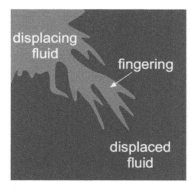 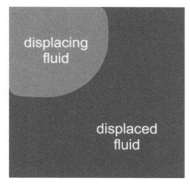

FIGURE 8.7 Qualitative pictures of a miscible displacement in which (A) viscous fingers form due to the high mobility ratio between the displacing and displaced fluid and (B) stable front by a low mobility ratio.

effective EOR technique but comes at higher costs than waterflooding. Injection rates can also be limited because the high viscosity results in higher well pressures.

8.9 Multicomponents, multidimensions, and additional forms

The equations developed in this chapter for compositional transport apply for any number of components; N_c, and $N_c - 1$ component equations are needed along with the pressure equation (alternatively, formulations involving N_c component equations without the pressure equation can be used). IMPEC is usually preferred over IMPIC or fully implicit for computational reasons, particularly, if the number of components is large. Coupling between component equations is often small or nonexistent which further motivates the use of an IMPEC scheme. However, for some applications, such as reactive transport, compositional equations can be strongly coupled which require the use of small timesteps or a fully implicit scheme. Although this chapter was mostly devoted to 1D flow, the matrix equations are applicable to 2D or 3D flow. The dispersion matrix, η, is pentadiagonal in 2D and heptadiagonal in 3D. For advection terms for an upwinded scheme, ζ_{up} is structured based on the upwinded velocities.

The equations were written in terms of concentrations (mass κ/pore volume), as was most of this chapter. Concentrations can be converted to mass fractions, ω_κ, or mole fractions, x_κ, using Eq. (1.3) (Chapter 1). In Chapter 10, multiphase, compositional models are presented in terms of mole fractions. The equations were presented using dimensionless coefficients, η, ζ, π, τ. However, the equations can also be written using dimensional coefficients (as was done with flow), which may have more physical meaning, by multiplying through by $A_i = V_t \varphi_i / \Delta t$. Furthermore, the upwinded concentrations can be lumped in with the transmissibility matrix, to create a new matrix $T_{\kappa, up}$, which leads to the matrix equations,

$$C_\kappa^{n+1} = \left\{ \mathbf{I} + \mathbf{A}^{-1} \left[\mathbf{U} + \vec{Q}_{prod} \right] \right\} C_\kappa^n + \mathbf{A}^{-1} \mathbf{T}_{\kappa,\mathbf{up}} \vec{P}^{n+1} + \mathbf{A}^{-1} \vec{Q}_{inj} \vec{C}_{\kappa,inj},$$

(8.20)

where $\mathbf{I}$ is the identity matrix, $\mathbf{A}$ is the accumulation matrix as defined in previous chapters, and Q_{inj} and Q_{prod} are the well rates, injection and production, respectively, in reservoir conditions (both constant rate and BHP wells allowed). $\mathbf{U}$ is a tridiagonal (in 1D) dispersivity matrix and $\mathbf{T}_{k,up}$ is the upwinded transmissibility,

$$U_{\kappa,i\pm\frac{1}{2}} = \left(\frac{\varphi D_\kappa A}{\Delta x} \right)_{i\pm\frac{1}{2}} ; T_{\kappa,up,i\pm\frac{1}{2}} = \left(\frac{kA}{\mu\Delta x} C_{\kappa,up} \right)_{i\pm\frac{1}{2}}.$$

(8.21)

The advantage of Eq. (8.20) is that the matrix equations can be solved without directly computing the velocity field first; however, the upwinded concentrations need to be included in the $T_{\kappa,up}$ matrix and cannot be extended easily to an implicit formulation. $T_{\kappa,up}$ is analogous to the phase transmissibilities which are used for multiphase flow in Chapter 9.

8.10 Pseudocode for component transport

The pseudocode for 1D IMPEC and IMPIC with no flux boundary conditions and wells is relatively straightforward for single-phase, single component component transport. First subroutines are created to compute interblock dimensionless dispersion coefficients (Fourier number) and Courant numbers. In another subroutine, *Transport Arrays*, **η**, **ζ**, **τ**, are developed by looping through all grids with special consideration at the boundaries and blocks with wells. *If* statements are used to determine the flow direction.

The main code involves marching through time and computing the (1) grid and well arrays, (2) solving the pressure equation implicitly, (3) looping through all components to solve component concentrations (or mass/mole fractions) explicitly or implicitly, and (4) updating the timestep. If flow is steady state, the pressure field may be given or can be solved outside the loop. Preprocessing (setting up the problem and defining input variables) and postprocessing (plotting and analyzing the results) subroutines can be developed similarly to previous chapters with some adjustments for the problem of compositional transport. The pseudocode can be generalized to multi-dimensions, multicomponents, and Dirichlet boundary conditions.

```
MAIN CODE IMPEC/IMPIC
CALL PREPROCESS
WHILE t < t_final
        CALL GRID ARRAYS
        CALL WELLS
        CALCULATE pressure implicitly using eqn 4.13b
        CALCULATE velocity field using Darcy's law, eqn 8.3
        CALCULATE dispersion field using eqn 7.7 and 7.8

        CALL TRANSPORT ARRAYS
        CALCULATE composition explicitly or implicitly (eqn. 8.12, 8.13)

        INCREMENT t = t + Δt
ENDWHILE
CALL POSTPROCESS
```

```
SUBROUTINE INTERBLOCK COURANT
Inputs: i, j, Δt, velocity field, reservoir properties
Outputs: Courant number (positive, scalar)

CALCULATE dimensionless Courant number b/w blocks i and j using EQN 8.5c
```

```
SUBROUTINE INTERBLOCK FOURIER
Inputs: i, j, Δt, dispersion field, reservoir properties
Outputs: Fourier number (positive, scalar)

CALCULATE dimensionless Fourier number b/w blocks i and j using EQN 8.5a
```

```
SUBROUTINE TRANSPORT ARRAYS
Inputs: velocity/dispersion fields, reservoir, fluid, well, numerical prop
Outputs: η, ζ, τ (NxN matrices)

FOR i = 1 to N (number of grids)
      IF block i contains an injector well
            SET τinj(i,i) using EQN 8.5d
      ELSEIF block i contains a producer well
            SET τp(i,i) using EQN 8.5d
      ENDIF

      IF i ≠ 1
            SET η(i,i-1)=-INTERBLOCK FOURIER(i,i-1)
            SET η(i,i)= η(i,i)- η(i,i-1)

            IF Φ(i-1)> Φ(i)
                  IF method is upwind
                        SET ζ(i,i-1)= ζ(i,i-1)-INTERBLOCK COURANT(i,i-1)
                        SET ζ(i,i)= ζ(i,i)
                  ELSEIF method is centered
                        SET ζ(i,i)=ζ(i,i)-0.5*INTERBLOCK COURANT(i,i-1)
                        SET ζ(i,i-1)=ζ(i,i-1)-0.5*INTERBLOCK COURANT(i,i-1)
                  END
            ELSE
                  IF method is upwind
                        SET ζ(i,i)= ζ(i,i)+INTERBLOCK COURANT(i,i-1)
                        SET ζ(i,i-1)=ζ(i,i-1)
                  ELSEIF method is centered
                        SET ζ(i,i)=ζ(i,i)+0.5*INTERBLOCK COURANT(i,i-1)
                        SET ζ(i,i-1)=ζ(i,i-1)+0.5*INTERBLOCK COURANT(i,i-1)
                  ENDIF
            ENDIF
      ENDIF

      IF i ≠ N
            SET η(i,i+1)=-INTERBLOCK FOURIER (i,i+1)
            SET η(i,i)= η(i,i)- η(i,i+1)

            IF Φ(i)> Φ(i+1)
                  IF method is upwind
                        SET ζ(i,i)= ζ(i,i)+INTERBLOCK COURANT(i,i+1)
                        SET ζ(i,i+1) =ζ(i,i+1)
                  ELSEIF method is centered
                        SET ζ(i,i)= ζ(i,i)+0.5*INTERBLOCK COURANT(i,i+1)
                        SET ζ(i,i+1)=ζ(i,i+1)+0.5*INTERBLOCK COURANT(i,i+1)
                  ENDIF
            ELSE
                  IF method is upwind
                        SET ζ(i,i+1)= ζ(i,i+1)-INTERBLOCK COURANT(i,i+1)
                        SET ζ(i,i)=ζ(i,i)
                  ELSEIF method = centered
                        SET ζ(i,i)= ζ(i,i)-0.5*INTERBLOCK COURANT(i,i+1)
                        SET ζ(i,i+1)=ζ(i,i+1)-0.5*INTERBLOCK COURANT(i,i+1)
                  ENDIF
            ENDIF
      ENDIF
ENDFOR
```

8.11 Exercises

Exercise 8.1. 2-point upwinding. Derive the discretized mass balance equations, analagous to Eq. (8.9), using the 2-point upwinding scheme proposed by Todd et al. (1972),

$$
C_{i+\frac{1}{2}} = \begin{cases} \frac{3}{2}C_i - \frac{1}{2}C_{i-1} & \text{if } \Phi_i > \Phi_{i+1} \\ \frac{3}{2}C_{i+1} - \frac{1}{2}C_{i+2} & \text{if } \Phi_i < \Phi_{i+1} \end{cases}
$$

Exercise 8.2. Numerical solution to ADE. Repeat Examples 8.1, 8.2, and 8.3 but the wells at $x = 0$ and $x = L$ are producers ($Q = 1000$ scf/day each) and the well at $x = L/2$ is an injector ($Q = 2000$ scf/day).

Exercise 8.3. Numerical solution to ADE with Dirichlet boundary condition. Repeat Examples 8.1, 8.2, and 8.3 but use a constant concentration boundary condition (1.0 lb$_m$/ft^3) at the injection wells.

Exercise 8.4. Numerical solution to ADE with 2-point upwining. Repeat Examples 8.1, 8.2, and 8.3 but use the 2-point upwinding scheme described in Exercise 8.1.

Exercise 8.5. Analytical solution for tracer pulse. Write a subroutine/function to calculate concentration versus space and time for the case of a pulse in which tracer is injected for a finite period time, t_{Ds}, using the analytical solution (valid for high Peclet numbers),

$$
c_D = \frac{1}{2}\left[\text{erf}\left\{ \frac{x_D - (t_D - t_{Ds})}{2\sqrt{\frac{(t_D - t_{Ds})}{N_{Pe}}}} \right\} - \text{erf}\left\{ \frac{x_D - t_D}{2\sqrt{\frac{t_D}{N_{Pe}}}} \right\} \right]
$$

Make plots of dimensionless concentration versus distance at $t_D = 0.25$, 0.5, and 0.75 and dimensionless concentration versus time at the exit, $x_D = 1$ for $N_{Pe} = 1000$.

Exercise 8.6. Component reservoir simulator. Develop a 1D reservoir simulator to solve single-phase, multicomponent transport in multidimensions using upwinding and allow for explicit and implicit method options. Allow for unsteady-state flow by calculating the pressure and velocity fields in each timestep. For additional flexibility, allow for the use of centered differences in addition to upwinding. Test your code against Example 8.3 and/or Exercise 8.2 and then increase the number of blocks to $N = 101$ and make plots of the concentration field at various times. For additional complexity, extend to 2D and 3D.

Exercise 8.7. Simulator validation against analytical solution. Using the simulator developed in Exercise 8.6, solve the concentration field for a 1D reservoir at steady-state flow if a tracer is injected at constant concentration for $t_{Ds} = 0.1$ pore volumes at $N_{Pe} = 1000$. After 0.1 pore-volumes, the injected concentration is 0. At the exit ($x_D = 1$), component is removed via a producer well at the same rate it is injected. Compare the numerical results to the analytical solution in Exercise 8.5.

Exercise 8.8. Component transport project. Use your reservoir simulator developed in Exercise 8.6 to solve for pressure and concentration in a reservoir with 2D, heterogeneous permeability and porosity of the Thomas reservoir first introduced in Exercise 1.8. The reservoir is saturated with water and an injector well is placed at ($x = 0$, $y = 0$) with injection rate 1000 STB/day and producer well at ($x = L$, $y = w$) with constant BHP 1000 psia. The initial pressure is 3000 psia.

First, determine the pressure and velocity field at steady state. Then, at steady state, tracer (1.0 lb_m/ft^3) is injected for 0.01 pore volumes. Make plots of the steady-state pressure field and concentration field at $t_D = 0.25, 0.5, 0.75,$ and 1.0. Also make plots of the producer concentration versus time and identify the breakthrough time.

References

Ataie-Ashtiani, B., Lockington, D.A., Volker, R.E., 1999. Truncation errors in finite difference models for solute transport equation with first-order reaction. Journal of Contaminant Hydrology 35 (4), 409–428.

Brandt, A., 1977. Multi-level adaptive solutions to boundary-value problems. Mathematics of Computation 31, 333–390.

Brenner, H., 1962. The diffusion model of longitudinal mixing in beds of finite length. Numerical Values. Chemical Engineering Science 17 (4), 229–243.

Hogarth, W.L., et al., 1990. A comparative study of finite difference methods for solving the one-dimensional transport equation with an initial-boundary value discontinuity. Computers and Mathematics with Applications 20 (11), 67–82.

Homsy, G.M., 1987-01-01. Viscous fingering in porous media. Annual Review of Fluid Mechanics 19 (1), 271–311. https://doi.org/10.1146/annurev.fl.19.010187.001415. Bibcode:1987AnRFM..19..271H. ISSN 0066-4189.

Jianchun, L., Pope, G.A., Sepehrnoori, K., 1995. A high-resolution finite-difference scheme for nonuniform grids. Applied Mathematical Modelling 19 (3), 162–172.

Marle, C., 1981. Multiphase flow in porous media. Éditions Technip.

Saad, N., Pope, G.A., Kamy, S., 1990. Application of higher-order methods in compositional simulation. SPE Reservoir Evaluation and Engineering 5, 623–630. https://doi.org/10.2118/18585-PA.

Saffman, P.G., Taylor, G.I., 1958. The penetration of a fluid into a porous medium or Hele-Shaw cell containing a more viscous liquid. Proceedings of the Royal Society of London. Series A. Mathematical and Physical Sciences 245 (1242), 312–329.

Schmidt, G.H., Jacobs, F.J., 1988. Adaptive local grid refinement and multi-grid in numerical reservoir simulation. Journal of Computational Physics 77 (1), 140–165.

Todd, M.R., O'Dell, P.M., Hirasaki, G.J., 1972. Methods for increased accuracy in numerical reservoir simulators. SPE Journal 12, 515–530. https://doi.org/10.2118/3516-PA.

Chapter 9

Numerical solution to the black oil model

9.1 Introduction

Many applications of flow and transport in subsurface media involve multiple flowing phases. For example, during secondary recovery, water or other fluids may be injected into the reservoir to increase the pressure and displace the oil. Thus at least two phases, aqueous and oleic, will be flowing. If the reservoir pressure decreases below the bubble point, a gaseous phase will also be present. Modeling several flowing phases requires the solution of coupled, nonlinear partial differential equations. These equations can rarely be solved analytically and advanced numerical techniques must be used.

In Chapter 10, we present the numerical approach to solve fully compositional flow for multiple phases. Here, in Chapter 9, we use a simplified compositional model: the black oil model, which assumes three pseudo-components (water, oil, and gas) and up to three phases (aqueous, oleic, and gaseous). Gas is present in the gaseous phase and can also be dissolved in the oleic phase with pressure-dependent composition, the solution gas ratio (R_s). Oil is only present in the oleic phase[1] and water only in the aqueous phase. The component balances can be combined to create a *pressure equation* and then coupled with N_c-1 component balances. Implicit in both pressure and saturation as well as implicit pressure and explicit saturation (IMPES) methods can be used to numerically solve the system.

9.2 The black oil model

The PDEs for three pseudocomponents (water, oil, and gas) and three phases (aqueous, oleic, and gaseous) flow were derived in Chapter 7 by performing component balances,

---

1. Some models use a volatilized oil ratio, R_v, to account for oil in the gaseous phase. However, we ignore R_v in this text.

An Introduction to Multiphase, Multicomponent Reservoir Simulation.
https://doi.org/10.1016/B978-0-323-99235-0.00004-X

$$\frac{\partial}{\partial t}\left(\frac{S_w\phi}{B_w}\right) = \nabla\cdot\left[\frac{\lambda_w}{B_w}\left(\nabla p_w - \rho_w g\nabla D\right)\right] + \widetilde{q}_w^{sc}$$

$$\frac{\partial}{\partial t}\left(\frac{S_o\phi}{B_o}\right) = \nabla\cdot\left[\frac{\lambda_o}{B_o}\left(\nabla p_o - \rho_o g\nabla D\right)\right] + \widetilde{q}_o^{sc}$$

$$\frac{\partial}{\partial t}\left[\phi\left(\frac{S_g}{B_g} + R_s\frac{S_o}{B_o}\right)\right] = \nabla\cdot\left[\frac{\lambda_g}{B_g}\left(\nabla p_g - \rho_o g\nabla D\right) + R_s\frac{\lambda_o}{B_o}\left(\nabla p_o - \rho_o g\nabla D\right)\right]$$

$$+\widetilde{q}_g^{sc} + R_s\widetilde{q}_o^{sc}, \tag{9.1a}$$

where the relative phase mobilities are defined as, $\lambda_\alpha = \frac{kk_{r,\alpha}}{\mu_\alpha}$.

In addition to the PDEs in Eq. (9.1a), auxiliary relationships can be written. First, the saturations of all phases must sum to unity,

$$S_w + S_o + S_g = 1 \tag{9.1b}$$

Capillary pressure and relative permeability are given as function of saturation,

$$p_{c,ow} = p_o - p_w = f(S_w); \; p_{c,og} = p_g - p_o = f\left(S_g\right)$$
$$k_{rw} = f(S_w); k_{ro} = f\left(S_w, S_g\right); k_{rg} = f\left(S_g\right) \tag{9.1c}$$

The saturation-dependent functions can be tabulated (and interpolation used) or the data fit to an empirical model such as those described in Chapter 1.

The three PDEs Eq. (9.1a) and six auxiliary Eqs. (9.1b and 9.1c) result in nine total equations and nine total unknowns (p_o, p_w, p_g, S_w, S_o, S_g, k_{rw}, k_{ro}, k_{rg}). Therefore, the system is well described when boundary and initial conditions are applied. However, the equations are also very nonlinear since many co-efficients are functions of unknown pressure and saturation.

The component balances in Eq. (9.1a) can be recombined to give a single overall balance. The result is equivalent to the combination of phase balance equations to obtain the diffusivity/pressure Eq. (2.10).

$$\phi c_t\frac{\partial p}{\partial t} = \sum_{\alpha=1}^{N_p}\frac{1}{\rho_\alpha}\nabla\cdot\left(\rho_\alpha\frac{k_{rl}\mathbf{k}}{\mu_\alpha}\nabla\Phi_\alpha\right) + \sum_{\alpha=1}^{N_p}\widetilde{q}_\alpha \tag{9.2a}$$

Or equivalently,

$$\phi c_t\frac{\partial p_o}{\partial t} = \sum_{\alpha=1}^{N_p}B_\alpha\nabla\cdot\left[\frac{\lambda_\alpha}{B_\alpha}\left(\nabla p_\alpha - \rho_\alpha g\nabla D\right)\right] - \sum_{\alpha=1}^{N_p}B_\alpha\widetilde{q}_\alpha^{sc} \tag{9.2b}$$

where N_p is the number of phases.

In Eq. (9.2b), the pressure is written in terms of oleic phase pressure, p_o, in the accumulation term, which is arbitrary but the usual convention if an oleic phase is present. Phase pressures in the advective transport terms should be converted to oleic phase pressure using capillary pressure, Eq. (9.1c). The equation is conveniently absent of phase saturations directly and thus often referred to as the *pressure equation*. However, saturations are included indirectly in the relative permeabilities, capillary pressures, and total compressibility, c_t. Any three of the four PDEs, Eqs. (9.1a) and (9.2b), can be used to solve for pressures and saturations in the reservoir with time. A common approach is to use the pressure Eq. (9.2b) with N_c-1 component balances Eq. (9.1a).

Eqs. (9.1) and (9.2) can be simplified for some common applications. For an undersaturated reservoir, there is no gaseous phase ($S_g = 0$) and the gas component Eq. (9.1a) is not unique and not needed. There is also no capillary pressure ($p_{c,og}$) between the oleic and gaseous phases, and no relative permeability (k_{rg}) of gaseous phase for this two-phase system. Another common simplification is that capillary pressure can be neglected. In many reservoirs, capillary pressure is small, $O(10^0$ psia) compared to the reservoir pressure $O(10^3$ psia), and does not significantly impact the solution. However, capillary pressure can be significant in reservoirs with low permeability/small pores (such as unconventional shale reservoirs). Importantly, even if the imbibition capillary pressure curve can be neglected during dynamic flow calculations, the reservoir saturations must still be initialized using the drainage capillary pressure curve as described in Chapter 1.

9.3 Finite difference equations for multiphase flow

Discretization of the reservoir results in algebraic conservation equations for each grid, replacing the complicated PDEs. The conservation equations can be derived either by applying finite differences to the PDEs (Eqs. 9.1 and 9.2) as was done in Chapter 3 or applying mass balances directly to the grids as was done in Chapter 4. Regardless of the approach used, Eqs. (9.6a)−(9.6d) are the finite difference equations for three pseudocomponents (water, oil, and gas) and the pressure equation, respectively,

$$\beta_w \Delta T_w \left(\Delta P_o^{n+1} - \Delta P_{c,ow} - \rho_w g \Delta D \right) = C_1 \Delta_t P_o + A \Delta_t S_w - \beta_w Q_w^{sc} \quad (9.6a)$$

$$\beta_o \Delta T_o \left(\Delta P_o^{n+1} - \rho_o g \Delta D \right) = C_2 \Delta_t P_o + A \Delta_t S_o - \beta_o Q_o^{sc} \quad (9.6b)$$

$$\beta_g \Delta T_g \left(\Delta P_o^{n+1} - \Delta P_{c,og} - \rho_g g \Delta D \right) + \beta_g \Delta R_s T_o \left(\Delta P_o^{n+1} - \rho_o g \Delta D \right)$$

$$= C_3 \Delta_t P_o + \beta_g A R_s \Delta_t S_o + A \Delta_t S_g - \beta_g \left(R_s Q_o^{sc} + Q_g^{sc} \right) \quad (9.6c)$$

$$\left[\beta_w \Delta T_w + \beta_o \Delta T_o + \beta_g \Delta T_g\right] \Delta P_o^{n+1} - \beta_w \Delta T_w \Delta P_{cow}^n + \beta_g \Delta T_g \Delta P_{cog}^n$$

$$- \left[\beta_w \rho_w g \Delta T_w + \beta_o \rho_o g \Delta T_o + \beta_g \rho_g g \Delta T_g\right] \Delta D = (Ac_t)\Delta_t P_o$$

$$- \left[\beta_w Q_w^{sc} + \beta_o Q_o^{sc} + \beta_g Q_g^{sc}\right] \tag{9.6d}$$

where $A_i = V_i/(\phi \Delta t)$ and the C coefficients for each block, i, on the accumulation terms are defined as,

$$C_{1,i} = \frac{\beta_{w,i}}{B_{w,i}} A_i S_{w,i}\left(c_{w,i} + c_{f,i}\right)$$

$$C_{2,i} = A_i S_{o,i}\left(c_{o,i}^* + c_{f,i}\right) \tag{9.7}$$

$$C_{3,i} = \frac{\beta_{g,i}}{B_{g,i}} A_i \left[S_{g,i}\left(c_{g,i} + c_{f,i}\right) + S_{o,i} R_{s,i}\left(c_{o,i} + c_{f,i}\right)\right]$$

The variables β_α are effectively formation volume factors with corrections for the derivative of capillary pressure with saturation,

$$\beta_{w,i} = \frac{B_{w,i}}{1 - S_{w,i} c_{w,i} P'_{cow,i}}; \quad \beta_{o,i} = B_{o,i}; \quad \beta_{g,i} = \frac{B_{g,i}}{1 - S_{g,i} c_{g,i} P'_{cow,i}} \tag{9.8}$$

If capillary pressure can be neglected, so can their derivatives with respect to saturation and β_α reduces to the formation volume factor, B_α. Even when capillary pressure is significant, derivatives are usually small except near residual saturations where both capillary pressure and its derivative are large and tend to infinity and/or negative infinity.

A new variable, c_o^*, is defined as

$$c_o^* = -\frac{1}{B_o} \frac{\partial B_o}{\partial p} = c_o - \frac{B_g}{B_o} \frac{\partial R_s}{\partial p} \tag{9.9}$$

and is only equal to the oleic phase compressibility, c_o, above the bubble point (i.e., no gaseous phase). The variable does not have physical significance below the bubble point but is defined here for convenience. The total compressibility, c_t, in Eq. (9.6d) is a weighted sum of phase and formation compressibilities,

$$c_{t,i} = \frac{\beta_{w,i}}{B_{w,i}} S_{w,i}\left(c_{w,i} + c_{f,i}\right) + S_{o,i}\left(c_{o,i} + c_{f,i}\right) + \frac{\beta_{g,i}}{B_{g,i}} S_{g,i}\left(c_{g,i} + c_{f,i}\right) \tag{9.10}$$

For the case where the derivative of capillary pressure can be neglected ($\beta_\alpha = B_\alpha$), the equation reduces the simpler definition originally presented in Eq. (1.18). The shorthand notation in Eq. (9.6) for differences was first introduced in Chapter 4 (Eq. 4.11),

$$\Delta_t P_o = P_{o,i}^{n+1} - P_{o,i}^n; \quad \Delta_t S_\alpha = S_{\alpha,i}^{n+1} - S_{\alpha,i}^n;$$

$$\Delta T_\alpha \Delta P_o = T_{\alpha,j-\frac{1}{2},k,l}\left(P_{o,j-1,k,l} - P_{o,j,k,l}\right) + T_{\alpha,j+\frac{1}{2},k,l}\left(P_{o,j+1,k,l} - P_{o,j,k,l}\right)$$

$$+T_{\alpha,j,k-\frac{1}{2},l}\left(P_{o,j,k-1,l} - P_{o,j,k,l}\right) + T_{\alpha,j,k+\frac{1}{2},l}\left(P_{o,j,k+1,l} - P_{o,j,k,l}\right)$$

$$+T_{\alpha,j,k,l-\frac{1}{2}}\left(P_{o,j,k,l-1} - P_{o,j,k,l}\right) + T_{\alpha,j,k,l+\frac{1}{2}}\left(P_{o,j,k,l+1} - P_{o,j,k,l}\right) \quad (9.11)$$

Eqs. (9.6a)−(9.6d) are the water, oil, and gas balances and pressure equation, respectively, for a grid block. For three-phase flow, only three of the four equations are needed (and for two-phase flow only two are needed) because the last equation is a combination of the others. Thus, for three-phase flow there are $3N$ equations that must be solved for N grid blocks in the finite difference model. The equations are coupled because they all include unknown pressure and saturation, although saturation is only included indirectly in Eq. (9.6d). The equations are very nonlinear because relative permeability and capillary pressure are nonlinear function of saturation. Example 9.1 demonstrates the appropriate balance equations for a two-phase (oleic and aqueous), 4-block reservoir model without capillary pressure and gravity.

Example 9.1. Finite difference equations.
Write the finite difference equations for oil, water, and overall for a 1D, two-phase (oleic and aqueous) four grid system, by expanding and simplifying Eq. (9.6). The boundaries are sealed to flow; there is a constant rate water injector well in block #1 and producer in block #4. Assume that gravity and capillary pressure can be neglected.

Solution
Block #1
    Expansion of the shorthand notation in only one dimension without gravity or capillary pressure results in the following balances for overall, water, and oil. Note that phase pressures are equal, therefore $P_o = P_w = P$.

$$\cancel{\left(B_{w,1}T_{w,\frac{1}{2}} + B_{o,1}T_{o,\frac{1}{2}}\right)\left(P_0^{n+1} - P_1^n\right)} + \left(B_{w,1}T_{w,\frac{3}{2}} + B_{o,1}T_{o,\frac{3}{2}}\right)\left(P_2^{n+1} - P_1^n\right) = A_1 c_{t,1}\left(P_1^{n+1} - P_1^n\right) - B_{w,1}Q_{w,1}^{sc} - \cancel{B_{o,1}Q_{o,}^{sc}}$$

$$= A_1 c_{t,1}\left(P_1^{n+1} - P_1^n\right) - B_{w,1}Q_{w,1}^{sc} - \cancel{B_{o,1}Q_{o,1}^{sc}}$$

$$\cancel{B_{o,1}T_{o,\frac{1}{2}}\left(P_0^{n+1} - P_1^{n+1}\right)} + B_{o,1}T_{o,\frac{3}{2}}\left(P_2^{n+1} - P_1^{n+1}\right) = A_1\left[S_{o,1}\left(c_{o,1} + c_{f,1}\right)\left(P_1^{n+1} - P_1^n\right) + \left(S_{o,1}^{n+1} - S_{o,1}^n\right)\right] - \cancel{B_{o,1}Q_{o,1}^{sc}}$$

$$\cancel{B_{w,1}T_{w,\frac{1}{2}}\left(P_0^{n+1} - P_1^{n+1}\right)} + B_{w,1}T_{w,\frac{3}{2}}\left(P_2^{n+1} - P_1^{n+1}\right) = A_1\left[S_{w,1}\left(c_{w,1} + c_{f,1}\right)\left(P_1^{n+1} - P_1^n\right) + \left(S_{w,1}^{n+1} - S_{w,1}^n\right)\right] - B_{w,1}Q_{w,1}^{sc}$$

    There is no flow at the boundary and no injection of oil (only water) into the well in block #1. The water equation can be rearranged to express water saturation explicitly,

Continued

Example 9.1. Finite difference equations.—cont'd

$$S_{w,1}^{n+1} = S_{w,1}^{n} + \frac{1}{A_1}\left[B_{w,1}T_{w,\frac{3}{2}}\left(P_2^{n+1} - P_1^{n+1}\right) + B_{w,1}Q_{w,1}^{sc}\right]$$

$$-S_{w,1}^{n}\left(c_{w,1} + c_{f,1}\right)\left(P_1^{n+1} - P_1^{n}\right)$$

The new water saturation is equal to the old water saturation plus the change due to flow of water in/out of the block plus a change due to water injection. The compressibility terms are usually negligible in the saturation equation.

Block #2

$$\left(B_{w,2}T_{w,\frac{3}{2}} + B_{o,2}T_{o,\frac{3}{2}}\right)\left(P_1^{n+1} - P_2^{n}\right) + \left(B_{w,2}T_{w,\frac{5}{2}} + B_{o,2}T_{o,\frac{5}{2}}\right)\left(P_3^{n+1} - P_2^{n}\right) = A_2 c_{t,2}\left(P_2^{n+1} - P_2^{n}\right) - \cancel{B_{w,2}Q_{w,2}^{sc}} - \cancel{B_{o,2}Q_{o,2}^{sc}}$$

$$B_{o,2}T_{o,\frac{3}{2}}\left(P_1^{n+1} - P_2^{n+1}\right) + B_{o,2}T_{o,\frac{5}{2}}\left(P_3^{n+1} - P_2^{n+1}\right) = A_2\left[S_{o,2}\left(c_{o,2} + c_{f,2}\right)\left(P_2^{n+1} - P_2^{n}\right) + \left(S_{o,2}^{n+1} - S_{o,2}^{n}\right)\right] - \cancel{B_{o,2}Q_{o,2}^{sc}}$$

$$B_{w,2}T_{w,\frac{3}{2}}\left(P_1^{n+1} - P_2^{n+1}\right) + B_{w,2}T_{w,\frac{5}{2}}\left(P_3^{n+1} - P_2^{n+1}\right) = A_2\left[S_{w,2}\left(c_{w,2} + c_{f,2}\right)\left(P_2^{n+1} - P_2^{n}\right) + \left(S_{w,2}^{n+1} - S_{w,2}^{n}\right)\right] - \cancel{B_{w,2}Q_{w,2}^{sc}}$$

There are no wells in block #2. The water equation can be rearranged to express water saturation explicitly,

$$S_{w,2}^{n+1} = S_{w,2}^{n} + \frac{1}{A_2}\left[B_{w,2}T_{w,\frac{3}{2}}\left(P_1^{n+1} - P_2^{n+1}\right) + B_{w,2}T_{w,\frac{5}{2}}\left(P_3^{n+1} - P_2^{n+1}\right) + \cancel{B_{w,2}Q_{w,2}^{sc}}\right] - S_{w,2}^{n}\left(c_{w,2} + c_{f,2}\right)\left(P_2^{n+1} - P_2^{n}\right)$$

Block #3

$$\left(B_{w,3}T_{w,\frac{5}{2}} + B_{o,3}T_{o,\frac{5}{2}}\right)\left(P_2^{n+1} - P_3^{n}\right) + \left(B_{w,3}T_{w,\frac{7}{2}} + B_{o,3}T_{o,\frac{7}{2}}\right)\left(P_4^{n+1} - P_3^{n}\right) = A_3 c_{t,3}\left(P_3^{n+1} - P_3^{n}\right) - \cancel{B_{w,3}Q_{w,3}^{sc}} - \cancel{B_{o,3}Q_{o,3}^{sc}}$$

$$B_{o,3}T_{o,\frac{5}{2}}\left(P_2^{n+1} - P_3^{n+1}\right) + B_{o,3}T_{o,\frac{7}{2}}\left(P_4^{n+1} - P_3^{n+1}\right) = A_3\left[S_{o,3}\left(c_{o,3} + c_{f,3}\right)\left(P_3^{n+1} - P_3^{n}\right) + \left(S_{o,3}^{n+1} - S_{o,3}^{n}\right)\right] - \cancel{B_{o,3}Q_{o,3}^{sc}}$$

$$B_{w,3}T_{w,\frac{5}{2}}\left(P_2^{n+1} - P_3^{n+1}\right) + B_{w,3}T_{w,\frac{5}{2}}\left(P_4^{n+1} - P_3^{n+1}\right) = A_3\left[S_{w,3}\left(c_{w,3} + c_{f,3}\right)\left(P_3^{n+1} - P_3^{n}\right) + \left(S_{w,3}^{n+1} - S_{w,3}^{n}\right)\right] - \cancel{B_{w,3}Q_{w,3}^{sc}}$$

Example 9.1. Finite difference equations.—cont'd

There are no wells in block #3. The water equation can be rearranged to express water saturation explicitly,

$$S_{w,3}^{n+1} = S_{w,3}^{n} + \frac{1}{A_3}\left[B_{w,3}T_{w,\frac{5}{2}}\left(P_2^{n+1}-P_3^{n+1}\right) + B_{w,3}T_{w,\frac{7}{2}}\left(P_4^{n+1}-P_3^{n+1}\right) + B_{w,3}Q_{w,3}^{sc}\right] - S_{w,3}^{n}\left(c_{w,3}+c_{f,3}\right)\left(P_3^{n+1}-P_3^{n}\right)$$

Block #4

$$\left(B_{w,4}T_{w,\frac{7}{2}}+B_{o,4}T_{o,\frac{7}{2}}\right)\left(P_3^{n+1}-P_4^{n}\right) + \left(B_{w,4}T_{w,\frac{9}{2}}+B_{o,4}T_{o,\frac{9}{2}}\right)\left(P_5^{n+1}-P_4^{n}\right) = A_4 c_{t,4}\left(P_4^{n+1}-P_4^{n}\right) - B_{w,4}Q_{w,4}^{sc} - B_{o,4}Q_{o,4}^{sc}$$

$$B_{o,4}T_{o,\frac{7}{2}}\left(P_3^{n+1}-P_4^{n+1}\right) + B_{o,4}T_{o,\frac{9}{2}}\left(P_5^{n+1}-P_4^{n+1}\right) = A_4\left[S_{o,4}\left(c_{o,4}+c_{f,4}\right)\left(P_4^{n+1}-P_4^{n}\right) + \left(S_{o,4}^{n+1}-S_{o,4}^{n}\right)\right] - B_{o,4}Q_{o,4}^{sc}$$

$$B_{w,4}T_{w,\frac{7}{2}}\left(P_3^{n+1}-P_4^{n+1}\right) + B_{w,4}T_{w,\frac{9}{2}}\left(P_5^{n+1}-P_4^{n+1}\right) = A_4\left[S_{w,4}\left(c_{w,4}+c_{f,4}\right)\left(P_4^{n+1}-P_4^{n}\right) + \left(S_{w,4}^{n+1}-S_{w,4}^{n}\right)\right] - B_{w,4}Q_{w,4}^{sc}$$

There is a producer well in block #4 which can produce both water and oil. The water equation can be rearranged to express water saturation explicitly,

$$S_{w,4}^{n+1} = S_{w,4}^{n} + \frac{1}{A_4}\left[B_{w,4}T_{w,\frac{7}{2}}\left(P_3^{n+1}-P_4^{n+1}\right) + B_{w,4}T_{w,\frac{9}{2}}\left(P_5^{n+1}-P_4^{n+1}\right) + B_{w,4}Q_{w,4}^{sc}\right]$$

$$- S_{w,4}^{n}\left(c_{w,4}+c_{f,4}\right)\left(P_4^{n+1}-P_4^{n}\right)$$

9.4 Solution methods

9.4.1 Implicit pressure, explicit saturation

A common approach for solving multiphase flow problems in numerical simulators is the IMPES method, which is an acronym for *implicit pressure, explicit saturation*. As the name implies, grid block pressures are first computed implicitly using the pressure Eq. (9.6d) and phase saturations can be computed explicitly by rearranging Eqs. (9.6a)–(9.6c) to solve for saturations, S_{α}^{n+1}. It is analogous to the IMPEC method introduced in Chapter 8. In matrix form, the implicit pressure equation can be expressed as,

$$\left(\mathbf{T} + \mathbf{J} + \mathbf{Ac}_t\right)\vec{P}_o^{\,n+1} = \mathbf{Ac}_t\,\vec{P}_o^{\,n} + \vec{Q} + \vec{G}, \tag{9.12a}$$

and the explicit saturation equations,

$$\vec{S}_w^{\,n+1} = \vec{S}_w^{\,n} + \mathbf{A}^{-1}\left[\underbrace{-\,\boldsymbol{\beta}_w\mathbf{T}_w^n\left(\vec{P}_o^{\,n+1} - \vec{P}_{c,ow}^{\,n} - \rho_w^n g\vec{D}\right)}_{\text{flow}} + \underbrace{\boldsymbol{\beta}_w\vec{Q}_w^{\,sc} + \mathbf{J}_w\left(P_{wf} - \vec{P}_o^{\,n+1}\right)}_{\text{wells}}\right]$$

$$\underbrace{-\mathbf{B}_w^{-1}\boldsymbol{\beta}_w\vec{S}_w^{\,n}\left(\mathbf{c}_w + \mathbf{c}_f\right)\left(\vec{P}_o^{\,n+1} - \vec{P}_o^{\,n}\right)}_{\text{compressibility}}$$

$$\vec{S}_o^{\,n+1} = S_o^n + \mathbf{A}^{-1}\left[-\,\boldsymbol{\beta}_o\mathbf{T}_o^n\left(\vec{P}_o^{\,n+1} - \rho_o^n g\vec{D}\right) + \boldsymbol{\beta}_o\vec{Q}_o^{\,sc} + \mathbf{J}_o\left(P_{wf} - \vec{P}_o^{\,n+1}\right)\right]$$

$$-S_o^n\left(c_o + \mathbf{c}_f\right)\left(\vec{P}_o^{\,n+1} - \vec{P}_o^{\,n}\right)$$

$$\vec{S}_g^{\,n+1} = 1 - \vec{S}_w^{\,n+1} - \vec{S}_o^{\,n+1} \tag{9.12b}$$

In the saturation Eq. (9.12b), the new saturation equals the old saturation plus any change due to flow in/out from adjacent grid blocks, injection/production from wells, and compressibility due to the change in pressure. The compressibility term in the saturation equations is often small and can be neglected but is included for completeness. The equations can be further simplified if capillary pressure and/or the derivative of capillary pressure can be neglected. For an undersaturated reservoir (i.e., reservoir pressure above the bubble point), gas saturation is zero and calculation of S_o using the second equation is not needed since $S_o = 1 - S_w$. However, the equation may be used as an additional check for accuracy.

The matrices and vectors in Eq. (9.12) have the form,

$$\mathbf{T} = \boldsymbol{\beta}_w\mathbf{T}_w + \boldsymbol{\beta}_o\mathbf{T}_o + \boldsymbol{\beta}_g\mathbf{T}_g; \quad \mathbf{J} = \mathbf{J}_w + \mathbf{J}_o + \mathbf{J}_g;$$

$$\mathbf{c}_t = \text{diag}\left[\mathbf{B}_w^{-1}\boldsymbol{\beta}_w\left(\mathbf{c}_w + \mathbf{c}_f\right)\vec{S}_w^{\,n} + \left(\mathbf{c}_o + \mathbf{c}_f\right)\vec{S}_o^{\,n} + \mathbf{B}_g^{-1}\boldsymbol{\beta}_g\left(\mathbf{c}_g + \mathbf{c}_f\right)\vec{S}_g^{\,n}\right]$$

$$\vec{G} = \boldsymbol{\beta}_w\mathbf{T}_w\vec{P}_{c,ow} + \boldsymbol{\beta}_g\mathbf{T}_g\vec{P}_{c,og} + \left[\rho_w g\boldsymbol{\beta}_w\mathbf{T}_w + \rho_o g\boldsymbol{\beta}_o\mathbf{T}_o + \rho_g g\boldsymbol{\beta}_g\mathbf{T}_g\right]\vec{D};$$

$$\vec{Q} = \beta_w \vec{Q}_w^{sc} + \beta_o \vec{Q}_o^{sc} + \beta_g \vec{Q}_g^{sc} + \mathbf{J}\vec{P}_{wf},$$

where,

$$\mathbf{T}_\alpha = \begin{pmatrix} T_{\alpha,\frac{3}{2}} & -T_{\alpha,\frac{3}{2}} & 0 & 0 \\ -T_{\alpha,\frac{3}{2}} & T_{\alpha,\frac{3}{2}} + T_{\alpha,\frac{5}{2}} & -T_{\alpha,\frac{5}{2}} & 0 \\ 0 & \ddots & \ddots & \ddots \\ 0 & 0 & -T_{\alpha,N-\frac{1}{2}} & T_{\alpha,N-\frac{1}{2}} \end{pmatrix} ; \quad \vec{D} = \begin{pmatrix} D_1 \\ D_2 \\ \vdots \\ D_N \end{pmatrix} ;$$

$$\vec{P}_{c,ow} = \begin{pmatrix} P_{cow,1} \\ P_{cow,2} \\ \vdots \\ P_{cow,N} \end{pmatrix} ; \quad \vec{P}_{c,og} = \begin{pmatrix} P_{cog,1} \\ P_{cog,2} \\ \vdots \\ P_{cog,N} \end{pmatrix} ; \quad \vec{Q}_\alpha^{sc} = \begin{pmatrix} Q_{\alpha,1}^{sc} \\ Q_{\alpha,2} \\ \vdots \\ Q_{\alpha,N} \end{pmatrix}^{CR} \quad \mathbf{A} = \frac{1}{\Delta t} \text{diag} \begin{pmatrix} V_1 \phi_1 \\ V_2 \phi_2 \\ \vdots \\ V_N \phi_N \end{pmatrix} ;$$

$$\beta_w = \text{diag} \begin{pmatrix} \dfrac{B_{w,1}}{1 - S_{w,1} c_{w,1} P_{cow,1}} \\ \dfrac{B_{w,2}}{1 - S_{w,2} c_{w,2} P_{cow,2}} \\ \vdots \\ \dfrac{B_{w,N}}{1 - S_{w,N} c_{w,N} P_{cow,N}} \end{pmatrix} ; \quad \beta_o = \text{diag} \begin{pmatrix} B_{o,1} \\ B_{o,2} \\ \vdots \\ B_{o,N} \end{pmatrix} ; \quad \beta_g = \text{diag} \begin{pmatrix} \dfrac{B_{g,1}}{1 - S_{g,1} c_{g,1} P_{cog,1}} \\ \dfrac{B_{g,2}}{1 - S_{g,2} c_{g,2} P_{cog,2}} \\ \vdots \\ \dfrac{B_{g,N}}{1 - S_{g,N} c_{g,N} P_{cog,N}} \end{pmatrix} ;$$

$$\mathbf{J}_\alpha = \text{diag} \begin{pmatrix} J_{\alpha,1} \\ J_{\alpha,2} \\ \vdots \\ J_{\alpha,N} \end{pmatrix}$$

The phase, $\mathbf{T}_\alpha$ (standard conditions), and total transmissibility, $\mathbf{T}$ (reservoir conditions), matrices are tridiagonal in 1D, pentadiagonal in 2D, and heptadiagonal in 3D. The accumulation ($\mathbf{A}$) and compressibility (c_t) matrices are diagonal. Note that the phase transmissibility matrix is symmetric but the total transmissibility is only strictly symmetric if formation volume factor is uniform through all blocks. The G, P_c, D, P_{wf}, and Q arrays are all $N \times 1$ vectors. $\mathbf{J}$ is diagonal matrix of productivity indices for constant bottomhole pressure wells. The Q_α vectors for each phase include a contribution from constant rate (CR) wells, but constant bottomhole pressure (BHP) wells are included directly in Eq. (9.12b), using the vector of bottomhole pressures, P_{wf}. Wells are discussed in more detail in Section 9.8.

The IMPES method employs an assumption that saturations remain constant over the timestep, Δt. The method is only conditionally stable since saturations are updated explicitly. Therefore, relatively small timesteps must be used in IMPES to ensure stability. The general strategy for solving the time-dependent problem is as follows,

1. Determine initial conditions for pressure and saturation of each block in the reservoir by using phase densities and primary drainage capillary pressure curves as described in Chapter 1 (Section 1.7).
2. Using current values of block pressures and saturations (e.g., solution from previous timestep), calculate the relative permeabilities and capillary pressures and then phase/total interblock transmissibilities, accumulation terms, and source vectors.
3. Solve the pressure Eq. (9.12a) implicitly, by solving a system of equations.
4. Solve the saturation Eq. (9.12b) explicitly.
5. Advance to the next timestep and repeat steps 2–4.

Both the pressure and saturation equations are nonlinear and, if necessary, various iterative techniques can be employed (Chen, 2007). In some applications the phase saturations change more quickly than pressure. In these situations, the pressure equation may not need to be resolved in every timestep. Example 9.2 demonstrates the use of the IMPES method for the first timestep for a 1D reservoir with negligible capillary pressure and gravity.

Example 9.2. IMPES solution to two-phase flow in 1D with negligible capillary pressure and gravity.

Recall the 1D reservoir introduced in Example 3.4 which has reservoir length, L, 4000 ft, width, w, 1000 ft, and thickness, h, of 20 ft. The reservoir has an initial pressure of 3000 psia and the following homogenous reservoir properties: $\phi = 0.2$, $k = 100$ mD, $c_f = 10^{-6}$ psi^{-1}. The fluid properties are $\mu_o = 5$ cp, $c_o = 10^{-5}$ psi^{-1}, $B_o = 1$ RB/STB and $\mu_w = 0.5$ cp, $c_w = 10^{-6}$ psi^{-1}, $B_w = 1$ RB/STB. The initial water saturation ($S_{wi} = 0.20$) is larger than the residual water saturation ($S_{wr} = 0.10$). Capillary pressure and gravity are negligible and relative permeability can be described by a Brooks–Corey type relationship (Chapter 1) with $n_o = n_w = 2$, $k_{ro}^0 = 1.0$, and $k_{rw}^0 = 0.2$, and $S_{or} = 0.2$. The boundary conditions are no flow at both ends. There is a water injector well at $x = 0$ of 2000 scf/day and producer well of 2000 scf/day at $x = L$. Solve for the pressure and water saturation field after one timestep, $\Delta t = 1$ day, using 4 uniform grids and the IMPES method.

Solution

Pressure can be solved implicitly and water saturation explicitly using Eq. (9.7a and b). At $t = 0$ the water saturation is uniform throughout the reservoir,

Example 9.2. IMPES solution to two-phase flow in 1D with negligible capillary pressure and gravity.—cont'd

$$k_{ro} = k_{ro}^0 \left(1 - \frac{S_w - S_{wr}}{1 - S_{wr} - S_{or}}\right)^{n_o} = 1.0 \left(1 - \frac{0.2 - 0.1}{1 - 0.1 - 0.2}\right)^2 = 0.735$$

$$k_{rw} = k_{rw}^0 \left(\frac{S_w - S_{wr}}{1 - S_{wr} - S_{or}}\right)^{n_w} = 0.2 \left(\frac{0.2 - 0.1}{1 - 0.1 - 0.2}\right)^2 = 0.004081$$

Calculating the water and oil transmissibilities (which are constant throughout the reservoir because reservoir/fluid properties as well as saturation are uniform at $t = 0$).

$$T_w = \frac{ka}{\mu_w B_w \Delta x} k_{rw} = \frac{(100 \text{ mD})(1000 \text{ ft} \cdot 20 \text{ ft})}{0.5 \text{ cp} \cdot 1 \cdot 1000 \text{ ft}} 0.00408$$

$$= 16.32 \frac{\text{mD} - \text{ft} - \text{scf}}{\text{cp} - \text{ft}^3} = 0.0133 \frac{\text{scf}}{\text{psi} - \text{day}}$$

$$T_o = \frac{ka}{\mu_o B_o \Delta x} k_{ro} = \frac{(100 \text{ mD})(1000 \text{ ft} \cdot 20 \text{ ft})}{5 \text{ cp} \cdot 1 \cdot 1000 \text{ ft}} 0.735$$

$$= 294 \frac{\text{mD} - \text{ft} - \text{scf}}{\text{cp} - \text{ft}^3} = 1.860 \frac{\text{scf}}{\text{psi} - \text{day}}$$

In matrix form, the water, oil, and total transmissibilities can be written as:

$$\mathbf{T}_w = 0.1033 \begin{pmatrix} 1 & -1 & & \\ -1 & 2 & -1 & \\ & -1 & 2 & -1 \\ & & -1 & 1 \end{pmatrix} \frac{\text{scf}}{\text{psi} - \text{day}};$$

$$\mathbf{T}_o = 1.860 \begin{pmatrix} 1 & -1 & & \\ -1 & 2 & -1 & \\ & -1 & 2 & -1 \\ & & -1 & 1 \end{pmatrix} \frac{\text{scf}}{\text{psi} - \text{day}};$$

$$\mathbf{T} = \beta_w \mathbf{T}_w + \beta_o \mathbf{T}_o = 1.964 \begin{pmatrix} 1 & -1 & & \\ -1 & 2 & -1 & \\ & -1 & 2 & -1 \\ & & -1 & 1 \end{pmatrix} \frac{\text{ft}^3}{\text{psi} - \text{day}}$$

where $\beta_w = B_w = 1$ RB/STB and $\beta_o = B_o = 1$ RB/STB. The total compressibility is a weighted average of the phase and formation compressibilities. Since saturation is uniform at $t = 0$, total compressibility is also uniform. Accumulation terms are uniform in the first timestep as well,

Continued

Example 9.2. IMPES solution to two-phase flow in 1D with negligible capillary pressure and gravity.—cont'd

$$c_t = c_f + S_o c_o + S_w c_w = 1.0 \times 10^{-6} + 0.8 \cdot 1.0 \times 10^{-5} + 0.2 \cdot 1.0 \times 10^{-6} = 9.20 \times 10^{-6} \text{psi}^{-1}$$

$$A_i = \frac{V_i \phi_i}{\Delta t} = \frac{1000 \text{ ft} \cdot 1000 \text{ ft} \cdot 20 \text{ ft} \cdot 0.2}{1 \text{ day}} = 4.0 \times 10^6 \frac{\text{ft}^3}{\text{day}}$$

$$c_t = 9.20 \times 10^{-6} \begin{pmatrix} 1 & & & \\ & 1 & & \\ & & 1 & \\ & & & 1 \end{pmatrix} \text{psi}^{-1}; \mathbf{A} = 4.0 \times 10^6 \begin{pmatrix} 1 & & & \\ & 1 & & \\ & & 1 & \\ & & & 1 \end{pmatrix} \frac{\text{ft}^3}{\text{day}}$$

The initial pressure is 3000 psia. The injector well (2000 scf/day) is only water. The producer well is computed from fractional flow

$$Q_{w,4}^{sc} = \frac{k_{rw}/\mu_w B_w}{k_{rw}/\mu_w B_w + k_{ro}/\mu_o B_o} \cdot Q_{well}^{sc} = \frac{0.00408/0.5 \cdot 1}{0.00408/0.5 \cdot 1 + 0.735/5 \cdot 1} \cdot -2000$$

$$\frac{\text{scf}}{\text{day}} = -105.2 \frac{\text{scf}}{\text{day}} \qquad Q_{o,4}^{sc} = Q^{sc} - Q_{w,4}^{sc} = 2000 - 105.2 = 1894.8 \frac{\text{scf}}{\text{day}}$$

$$P^0 = \begin{pmatrix} 3000 \\ 3000 \\ 3000 \\ 3000 \end{pmatrix} \text{psia}; \quad S_w^0 = \begin{pmatrix} 0.2 \\ 0.2 \\ 0.2 \\ 0.2 \end{pmatrix} \quad \vec{Q}_w^{sc} = \begin{pmatrix} 2000 \\ \\ \\ -105.2 \end{pmatrix} \frac{\text{scf}}{\text{day}};$$

$$\vec{Q}_o^{sc} = \begin{pmatrix} 0 \\ \\ \\ -1894.8 \end{pmatrix} \frac{\text{scf}}{\text{day}}; \quad \vec{Q} = \beta_w \vec{Q}_w^{sc} + \beta_o \vec{Q}_o^{sc} = \begin{pmatrix} 2000 \\ \\ \\ -2000 \end{pmatrix} \frac{\text{ft}^3}{\text{day}}$$

Solving for the pressure implicitly using Eq. (9.7a) we get:

$$(\mathbf{T} + \mathbf{J} + \mathbf{A}c_t)\vec{P}_o^{n+1} = \mathbf{A}c_t \vec{P}_o^n + \vec{Q} + \vec{\mathcal{C}}$$

$$\vec{P}^1 = \begin{pmatrix} 3051.7 \\ 3002.4 \\ 2997.6 \\ 2948.2 \end{pmatrix} \text{psia}$$

The saturation can then be found explicitly using Eq. (9.7b). Capillary pressure is negligible so $\beta_w = B_w = 1.0$ in all blocks.

$$\vec{S}_w^{n+1} = \vec{S}_w^n + \mathbf{A}^{-1} \left[-\beta_w \mathbf{T}_w^n \left(\vec{P}_o^{n+1} - \vec{P}_{cow}^n - \rho_w' g \vec{D} \right) + \beta_w \vec{Q}_w^{sc} + \mathbf{J}_w \left(P_{wf} - \vec{P}_o^{n+1} \right) \right] - \mathbf{B}_w^{-1} \beta_w \vec{S}_w^n \left(c_w + c_f \right) \left(\vec{P}_o^{n+1} - \vec{P}_o^n \right)$$

Example 9.2. IMPES solution to two-phase flow in 1D with negligible capillary pressure and gravity.—cont'd

$$\vec{S}_w^{\,1} = \begin{pmatrix} 0.2005 \\ 0.2000 \\ 0.2000 \\ 0.2000 \end{pmatrix}$$

The pressure increases in block #1 due to the injector well and decreases in block #4 due to the producer. Since water is injected in block #1, the S_w increases in that block. There is very little change in the other blocks.

9.4.2 Simultaneous solution method

An alternative numerical approach to IMPES is the *simultaneous solution* (SS) method (Douglas et al., 1959) which, as the name suggests, requires solution of the component balances (Eqs. 9.6a–c) simultaneously, without the use of the combined (pressure) Eq. (9.6d). There are many different ways to formulate the SS equations including the use of all phase pressures (P_w, P_o, and P_g) as the unknown variables. Saturations can then be found from inversion of capillary pressure functions. Here, we show one SS approach in which the unknowns are pressure of one phase (e.g., P_o) and saturation of the other phases (e.g., S_w). For simplicity, we only show the equations here for two-phase undersaturated (oleic and aqueous) flow but they can be easily extended to three-phase flow.

The block equations can be derived by rearranging Eqs. 9.6a and 9.6b and using the approximation that $\Delta P_{c,ow} \sim P_c' \Delta S_w$, where $P_c' = dP_c/dS_w$.

$$-\beta_w \Delta T_w \Delta P_o^{n+1} - \beta_w \Delta \left(T_w P_c' \right) \Delta S_w^{n+1} + C_1 P_o^{n+1} + A S_w^{n+1}$$

$$= C_1 P_o^n + A S_w^n - \beta_w \Delta T_w (\rho_w g \Delta D) + \beta_w Q_w^{sc} \qquad (9.13a)$$

$$-\beta_o \Delta T_o \Delta P_o^{n+1} + C_2 P_o^{n+1} - A S_w^{n+1} = C_2 P_o^n - A S_w^n + \beta_o \Delta T_o \rho_o g \Delta D + \beta_o Q_o^{rc} \qquad (9.13b)$$

In the above block equations, the unknown variables are block oil pressures, P_o^{n+1}, and water saturations, S_w^{n+1}, which appear in both the water and oil balances. Therefore, there are two unknowns per block (pressure, P_o, and water saturation, S_w). This results in a system of $2N$ equations with $2N$ unknowns which can be solved simultaneously,

$$\left(\widehat{\mathbf{T}} + \widehat{\mathbf{J}} + \widehat{\mathbf{C}} \right) \vec{X}^{\,n+1} = \widehat{\mathbf{C}} \vec{X}^{\,n} + \widehat{Q} + \widehat{G} \qquad (9.14)$$

where the new matrices and vectors are defined in 1D as:

$$
\widehat{\mathbf{T}} = \widehat{\boldsymbol{\beta}}
\begin{pmatrix}
\left(T_{w,\frac{1}{2}} + T_{w,\frac{3}{2}}\right) & -\left[(T_w P_c')_{\frac{1}{2}} + (T_w P_c')_{\frac{3}{2}}\right] & -T_{w,\frac{3}{2}} & (T_w P_c')_{\frac{3}{2}} & & & \\
\left(T_{o,\frac{1}{2}} + T_{o,\frac{3}{2}}\right) & 0 & -T_{o,\frac{3}{2}} & 0 & & & \\
-T_{w,\frac{3}{2}} & (T_w P_c')_{\frac{3}{2}} & \left(T_{w,\frac{3}{2}} + T_{w,\frac{5}{2}}\right) & -\left[(T_w P_c')_{\frac{3}{2}} + (T_w P_c')_{\frac{5}{2}}\right] & -T_{w,\frac{5}{2}} & (T_w P_c')_{\frac{5}{2}} & \\
-T_{o,\frac{3}{2}} & 0 & \left(T_{o,\frac{3}{2}} + T_{o,\frac{5}{2}}\right) & 0 & -T_{o,\frac{5}{2}} & 0 & \\
& & & \ddots & \ddots & \ddots & \ddots \\
& & & & -T_{w,N-\frac{1}{2}} & (T_w P_c')_{N-\frac{1}{2}} & \left(T_{w,N-\frac{1}{2}} + T_{w,N+\frac{1}{2}}\right) & -\left[(T_w P_c')_{N-\frac{1}{2}} + (T_w P_c')_{N+\frac{1}{2}}\right] \\
& & & & -T_{o,N-\frac{1}{2}} & 0 & \left(T_{o,N-\frac{1}{2}} + T_{o,N+\frac{1}{2}}\right) & 0
\end{pmatrix}
$$

$$
\widehat{\mathbf{C}} =
\begin{pmatrix}
C_{1,1} & A_1 & & & & \\
C_{2,1} & -A_1 & & & & \\
& & C_{1,2} & A_2 & & \\
& & C_{2,2} & -A_2 & & \\
& & & & \ddots & \ddots \\
& & & & C_{1,N} & A_N \\
& & & & C_{2,N} & -A_N
\end{pmatrix}
\; ; \;
\widehat{\mathbf{J}} =
\begin{pmatrix}
J_{w,1} & 0 & & \\
J_{o,1} & 0 & & \\
& & J_{w,2} & 0 \\
& & J_{o,2} & 0 \\
& & & \ddots & \ddots \\
& & & & J_{w,N} & 0 \\
& & & & J_{o,N} & 0
\end{pmatrix}
\; ; \; \widehat{\boldsymbol{\beta}} = diag
\begin{pmatrix}
\beta_{w,1} \\ \beta_{o,1} \\ \beta_{w,2} \\ \beta_{o,2} \\ \vdots \\ \beta_{w,N} \\ \beta_{o,N}
\end{pmatrix}
$$

$$
\overline{X} =
\begin{pmatrix}
P_1 \\ S_{w,1} \\ P_2 \\ S_{w,2} \\ \vdots \\ \vdots \\ P_N \\ S_{w,N}
\end{pmatrix}
\; ; \;
\widehat{Q} = \widehat{\boldsymbol{\beta}}
\begin{pmatrix}
Q_{w,1} \\ Q_{o,1} \\ Q_{w,2} \\ Q_{o,2} \\ \vdots \\ \vdots \\ Q_{w,4} \\ Q_{o,4}
\end{pmatrix}^{sc}
+ \widehat{\mathbf{J}}
\begin{pmatrix}
P_{wf,1} \\ P_{wf,1} \\ P_{wf,2} \\ P_{wf,2} \\ \vdots \\ \vdots \\ P_{wf,N} \\ P_{wf,N}
\end{pmatrix}
\; ; \; \widehat{\rho g} = diag
\begin{pmatrix}
\rho_{w,1} g \\ \rho_{o,1} g \\ \rho_{w,2} g \\ \rho_{w,2} g \\ \vdots \\ \vdots \\ \rho_{w,1} g \\ \rho_{o,1} g
\end{pmatrix}
\; ; \;
\widehat{D} =
\begin{pmatrix}
D_1 \\ 0 \\ D_2 \\ 0 \\ \vdots \\ \vdots \\ D_N \\ 0
\end{pmatrix}
\; ; \; \widehat{G} = \widehat{\rho g} \widehat{\mathbf{T}} \widehat{D}
$$

All arrays in the SS method are defined differently than in the IMPES method and now have $2N$ rows, since there are two unknowns (P and S_W) per grid block. The matrix $\widehat{\mathbf{C}}$ is "block-diagonal" and the matrix $\widehat{\mathbf{T}}$ is "block tridiagonal" in 1D. The "hat," or circumflex, is added above the arrays to avoid confusion with arrays in the IMPES method. The X solution vector includes both block oleic phase pressures and aqueous phase saturations. The interblock capillary pressure derivatives in the $\widehat{\mathbf{T}}$ matrix can be determined by upwinding from the block of higher capillary pressure.

In the simplest form of SS, relative permeability (and therefore phase transmissibility) is determined at the beginning of the timestep, using the saturation at time $= n$. Variables such as relative permeability can, and do, change during the timestep, so there remains a degree of "explicitness" in the problem; therefore, instabilities may still arise if too large of a timestep is used. More advanced methods (see Section 9.4.3; fully implicit) utilize an iterative approach at each timestep, so that the problem is more stable.

The *SS* method is computationally slow per timestep compared to IMPES because $2N$ simultaneous equations must be solved implicitly for two-phase flow. Since solution methods for linear systems of equations (e.g., gauss elimination, conjugate gradient) do not generally scale linearly with the number of equations (sometimes $\sim N^3$ for direct methods), the computation time can be up to eight times slower per time step for two-phase flow when compared to IMPES. Example 9.3 demonstrates the use of the IMPES method for the first timestep without capillary pressure or gravity.

Example 9.3. SS solution to two-phase flow in 1D without capillary pressure.
Repeat Example 9.2, but use the SS method.

Solution
The phase relative permeabilities and phase transmissibilities for all blocks were computed in Example 9.2. In matrix form the transmissibilities can be written as the following (recall the block-nature in the SS method):

$$\hat{\mathbf{T}} = \begin{pmatrix} \left(T_{o,\frac{1}{2}}+T_{o,\frac{3}{2}}\right) -\left[\left(T_oP_c'\right)_{\frac{1}{2}}+\left(T_oP_c'\right)_{\frac{3}{2}}\right] & -T_{o,\frac{3}{2}} & \left(T_oP_c'\right)_{\frac{3}{2}} & & & & \\ \left(T_{o,\frac{1}{2}}+T_{o,\frac{3}{2}}\right) & 0 & -T_{o,\frac{3}{2}} & 0 & & & \\ -T_{o,\frac{3}{2}} & \left(T_oP_c'\right)_{\frac{3}{2}} & \left(T_{o,\frac{3}{2}}+T_{o,\frac{5}{2}}\right) -\left[\left(T_oP_c'\right)_{\frac{3}{2}}+\left(T_oP_c'\right)_{\frac{5}{2}}\right] & -T_{o,\frac{5}{2}} & \left(T_oP_c'\right)_{\frac{5}{2}} & & \\ -T_{o,\frac{3}{2}} & 0 & \left(T_{o,\frac{3}{2}}+T_{o,\frac{5}{2}}\right) & 0 & -T_{o,\frac{5}{2}} & 0 & \\ & & -T_{o,\frac{5}{2}} & \left(T_oP_c'\right)_{\frac{5}{2}} & \left(T_{o,\frac{5}{2}}+T_{o,\frac{7}{2}}\right) -\left[\left(T_oP_c'\right)_{\frac{5}{2}}+\left(T_oP_c'\right)_{\frac{7}{2}}\right] & -T_{o,\frac{7}{2}} & \left(T_oP_c'\right)_{\frac{7}{2}} \\ & & -T_{o,\frac{5}{2}} & 0 & \left(T_{o,\frac{5}{2}}+T_{o,\frac{7}{2}}\right) & 0 & -T_{o,\frac{7}{2}} & 0 \\ & & & & -T_{o,\frac{7}{2}} & \left(T_oP_c'\right)_{\frac{7}{2}} & \left(T_{o,\frac{7}{2}}+T_{o,\frac{9}{2}}\right) -\left[\left(T_oP_c'\right)_{\frac{7}{2}}+\left(T_oP_c'\right)_{\frac{9}{2}}\right] \\ & & & & -T_{o,\frac{7}{2}} & 0 & \left(T_{o,\frac{7}{2}}+T_{o,\frac{9}{2}}\right) & 0 \end{pmatrix}$$

Capillary pressure (and its derivative) is neglected in the problem,

$$\widehat{\mathbf{T}} = \begin{pmatrix} 0.1034 & 0 & -0.1034 & 0 & & & & \\ 1.860 & 0 & -1.860 & 0 & & & & \\ -0.1034 & 0 & 0.2067 & 0 & -0.1034 & 0 & & \\ -1.860 & 0 & 3.721 & 0 & -1.860 & 0 & & \\ & & -0.1034 & 0 & 0.2067 & 0 & -0.1034 & 0 \\ & & -1.860 & 0 & 3.721 & 0 & -1.860 & 0 \\ & & & & -0.1034 & 0 & 0.1033 & 0 \\ & & & & -1.860 & 0 & 1.860 & 0 \end{pmatrix} \frac{\text{ft}^3}{\text{psi} - \text{day}}$$

Continued

Example 9.3. SS solution to two-phase flow in 1D without capillary pressure.—cont'd

The accumulation terms are the same in all blocks because all properties are uniform at $t = 0$

$$A_i = \frac{V_i \phi_i}{\Delta t} = 4.0 \times 10^6 \frac{ft^3}{day}$$

$$C_{1,i} = \frac{\beta_{w.i}}{B_{w.i}} A_i S_{w,i} (c_{w,i} + c_{f,i})$$

$$= \frac{1.0 \frac{RB}{STB} 1000\,ft \cdot 1000\,ft \cdot 20\,ft \cdot 0.2}{1.0 \frac{RB}{STB} 1\,day} \cdot 0.2 \cdot 2E - 06$$

$$= 1.6 \frac{ft^3}{psi - day}$$

$$C_{2,i} = A_i S_{o,i} (c_{o,i} + c_{f,i})$$

$$= 4.0 \times 10^6 \cdot 0.8 \cdot 1.1E - 05 = 35.2 \frac{ft^3}{psi - day}$$

which gives,

$$\hat{C} = \begin{pmatrix} 1.6 & 4.0E06 & & & & & & \\ 35.2 & -4.0E06 & & & & & & \\ & & 1.6 & 4.0E06 & & & & \\ & & 35.2 & -4.0E06 & & & & \\ & & & & 1.6 & 4.0E06 & & \\ & & & & 35.2 & -4.0E06 & & \\ & & & & & & 1.6 & 4.0E06 \\ & & & & & & 35.2 & -4.0E06 \end{pmatrix} \frac{ft^3\text{-psi}}{day}$$

$$\vec{X}^0 = \begin{pmatrix} 3000 \\ 0.2 \\ 3000 \\ 0.2 \\ 3000 \\ 0.2 \\ 3000 \\ 0.2 \end{pmatrix} ; \quad \hat{Q} = \begin{pmatrix} 2000 \\ 0 \\ \\ \\ \frac{ft^3}{day} \\ \\ -105.3 \\ 1894.7 \end{pmatrix}$$

Calculating the solution: $(\hat{T} + \hat{J} + \hat{C})\vec{X}^{n+1} = \hat{C}\vec{X}^n + \hat{Q} + \hat{\mathcal{E}}$

Example 9.3. SS solution to two-phase flow in 1D without capillary pressure.—cont'd

$$\vec{X}^1 = \begin{pmatrix} 3051.7 \\ 0.2005 \\ 3002.4 \\ 0.2 \\ 2997.6 \\ 0.2 \\ 2948.3 \\ 0.2 \end{pmatrix}$$

The pressures (psia) are the odd-numbered elements and the water saturations the even numbered elements of vector X. The solution is extremely close to the solution obtained from IMPES.

9.4.3 Fully implicit method

The mass balance equations are nonlinear because several variables in the multiphase flow equations are functions of state variables S_w and P_o, and therefore time. Nonlinearities are classified as being weak or strong. Fluid (e.g., viscosity, formation volume factor) and rock properties (e.g., porosity, permeability) can be functions of pressure, but these dependences are usually weak (exceptions were discussed in Chapter 6). As a result, explicit calculation of these variables using the pressure at time $= n$ is often sufficient. However, saturation-dependent variables, such as relative permeability and capillary pressure, are usually strong, nonlinear functions. These variables are so strong that they may change significantly between time n and $n + 1$. In order to maintain stability, the timestep can be severely limited by the explicit calculation of these variables if the simple SS method is used.

Saturation and pressure-dependent variables can be updated iteratively within a timestep. As discussed in Chapter 6, the simplest iterative approach involves direct updates (Picard iteration); pressures and saturations are guessed, nonlinear variables (e.g., relative permeability) are computed, the linear system of equations is solved for X, and the nonlinear variables are recomputed based on the improved estimate of X. The iterations continue until convergence is obtained and then the algorithm proceeds to the next timestep. These internal iterations allow for larger timesteps than the simple SS method but are still restrictive in that the solution may not converge, and many iterations may be necessary.

A more robust iterative method is the *fully implicit* method, which is defined here as any SS formulation with Newton–Raphson iterations on the nonlinear multiphase equations. The problem can be formulated as a nonlinear system of equations:

$$\vec{F}\left(\vec{X}^{n+1}\right) = -\left(\widehat{\mathbf{T}}^{n+1} + \widehat{\mathbf{J}}^{n+1} + \widehat{\mathbf{C}}^{n+1}\right)\vec{X}^{n+1} + \widehat{\mathbf{C}}^{n+1}\vec{X}^{n} + \widehat{Q}^{n+1} + \widehat{G}^{n+1} = 0$$

(9.15)

In the above nonlinear matrix equations, it is assumed that some or all of the arrays are functions of the unknown solution vector, X^{n+1}. The system of $2N$ nonlinear equations and $2N$ unknowns can be solved using the usual Newton–Raphson approach, which was described in Chapter 6. The partial derivatives for the Jacobian are computed analytically if possible because it is more computationally efficient and includes less roundoff error. If not, they can be computed numerically using finite differences. The Jacobian can be computed in parts as a summation if it makes calculations easier/more amenable to an analytical solution,

$$\Im_{ij} = \frac{\partial \vec{F}_i}{\partial \vec{X}_j} = -\frac{\partial\left(\widehat{\mathbf{T}}\vec{X}^{n+1}\right)_i}{\partial \vec{X}_j} - \frac{\partial\left(\widehat{\mathbf{J}}\vec{X}^{n+1}\right)_i}{\partial \vec{X}_j} - \frac{\partial\left(\widehat{\mathbf{C}}\vec{X}^{n+1}\right)_i}{\partial \vec{X}_j} + \frac{\partial\left(\widehat{\mathbf{C}}\vec{X}^{n}\right)_i}{\partial \vec{X}_j} - \frac{\partial\left(\widehat{G}_i^{n+1}\right)}{\partial \vec{X}_j} + \frac{\partial\left(\widehat{Q}_i^{n+1}\right)}{\partial \vec{X}_j}$$

(9.16)

In doing so, some terms in Eq. (9.11) may vanish if some arrays are independent of X.

The fully implicit method is *unconditionally stable* and as the name implies is implicit in all variables. However, it is less accurate (more truncation error) than methods with some degree of explicitness (IMPES and SS). As stated by Aziz and Settari (1979) (page 167), *"as the implicitness of the method increases, stability improves but the truncation errors also increases."*

The fully implicit method is very computationally demanding. Not only is it necessary to solve $2N$ simultaneous equations in two-phase flow ($3N$ for three phases), but iterations are also required within the timestep. The advantage is that larger timesteps can be utilized and stability is maintained, but small enough timesteps must still be used to maintain accuracy. Almost all reservoir simulators have IMPES as one solution option, but most modern simulators also include the fully implicit method. Advanced simulators may adaptively switch between IMPES and Fully Implicit when appropriate and some may even use one formulation in parts of the reservoir domain and other formulations in other parts (Peszyńska et al., 2002).

9.5 Interblock transmissibilities and upwinding

In all solution methods, interblock transmissibilities for each phase must be calculated prior to solving the pressure and saturation field. As was done for single-phase flow, the geometric properties (permeability, area, grid size) should be computed using the harmonic mean as described in Chapter 4. For example, in the x-direction between blocks i and $i+1$,

$$\left(\frac{ka}{\Delta x}\right)_{i+\frac{1}{2}} = \frac{2k_i a_i k_{i+1} a_{i+1}}{k_i a_i \Delta x_{i+1} + k_{i+1} a_{i+1} \Delta x_i}. \tag{9.17a}$$

However, the interblock fluid properties, including relative permeabilities and viscosities, are computed differently. These variables must be determined with great care because relative permeability is a very nonlinear function of saturation which can vary significantly between grid blocks. An option for computing interblock fluid properties is to use some sort of mean, for example, an arithmetic average of the block saturations, and then evaluate relative mobility (or alternatively evaluate relative permeability in both blocks and average them). However, one can show that this is not accurate and employing a harmonic mean is no better.

The proper approach for computing the interblock fluid properties is *upwinding* which was introduced in Chapter 8 for component transport. In upwinding the relative mobility is evaluated in the block upstream of flow, i.e., in the block that flow is coming from. This is the block of higher phase pressure, or more generally if gravity is present, of higher phase potential.

$$\left(\frac{k_r}{\mu B}\right)_{\alpha, i+1/2} = \begin{cases} \left(\dfrac{k_r}{\mu B}\right)_{\alpha, i} & \text{if} \quad \Phi_{\alpha, i} > \Phi_{\alpha, i+1} \\[4mm] \left(\dfrac{k_r}{\mu B}\right)_{\alpha, i+1} & \text{if} \quad \Phi_{\alpha, i+1} > \Phi_{\alpha, i} \end{cases} \tag{9.17b}$$

If capillary pressure is important, the potentials of each phase will be different. It is possible the flow direction will be in opposition for the different phases; this is referred to as *countercurrent flow*. The interblock phase transmissibility is computed as:

$$T_{\alpha, i\pm\frac{1}{2}} = \left(\frac{k_r}{B\mu}\right)_{\alpha, i\pm\frac{1}{2}} \left(\frac{ka}{\Delta x}\right)_{i\pm\frac{1}{2}} \tag{9.17c}$$

Example 9.4 demonstrates the use of upwinding Example 9.1. For the 1D displacement process, Figs. 9.1 and 9.2 show the IMPES numerical solution for pressure and water saturation, respectively, versus distance at $t_D = 0.1$, 0.25, and 0.4 pore volumes injected. 50 uniform grids and upwinding were utilized. As expected, the pressure (Fig. 9.1) increases with time near the injector well and decreases with time near the producer well. An inflection point occurs at the *shock front*, the location behind which contains injected water. The displacement of oleic phase by aqueous phase, and the shock front, is apparent in Fig. 9.2. Good agreement is observed with the Buckley–Leverett semianalytical solution (Chapter 7); however, there is some smearing of the shock front in the numerical solution due to using a finite

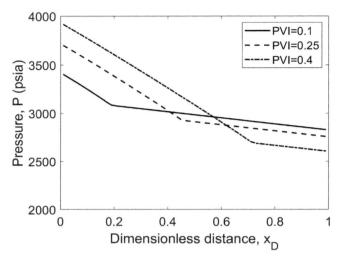

FIGURE 9.1   Pressure versus distance using $N = 50$ grids in the IMPES numerical solution for 1D displacement of oleic phase by aqueous phase without capillary pressure. The reservoir has an initial pressure of 3000 psia, porosity of 20%, and permeability of 100 mD. The viscosities of both phases are 1.0 cp. The initial water saturation ($S_{wi} = 0.20$) is the residual water saturation and Corey-constants $n_o = n_w = 2$, $k_{ro}^0 = 1.0$, and $k_{rw}^0 = 0.2$, and $S_{or} = 0.2$. There is a water injector well at $x = 0$ and producer well at $x = L$, both 1000 scf/day.

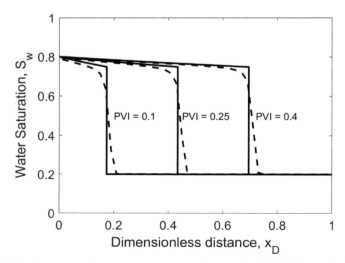

FIGURE 9.2   Comparison of water saturation versus distance for the IMPES numerical solution using $N = 50$ grids and Buckley—Leverett semianalytical solution for 1D displacement of oleic phase by aqueous phase without capillary pressure using properties described in Fig. 9.1.

number of grid blocks. This is analogous to the numerical dispersion observed in Chapter 8.

Fig. 9.3 A shows that the amount of smearing at the shock front is dependent on the number of grids employed. Increasing the number of grids can reduce, but not eliminate, the amount of smearing of the front. More advanced techniques, such as higher order discretization (Saad et al., 1990; Liu et al., 1995) and adaptive gridding (Brandt, 1977; Schmidt and Jacobs, 1988), can also be used to

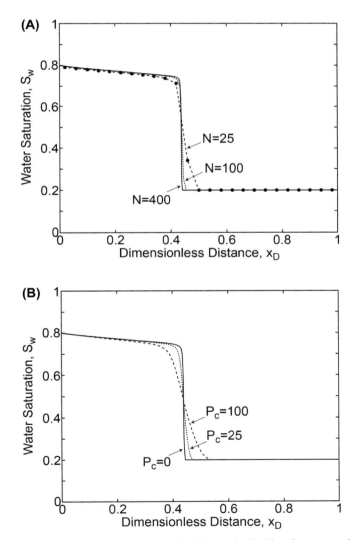

FIGURE 9.3 Comparison of IMPES numerical solution to the Buckley–Leverett semianalytical solution for 1D displacement of oil by water at $t_D = 0.25$ PVI for reservoir, fluid, and well properties in Example 9.1 using (a) $N = 25$, 100, and 400 without capillary pressure and (b) $N_x = 400$ grids with a capillary entry pressure, $p_e = 100$ psi, 25 psi, and 0 psi.

improve accuracy. Fig. 9.3B compares the numerical solutions for varying capillary entry pressures ($p_e = 100$ psi, 25 psi, and 0 psi) but a relatively large number of grids, $N = 400$. Capillary pressure is analogous to physical dispersion (Chapter 8) and increasing capillary pressure has a similar effect (smearing the shock front) on the numerical solution as reducing the number of grid blocks.

Example 9.4. IMPES solution to two-phase flow in 1D with capillary pressure.

The reservoir described in Example 9.2 has been waterflooded for 1000 days and the current pressure and saturation in the four gridblocks are given below.

$$P = [4620.2 \quad 3624.6 \quad 2441.9 \quad 1340.1] \text{ psia}$$

$$S_w = [0.4774 \quad 0.3450 \quad 0.2440 \quad 0.2065]$$

Solve for the pressure and water saturation using $\Delta t = 100$ days and the IMPES method. Include capillary pressure for imbibition and assume that it can be described by a Brooks−Corey model with $p_e = 10$ psi and $\lambda = 2.0$.

Solution

Saturations are not uniform and the relative permeability must be upwinded from the block of higher potential (or pressure because there are no gravitational effects).

The block relative permeabilities can be computed using the Brooks-Corey model which gives,

$$k_{rw} = [0.0581 \quad 0.0245 \quad 0.00845 \quad 0.00463]$$

$$k_{ro} = [0.213 \quad 0.423 \quad 0.631 \quad 0.719]$$

Interblock relative permeabilities require upwinding. Since, $P_1 > P_2 > P_3 > P_4$, interblock relative permeabilities for the $i - 1/2$ interface are upwinded from block $i - 1$ and for the $i + 1/2$ interblock are upwinded from block i.

$$k_{rw,\frac{1}{2}} = 0.0581; \quad k_{rw,\frac{3}{2}} = 0.0245; k_{rw,\frac{5}{2}} = 0.00845$$

$$k_{ro,\frac{1}{2}} = 0.213; \quad k_{ro,\frac{3}{2}} = 0.423; k_{ro,\frac{5}{2}} = 0.631$$

The transmissibility matrices can be calculated from interblock relative permeabilities:

$$\mathbf{T_w} = \begin{pmatrix} 1.471 & -1.471 & & \\ -1.471 & 2.090 & -0.619 & \\ & -0.619 & 0.833 & -0.214 \\ & & -0.214 & 0.214 \end{pmatrix} \frac{\text{scf}}{\text{psi} - \text{day}};$$

$$\mathbf{T_o} = \begin{pmatrix} 0.538 & -0.538 & & \\ -0.538 & 1.609 & -1.071 & \\ & -1.071 & 2.669 & -1.598 \\ & & -1.598 & 1.598 \end{pmatrix} \frac{\text{scf}}{\text{psi} - \text{day}};$$

Example 9.4. IMPES solution to two-phase flow in 1D with capillary pressure.—cont'd

$$\mathbf{T} = \beta_w \mathbf{T}_w + \beta_o \mathbf{T}_o = 1.964 \begin{pmatrix} 2.009 & -2.009 & & \\ -2.009 & 3.699 & -1.69 & \\ & -1.69 & 3.502 & -1.811 \\ & & -1.811 & 1.811 \end{pmatrix} \frac{\text{ft}^3}{\text{psi} - \text{day}}$$

Capillary pressure and the derivative of capillary pressure can be computed in all blocks. For example, in block #1,

$$S_e = \frac{S_w - S_{wr}}{1 - S_{wr} - S_{or}} = \frac{0.4774 - 0.1}{1 - 0.2 - 0.1} = 0.539$$

$$p_{c,1} = p_d \left(S_e^{-\frac{1}{\lambda}} - 1 \right) = 10 \left(0.539^{-\frac{1}{2}} - 1 \right) = 3.622 \text{ psi}$$

$$p'_{c,1} = \frac{-p_d}{\lambda} \left(S_e^{-(1+\lambda)/\lambda} \right) = \frac{-10}{2} \left(S_e^{-\frac{3}{2}} \right) = -12.63 \text{ psi}$$

$$\beta_{w,1} = \frac{1}{1 - S_{w,1} c_{w,1} p'_{cow,1}} = \frac{1}{1 - 0.47739 \cdot 1E - 6 \cdot (-12.63)} \approx 1.0$$

Computing for all blocks

$$\bar{P}_{c,ow} = \begin{pmatrix} 3.622 \\ 6.911 \\ 12.057 \\ 15.657 \end{pmatrix} \text{psi}; \quad \bar{P}'_{c,ow} = \begin{pmatrix} -12.63 \\ -24.15 \\ -53.57 \\ -84.27 \end{pmatrix} \text{psi}; \quad \beta_w \approx diag \begin{pmatrix} 1 \\ 1 \\ 1 \\ 1 \end{pmatrix};$$

$$\vec{G} = \beta_w \mathbf{T}_w \bar{P}_{c,ow} + [\rho_w g \beta_w \mathbf{T}_w + \rho_o g \beta_o \mathbf{T}_o] \vec{D} = \begin{pmatrix} 1 & & & \\ & 1 & & \\ & & 1 & \\ & & & 1 \end{pmatrix} \begin{pmatrix} 1.471 & -1.471 & & \\ -1.471 & 2.090 & -0.619 & \\ & -0.619 & 0.833 & -0.214 \\ & & -0.214 & 0.214 \end{pmatrix} \begin{pmatrix} 3.622 \\ 6.911 \\ 12.057 \\ 15.657 \end{pmatrix} = \begin{pmatrix} -4.837 \\ 1.651 \\ 2.416 \\ 0.7707 \end{pmatrix}$$

The **A** matrix remains unchanged from Example 9.2, but total compressibility is a function of saturation, and therefore block dependent

Continued

Example 9.4. IMPES solution to two-phase flow in 1D with capillary pressure.—cont'd

$$c_t = 1 \times 10^{-6} \begin{pmatrix} 6.70 & & & \\ & 7.90 & & \\ & & 8.80 & \\ & & & 9.14 \end{pmatrix} \text{psi}^{-1}; \mathbf{A} = 4.0 \times 10^{7} \begin{pmatrix} 1 & & & \\ & 1 & & \\ & & 1 & \\ & & & 1 \end{pmatrix} \frac{\text{ft}^3}{\text{day}}$$

The water cut of the production well depends on the relative permeability of block #4,

$$\vec{Q}_w = \begin{pmatrix} 2000 \\ \\ \\ -121.01 \end{pmatrix} \frac{\text{scf}}{\text{day}}; \vec{Q}_o = \begin{pmatrix} 0 \\ \\ \\ -1879.0 \end{pmatrix} \frac{\text{scf}}{\text{day}};$$

$$\vec{Q} = \beta_w \vec{Q}_w + \beta_o \vec{Q}_o = \begin{pmatrix} 2000 \\ \\ \\ -2000 \end{pmatrix} \frac{\text{ft}^3}{\text{day}}$$

Our new solution at 1100 days ($n = 11$) is given by:

$$(\mathbf{T} + \mathbf{J} + \mathbf{A}c_t)\vec{P}_o^{\,n+1} = \mathbf{A}c_t \vec{P}_o^{\,n} + \vec{Q} + \vec{G}$$

$$\vec{S}_w^{\,n+1} = \vec{S}_w^{\,n} + \mathbf{A}^{-1}\left[-\beta_w \mathbf{T}_w^n \left(\vec{P}_o^{\,n+1} - \vec{P}_{c,ow}^{\,n} - \rho_w^n g \vec{D} \right) + \beta_w \vec{Q}_w^{\,sc} \right.$$

$$\left. + \mathbf{J}_w \left(\vec{P}_{wf}^{\,n} - \vec{P}_o^{\,n+1} \right) \right] - \mathbf{B}_w^{-1} \beta_w \vec{S}_w^{\,n}(c_w + c_f)\left(\vec{P}_o^{\,n+1} - \vec{P}_o^{\,n} \right)$$

$$P_o^{11} = \begin{pmatrix} 4618.1 \\ 3625.0 \\ 2443.3 \\ 1339.8 \end{pmatrix} \text{psia}; S_w^{11} = \begin{pmatrix} 0.49073 \\ 0.36323 \\ 0.25651 \\ 0.20944 \end{pmatrix}$$

The time stepping continues until steady state is reached. Table 9.1 summarizes the results.

Example 9.4. IMPES solution to two-phase flow in 1D with capillary pressure.—cont'd

TABLE 9.1 Summary of block pressures and saturations.

	Block #1 500 ft		Block #2 1500 ft		Block #3 2500 ft		Block #4 3500 ft	
	P_o, psia	S_w	P_o, psia	S_w	P_o, psia	S_w	P_o, psia	S_w
Initial	3000	0.2	3000	0.2	3000	0.2	3000	0.2
1000 days	4620.2	0.4774	3624.6	0.345	2441.9	0.244	1340.1	0.2065
1100 days	4618.1	0.49073	3625	0.36323	2443.3	0.25651	1339.8	0.20944
1200 days	4597.8	0.50254	3632.2	0.38031	2457.3	0.2705	1334.9	0.21336
1300 days	4574.9	0.51315	3636.1	0.39604	2474.1	0.28555	1332.1	0.2185
1400 days	4551	0.52274	3636.5	0.41046	2491.7	0.3012	1332.2	0.22502
Steady state	3539.9	0.8	3140.7	0.8	2739	0.8	2334.9	0.8

9.6 Stability

Both the IMPES and SS method are conditionally stable because of the explicitness in calculation of relative permeabilities. The CFL condition, first introduced in Chapter 8 must be met in order to maintain stability. The CFL condition for two-phase IMPES is more complicated than single-phase component transport (Coats, 1968, 2003; Todd et al., 1972; Aziz and Setari, 1979; Franc et al., 2016) because it includes the derivative of fractional flow. For 1D flow with homogenous properties and uniform grids, a timestep must be chosen such that,

$$\Delta t \leq \frac{\phi \Delta x}{\max\left(f'_{w,i} u_i\right)} \quad i = 1 \text{ to } N, \tag{9.18a}$$

where the derivative of fractional flow ($f'_{w,i}$) is evaluated using the block saturation and u_i is the total velocity in the block. A safe timestep is to use the maximum possible derivative at all saturations.[2] The timestep can be optimized and adapted by updating at the end of each timestep and solving Eq. (9.18a). The timestep required for stability is the minimum of value of Eq. (9.13a) for all grids. An additional stability criterion is required for IMPES if capillary pressure is included (Aziz & Settari, 1979),

$$\Delta t_{p_c,i} \leq \frac{\phi_i V_i}{\max\left(\left|p'_{c,i}\right| \left|T_f \psi\right|\right)} \quad i = 1, N \tag{9.18b}$$

where $T_f = ka/\Delta x$ and $\Psi = 2\lambda_w \lambda_o/(\lambda_w + \lambda_o)$. For the IMPES method, the smaller of the two criteria should be used. Fig. 9.4 shows the IMPES solution of Example 9.12 with 100 grids but a timestep is chosen so that the solution is unstable.

9.7 Wells and well models

Wells and well models work similarly in multiphase flow as they did in single-phase flow. Wells can be operated at constant rate or constant bottomhole pressure, be injectors or producers, and be vertical or horizontal. Producer wells may produce multiple phases, with phase cuts determined by the mobility of the phase in the perforated block. Injector wells have a composition and/or saturation specified by the operator. Oil is rarely (or never) injected into the reservoir as this is not profitable. Water and gas injection are common. For our purposes, we will assume only one phase (aqueous or

---

2. It might seem that the maximum possible derivative for the 1D Buckley—Leverett problem is at the front saturation, S_{wf}. However, S_w can be between S_{wi} and S_{wf} due to numerical smearing of the shock front. The safest maximum timestep evaluates Eq. (9.18a) at the peak of the f_w' versus S_w curve (Fig. 7.6).

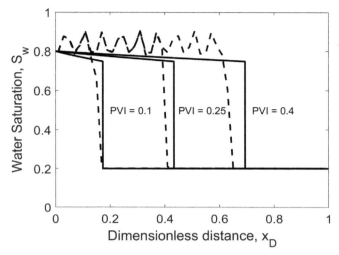

FIGURE 9.4 Comparison of water saturation versus distance for the IMPES numerical solution using $N = 50$ grids and Buckley–Leverett semianalytical solution for 1D displacement of oleic phase by aqueous phase without capillary pressure using reservoir, fluid, and well properties (Fig. 9.1). Plots are made at dimensionless times of $t_D = 0.1$, 0.25, and 0.4 pore volumes injected (PVI). The timestep is chosen so that the solution is unstable.

gaseous) is injected in a single well at once, but water could be injected in some wells, and gas in others. Also, the injected fluid in a well can change over time. For example, a common enhanced oil recovery strategy is water alternating gas (WAG) injection.

The well rate (reservoir conditions) of each phase in a block, $Q_{\alpha,i}$, is related to the bottomhole pressure, P_{wf}, by a well model,

$$Q_{\alpha,i} = J_{\alpha,i}\left(P_{wf} - P_i\right) \tag{9.19}$$

where $J_{\alpha,i}$ is the phase productivity index in block i and can be computed for each perforated block of each well. Each grid block that is intercepted by a well has phase productivity indices ($J_{\alpha,i} = 0$ for blocks without wells). The equation is based on Peaceman's correction as was done for single-phase flow (Chapter 5),

$$J_{\alpha,i} = \frac{2\pi k_{H,i} k_{r\alpha,i} h_i}{\mu_{\alpha,i}\left[\ln\left(\frac{r_{eq,i}}{r_{w,i}}\right) + s_i\right]}, \tag{9.20}$$

where $k_{H,i}$ is the geometric mean of grid block permeability of the two directions not perpendicular to the wellbore, h_i is the thickness of the block, $\mu_{\alpha,i}$ is the phase viscosity, $r_{w,i}$ is the wellbore radius, $r_{eq,i}$ is the equivalent radius of the block as defined in Chapter 5, and s_i is the skin factor in the block. The phase relative permeability of the block, $k_{r\alpha,i}$, is computed using the saturation of the phase in the block for producer wells. For injector wells, the relative permeability of the injected phase is unity and is zero for phases not injected. Eq. (9.20) is appropriate for both vertical and horizontal wells but calculation of k_H, h, and r_{eq} should be performed carefully based on the direction of the well, e.g., vertical (z), horizontal-x, or horizontal-y, as described in Chapter 5.

Wells are included in the matrix Eq. (9.12 for IMPES and 9.14 for SS) as source terms. For constant rate wells, the rate of each phase is specified and therefore can be directly substituted into the matrix equations as a vector Q_α (IMPES, 9.7b) or $\hat{Q}$ (SS, 9.9). For the pressure equation in IMPES (9.7a), a total rate vector, Q, is needed which is a weighted sum of phase rates and bottomhole pressures, $Q = \beta_w Q_w^{sc} + \beta_o Q_o^{sc} + \beta_g Q_g^{sc} + \mathbf{J}P_{wf}$. The phase rates are in standard conditions (scf/day) and total rate vectors must be in reservoir rates (e.g., ft³/day) for calculation purposes. Note that for constant rate wells, the productivity index is not substituted into the matrix equations but are used for postprocessing to calculate the bottomhole pressure, P_{wf}, using Eq. (9.16). For bottomhole pressure wells, the phase, $\mathbf{J}_\alpha$, and total, $\mathbf{J} = \mathbf{J}_w + \mathbf{J}_o + \mathbf{J}_g$, productivity indices are used directly in the matrix equations, instead of the rates. Eq. (9.19) can then be used in postprocessing to compute the well rates. Four specific scenarios, constant rate injector, constant rate producer, constant bottomhole pressure injector, and constant bottomhole producer, are described in more detail below.

9.7.1 Constant rate injector wells

For a constant rate injector well, one phase (e.g., aqueous or gaseous) is usually injected in a well at a given time and that injection rate (Q_α^{sc}) is specified. The injection rates of all other phases are zero. Usually the injected rate is specified in standard conditions (STB/day or scf/day) and must be converted to reservoir conditions, for use in the Q vectors in Eq. (9.12a) (IMPES) and $\hat{Q}$ in Eq. (9.14) (SS). The rate is always positive for injection wells when included in the matrix equations.

A well may be perforated in multiple layers or grids in which case the specified injection rate must be distributed among all perforated blocks with those of higher permeability, thickness, etc., accounting for greater percentages of flow. A simple option is to distribute the well injection by the productivity index of perforated grids of the well,

$$Q_{\alpha,i}^{sc} = \frac{J_{\alpha,i}/B_{\alpha,i}}{\sum\limits_{j=1}^{Nperf} J_{\alpha,i}/B_{\alpha,i}} Q_{\alpha,inj}^{sc}. \tag{9.21}$$

A more complicated, but potentially more accurate, option is to assume the well bottomhole pressure, P_{wf}, is uniform in all perforated blocks and then model the well as a constant BHP well (as described on Section 9.8.3). However, this would then require iteration in each timestep to determine the bottomhole pressure that results in the specified well injection rate.

9.7.2 Constant rate producer wells

For constant rate producer wells, the oil rate, total liquid (water and oil) rate, or total production rate (water, oil, and gas), at standard conditions, will usually be specified. Here, we discuss the case of a total production rate constraint, but other constraints easily follow this approach. The relative production rates of aqueous, oleic, and gaseous phase in a perforated grid block are determined from the phase cut of the grid block which must be computed using the relative permeabilities and viscosities of the phase. The relative production rates in a perforated grid block of each phase ($Q_{\alpha,i}$) is then computed as,

$$Q_{\alpha,i}^{sc} = \frac{J_{\alpha,i}/B_{\alpha,i}}{\sum\limits_{\alpha=1}^{N_p} \sum\limits_{j=1}^{N_{perf}} J_{\alpha,i}/B_{\alpha,i}} \cdot Q_{i,prod}^{sc} \tag{9.22}$$

Alternatively, the well can be modeled as a constant BHP well (Section 9.7.4) and P_{wf} can be iterated on to achieve the specified production rate.

Rates are negative in the matrix equations for constant rate producers. During postprocessing, P_{wf} can be computed using Eq. (9.19) and the phase productivity index. Also, during postprocessing the water cut (percent of liquids produced that are water), water-oil ratio (WOR), and gas-oil ratio (GOR) can be computed,

$$\text{water cut}(\%) = 100 \times \frac{\sum\limits_{j=1}^{N_{perfs}} Q_{w,i}^{sc}}{\sum\limits_{j=1}^{N_{perfs}} \left(Q_{w,i}^{sc} + Q_{o,i}^{sc} \right)} \tag{9.23a}$$

$$\text{WOR} = \frac{\sum_{j=1}^{N_{\text{perfs}}} Q_{w,i}^{sc}}{\sum_{j=1}^{N_{\text{perfs}}} Q_{o,i}^{sc}} \tag{9.23b}$$

$$\text{GOR} = \frac{\sum_{j=1}^{N_{\text{perfs}}} \left(Q_{g,i}^{rc} + R_{s,i} Q_{o,i}^{rc} \right) \Big/ B_{g,i}}{\sum_{j=1}^{N_{\text{perfs}}} Q_{o,i}^{sc}} \tag{9.23c}$$

The water cut is bound between 0% and 100%. The WOR and GOR do not have an upper bound because the denominator (oil produced) can approach zero. The units of WOR and GOR are usually STB/STB and scf/STB, respectively.

Fig. 9.5 shows the water cut and bottomhole pressure versus dimensionless time (pore volumes injected) for the reservoir and fluid properties provided in Example 9.2 using $N = 100$ grids and the IMPES method. However, the initial water saturation is the residual value, S_{wr}. Prior to breakthrough, no water is produced because the only water near the producer well is immobile ($S_w = S_{wr}$). At the breakthrough time ($t_D = 0.56$), some of the water near the producer well becomes mobile ($S_w > S_{wr}$) and the water cut increases sharply.

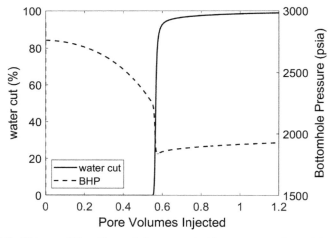

FIGURE 9.5 Water cut (%) and bottomhole pressure (psia) versus pore volumes injected for the producer well using reservoir, fluid, and well properties described in Ex. 9.2, $S_{wi} = S_{wr}$, and $N = 100$ grids.

The bottomhole pressure decreases until breakthrough because the pressure of block that has the producer well is decreasing. At breakthrough, the water saturation increases sharply and the total mobility decreases; a lower bottomhole pressure and greater drawdown pressure is needed to produce the specified production rate. After breakthrough, total mobility gradually increases and the bottomhole pressure increases.

For a constant rate producer well, the BHP of the well may eventually decrease below the bubble point and result in a gaseous phase in the reservoir. In order to avoid the adverse effects on relative mobility by a gaseous phase, one may choose to convert the constant rate producer well to a constant BHP well once it reaches the bubble point.

9.7.3 Constant BHP injector wells

For constant BHP wells, P_{wf} is specified. If the well is an injector then only one phase (aqueous or gaseous) is injected at a given time and the phase productivity index, J_α, and P_{wf} of the perforated blocks are included in the matrix equations. The relative permeability of the injected phase is 1.0. During postprocessing, well injection rate versus time can be computed using Eq. (9.19).

9.7.4 Constant BHP producer wells

Constant BHP producers are modeled similarly to injectors. However, the relative permeabilities (and therefore productivity indices) of each phase are computed using the phase saturation in the perforated grid block. During postprocessing, well production rate versus time can be computed using Eq. (9.19).

9.7.5 Time-dependent well constraints

The constraints on wells can change over time; for example, wells can be shut in (zero rate) or opened, producers can be converted to injectors, constant rate wells can be converted to BHP wells, or the value of the constraint (Q or P_{wf}) can be time dependent. In some cases, these changes follow a predetermined schedule. For example, a well may not begin producing until some predetermined date (after the well is drilled and completed). In other cases, the well constraint is changed when some reservoir condition is met. For example, if the bottomhole pressure of the well reaches a specified limit (e.g., the bubble point), the well might be converted to a constant BHP well equal to that limit. If the water cut of a producer well exceeds some maximum value, it might be shut in or switched to an injector.

It is possible that, during the simulation, the grid block pressure will become lower than P_{wf} of a producer well or vice versa, the block pressure becomes higher than P_{wf} for an injector. If this occurs, then mathematically the sign of the drawdown pressure changes and the producer becomes an injector (or vice versa). However, this is not practical as a producer well will not begin injecting fluid without a decision from the operator. A check in the simulator is needed to shut the well in when the drawdown switch occurs and reopen when a prespecified criterion is met. Example 9.5 demonstrates the IMPES method for a 2D reservoir with both vertical and horizontal wells and injectors/producers.

Example 9.5. IMPES solution to two-phase flow in 2D without capillary pressure.

Recall Example 4.2 (and Example 5.2) which introduced a heterogeneous 2D reservoir ($L = 3000$ ft, $w = 3000$ ft, $h = 20$ ft). The oleic and aqueous phase viscosities are 5 cp and 0.5 cp, respectively, and the formation volume factor of both phases is 1.0. The compressibilities of oleic phase, aqueous phase, and the formation are 1E-5 psi^{-1}, 1E-6 psi^{-1}, and 1E-6 psi^{-1}, respectively. The reservoir is discretized into $N_x = 3$, $N_y = 3$, $N_z = 1$ nonuniform grids, $\Delta x = [750\ 1000\ 1250]$ ft and $\Delta y = [750\ 1000\ 1250]$ ft. The permeability is anisotropic, $k_y = 0.5 \times k_x$, $k_z = 0.1 \times k_x$, and both permeability and porosity are heterogeneous as given below. The depth is uniform and the initial pressure is 3000 psia.

$$k_x = [50\quad 100\quad 200\quad 100\quad 150\quad 250\quad 150\quad 200\quad 300]\,mD$$

$$\phi = [0.15\quad 0.18\quad 0.20\quad 0.17\quad 0.20\quad 0.22\quad 0.22\quad 0.25\quad 0.26]$$

The reservoir now has an initial water saturation ($S_{wi} = 0.2$) greater than the residual water saturation ($S_{wr} = 0.1$). Capillary pressure can be ignored and relative permeability can be described by a Brooks–Corey type relationship (Chapter 1) with $n_o = n_w = 2$, $k_{ro}^0 = 1.0$, and $k_{rw}^0 = 0.2$, and $S_{or} = 0.2$.

There is a constant rate, vertical, injector well in block #1 ($Q = 3000$ scf/day), and constant BHP horizontal producer perforated in blocks #8 and #9 ($P_{wf} = 2000$ psia = bubble point). The wellbore radius of both wells is 0.25 ft and there is no skin.

a. Calculate the pressure and water saturation after one timestep using the IMPES method and $\Delta t = 1$ day.

b. The pressure and water saturation after 1000 days of waterflooding is given by,

$$P = [4338.2\quad 3349.4\quad 2844.3\quad 2967.2\quad 2720.4\quad 2523\quad 2161.2\quad 2032.1\quad 2018.3]psia$$

$$S_w = [0.62323\quad 0.46393\quad 0.26893\quad 0.42397\quad 0.28922\quad 0.2091\quad 0.2313\quad 0.207\quad 0.20034]$$

Example 9.5. IMPES solution to two-phase flow in 2D without capillary pressure.—cont'd

Calculate the pressure at water saturation at 1100 days using the IMPES method and $\Delta t = 100$ days.

Solution
a. Interblock transmissibilities were computed in Example 4.2 for single-phase flow using a harmonic mean on the geometric properties. Here, the phase relative permeabilities are both less than one (0.735 and 0.004081 for oleic and aqueous, respectively) but uniform at $t = 0$ since $S_w = 0.2$ in all blocks. The transmissibility matrices are thus:

$$T_w = \begin{bmatrix} 0.093012 & -0.062008 & & -0.031004 \\ -0.062008 & 0.23036 & -0.095397 & & -0.072951 \\ & -0.095397 & 0.26209 & & & -0.16669 \\ -0.031004 & & & 0.21918 & -0.1459 & & -0.042278 \\ & -0.072951 & & -0.1459 & 0.47603 & -0.17717 & & -0.080011 \\ & & -0.16669 & & -0.17717 & 0.50204 & & & -0.15818 \\ & & & -0.042278 & & & 0.30065 & -0.25837 \\ & & & & -0.080011 & & -0.25837 & 0.62023 & -0.28186 \\ & & & & & -0.15818 & & -0.28186 & 0.44004 \end{bmatrix} \frac{scf}{psi - day}$$

$$T_o = \begin{bmatrix} 1.6742 & -1.1161 & & -0.55807 \\ -1.1161 & 4.1464 & -1.7171 & & -1.3131 \\ & -1.7171 & 4.7175 & & & -3.0004 \\ -0.55807 & & & 3.9453 & -2.6262 & & -0.76101 \\ & -1.3131 & & -2.6262 & 8.5685 & -3.189 & & -1.4402 \\ & & -3.0004 & & -3.189 & 9.0367 & & & -2.8473 \\ & & & -0.76101 & & & 5.4116 & -4.6506 \\ & & & & -1.4402 & & -4.6506 & 11.164 & -5.0734 \\ & & & & & -2.8473 & & -5.0734 & 7.9207 \end{bmatrix} \frac{scf}{psi - day}$$

$$T = \beta_w T_w + \beta_o T_o = \begin{bmatrix} 1.7672 & -1.1782 & & -0.58908 \\ -1.1782 & 4.3768 & -1.8125 & & -1.3861 \\ & -1.8125 & 4.9796 & & & -3.1671 \\ -0.58908 & & & 4.1645 & -2.7721 & & -0.80329 \\ & -1.3861 & & -2.7721 & 9.0446 & -3.3662 & & -1.5202 \\ & & -3.1671 & & -3.3662 & 9.5387 & & & -3.0055 \\ & & & -0.80329 & & & 5.7123 & -4.909 \\ & & & & -1.5202 & & -4.909 & 11.784 & -5.3553 \\ & & & & & -3.0055 & & -5.3553 & 8.3607 \end{bmatrix} \frac{ft^3}{psi\text{-}day}$$

Continued

Example 9.5. IMPES solution to two-phase flow in 2D without capillary pressure.—cont'd

where, $\beta_w = B_w = 1$ RB/STB and $\beta_o = B_o = 1$ RB/STB in all blocks. The **A** matrix can be computed using Eqn (4.20). The total compressibility is uniform since saturations are uniform at $t = 0$:

$$c_t = c_f + S_o c_o + S_w c_w = 1.0 \times 10^{-6} + 0.8 \cdot 1.0 \times 10^{-5} + 0.2 \cdot 1.0 \times 10^{-6}$$

$$= 9.20 \times 10^{-6} \text{psi}^{-1}$$

$$\mathbf{A} = \begin{pmatrix} 1.688 & & & & & & & & \\ & 2.70 & & & & & & & \\ & & 3.75 & & & & & & \\ & & & 2.55 & & & & & \\ & & & & 4.00 & & & & \\ & & & & & 5.50 & & & \\ & & & & & & 4.125 & & \\ & & & & & & & 6.25 & \\ & & & & & & & & 8.125 \end{pmatrix} \times 10^6 \frac{\text{ft}^3}{\text{psi} - \text{day}}$$

$$\mathbf{c}_t = 9.20E - 06 \begin{pmatrix} 1 & & & & & & & & \\ & 1 & & & & & & & \\ & & 1 & & & & & & \\ & & & 1 & & & & & \\ & & & & 1 & & & & \\ & & & & & 1 & & & \\ & & & & & & 1 & & \\ & & & & & & & 1 & \\ & & & & & & & & 1 \end{pmatrix} \text{psi}^{-1}$$

In Example 5.2, single-phase productivity indices were computed in blocks #8 and #9 for the horizontal BHP well but these must be recomputed for multiphase flow. The equivalent radius for both blocks was 108.2 ft.

Example 9.5. IMPES solution to two-phase flow in 2D without capillary pressure.—cont'd

Block #8

$$J_{o,8} = \frac{2\pi\Delta x_8 k_{ro,8}\sqrt{k_{y,8}k_{z,8}}}{\mu_o\left[\ln\left(\frac{r_{eq}}{r_w}\right) + s\right]} = \frac{2\pi\cdot0.735\cdot1000\sqrt{100\cdot20}}{5\left[\ln\left(\frac{108.2}{0.25}\right) + 0\right]}$$

$$= 0.735\cdot9258.0\frac{mD-ft}{cp} = 43.1\frac{ft^3}{psi-day}$$

$$J_{w,8} = \frac{2\pi\Delta x_8 k_{rw,8}\sqrt{k_{y,8}k_{z,8}}}{\mu_w\left[\ln\left(\frac{r_{eq}}{r_w}\right) + s\right]} = \frac{2\pi\cdot0.00408\cdot1000\sqrt{100\cdot20}}{0.5\left[\ln\left(\frac{108.2}{0.25}\right) + 0\right]}$$

$$= 0.00408\cdot92580\frac{mD-ft}{cp} = 2.392\frac{ft^3}{psi-day}$$

Block #9

$$J_{o,9} = \frac{2\pi\Delta x_9 k_{ro,9}\sqrt{k_{y,9}k_{z,9}}}{\mu_o\left[\ln\left(\frac{r_{eq}}{r_w}\right) + s\right]} = \frac{2\pi\cdot0.735\cdot1250\sqrt{150\cdot30}}{5\left[\ln\left(\frac{108.2}{0.25}\right) + 0\right]}$$

$$= 0.735\cdot17,358\frac{mD-ft}{cp} = 80.73\frac{ft^3}{psi-day}$$

$$J_{w,9} = \frac{2\pi\Delta x_9 k_{rw,9}\sqrt{k_{y,9}k_{z,9}}}{\mu_w\left[\ln\left(\frac{r_{eq}}{r_w}\right) + s\right]} = \frac{2\pi\cdot0.00408\cdot1250\sqrt{150\cdot30}}{0.5\left[\ln\left(\frac{108.2}{0.25}\right) + 0\right]}$$

$$= 0.00408\cdot173,580\frac{mD-ft}{cp} = 4.48\frac{ft^3}{psi-day}$$

The productivity matrices and source vectors are then

Continued

Example 9.5. IMPES solution to two-phase flow in 2D without capillary pressure.—cont'd

$$
\mathbf{J}_w = \begin{pmatrix} & & \\ & & \\ & 2.39 & \\ & & 4.48 \end{pmatrix} \quad
\mathbf{J}_o = \begin{pmatrix} & & \\ & & \\ & 43.1 & \\ & & 80.73 \end{pmatrix}
$$

$$
\mathbf{J} = \begin{pmatrix} & & \\ & & \\ 45.45 & \\ & 85.21 \end{pmatrix} \dfrac{ft^3}{psi - day}
$$

Example 9.5. IMPES solution to two-phase flow in 2D without capillary pressure.—cont'd

$$\vec{Q}_w = \begin{pmatrix} Q_{w,1} \\ \\ \\ \\ \\ \end{pmatrix} = \begin{pmatrix} 3000 \\ \\ \\ \\ \\ \end{pmatrix}; \quad \vec{Q}_o = \begin{pmatrix} Q_{o,1} \\ \\ \\ \\ \\ \end{pmatrix} = \begin{pmatrix} 0 \\ \\ \\ \\ \\ \end{pmatrix};$$

$$\vec{Q} = \begin{pmatrix} Q_1 \\ \\ \\ J_8 P_{wf,8} \\ J_9 P_{wf,9} \end{pmatrix} = \begin{pmatrix} 0 \\ \\ \\ 90,891 \\ 170,420 \end{pmatrix} \frac{ft^3}{psi-day}$$

The pressure and saturation can then be computed,

$$(\mathbf{T} + \mathbf{J} + \mathbf{A}c_t)\vec{P}_o^{\,n+1} = \mathbf{A}c_t \vec{P}_o^{\,n} + \vec{Q} + \vec{G}$$

$$\vec{S}_w^{\,n+1} = \vec{S}_w^{\,n} + \mathbf{A}^{-1}\left[-\beta_w \mathbf{T}_w^n \left(\vec{P}_o^{\,n+1} - \vec{P}_{c,ow}^{\,n} - \rho_w^n g \vec{D} \right) \right.$$

$$\left. + \beta_w \vec{Q}_w^{sc} + \mathbf{J}_w \left(\vec{P}_{wf} - \vec{P}_o^{\,n+1} \right) \right] - \mathbf{B}_w^{-1} \beta_w \vec{S}_w^{\,n} (c_w + c_f) \left(\vec{P}_o^{\,n+1} - \vec{P}_o^{\,n} \right)$$

Continued

Example 9.5. IMPES solution to two-phase flow in 2D without capillary pressure.—cont'd

$$\vec{P} = \begin{bmatrix} 3173.9 \\ 3006.1 \\ 2998.1 \\ 3000.7 \\ 2984.2 \\ 2973 \\ 2952.5 \\ 2577.4 \\ 2479.8 \end{bmatrix} \text{psia} \quad \vec{S}_w = \begin{bmatrix} 0.2017 \\ 0.2 \\ 0.2 \\ 0.2 \\ 0.2 \\ 0.2 \\ 0.2 \\ 0.2 \\ 0.2 \end{bmatrix}$$

At $t = 2$ days

$$\vec{P} = \begin{bmatrix} 3331.1 \\ 3016.4 \\ 2994.0 \\ 3000.5 \\ 2961.7 \\ 2936.9 \\ 2885.8 \\ 2351.2 \\ 2241.0 \end{bmatrix} \text{psia} \quad \vec{S}_w = \begin{bmatrix} 0.20339 \\ 0.2 \\ 0.2 \\ 0.2 \\ 0.2 \\ 0.2 \\ 0.2 \\ 0.2 \\ 0.2 \end{bmatrix}$$

At $t = 3$ days

$$\vec{P} = \begin{bmatrix} 3473.5 \\ 3029.1 \\ 2988.1 \\ 2998.7 \\ 2937.1 \\ 2899.3 \\ 2814.1 \\ 2229.4 \\ 2130.4 \end{bmatrix} \text{psia} \quad \vec{S}_w = \begin{bmatrix} 0.20509 \\ 0.2 \\ 0.2 \\ 0.2 \\ 0.2 \\ 0.2 \\ 0.2 \\ 0.2 \\ 0.2 \end{bmatrix}$$

b. The saturations are not uniform at 1000 days as upwinding of interblock transmissibilities must be performed. The upwinded directions are shown in the figure below,

Example 9.5. IMPES solution to two-phase flow in 2D without capillary pressure.—cont'd

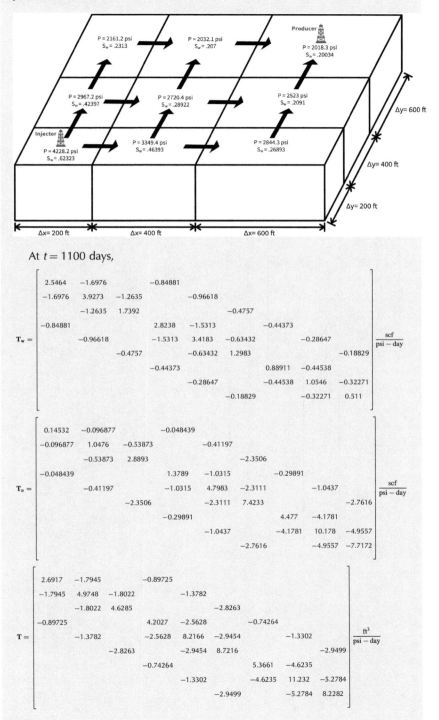

At $t = 1100$ days,

$$T_w = \begin{bmatrix} 2.5464 & -1.6976 & & -0.84881 & & & & & \\ -1.6976 & 3.9273 & -1.2635 & & -0.96618 & & & & \\ & -1.2635 & 1.7392 & & & -0.4757 & & & \\ -0.84881 & & & 2.8238 & -1.5313 & & -0.44373 & & \\ & -0.96618 & & -1.5313 & 3.4183 & -0.63432 & & -0.28647 & \\ & & -0.4757 & & -0.63432 & 1.2983 & & & -0.18829 \\ & & & -0.44373 & & & 0.88911 & -0.44538 & \\ & & & & -0.28647 & & -0.44538 & 1.0546 & -0.32271 \\ & & & & & -0.18829 & & -0.32271 & 0.511 \end{bmatrix} \frac{scf}{psi-day}$$

$$T_o = \begin{bmatrix} 0.14532 & -0.096877 & & -0.048439 & & & & & \\ -0.096877 & 1.0476 & -0.53873 & & -0.41197 & & & & \\ & -0.53873 & 2.8893 & & & -2.3506 & & & \\ -0.048439 & & & 1.3789 & -1.0315 & & -0.29891 & & \\ & -0.41197 & & -1.0315 & 4.7983 & -2.3111 & & -1.0437 & \\ & & -2.3506 & & -2.3111 & 7.4233 & & & -2.7616 \\ & & & -0.29891 & & & 4.477 & -4.1781 & \\ & & & & -1.0437 & & -4.1781 & 10.178 & -4.9557 \\ & & & & & -2.7616 & & -4.9557 & -7.7172 \end{bmatrix} \frac{scf}{psi-day}$$

$$T = \begin{bmatrix} 2.6917 & -1.7945 & & -0.89725 & & & & & \\ -1.7945 & 4.9748 & -1.8022 & & -1.3782 & & & & \\ & -1.8022 & 4.6285 & & & -2.8263 & & & \\ -0.89725 & & & 4.2027 & -2.5628 & & -0.74264 & & \\ & -1.3782 & & -2.5628 & 8.2166 & -2.9454 & & -1.3302 & \\ & & -2.8263 & & -2.9454 & 8.7216 & & & -2.9499 \\ & & & -0.74264 & & & 5.3661 & -4.6235 & \\ & & & & -1.3302 & & -4.6235 & 11.232 & -5.2784 \\ & & & & & -2.9499 & & -5.2784 & 8.2282 \end{bmatrix} \frac{ft^3}{psi-day}$$

Continued

Example 9.5. IMPES solution to two-phase flow in 2D without capillary pressure.—cont'd

$$J_w = \begin{bmatrix} & & \\ & & \\ & 2.7386 & \\ & & 4.5156 \end{bmatrix} \frac{ft^3}{psi-day}; J_o = \begin{bmatrix} & & \\ & & \\ & 42.055 & \\ & & 80.633 \end{bmatrix} \frac{ft^3}{psi-day};$$

$$J = \begin{bmatrix} & & \\ & & \\ 44.793 & \\ & 85.149 \end{bmatrix} \frac{ft^3}{psi-day}$$

$$\vec{Q}_w = \begin{bmatrix} 3000 \\ \\ \\ \end{bmatrix} \frac{scf}{day}; \quad \vec{Q}_o = \begin{bmatrix} 0 \\ \\ \\ \end{bmatrix} \frac{scf}{day}; \quad \vec{Q} = \begin{bmatrix} 3000 \\ \\ 89586 \\ 1.703E5 \end{bmatrix} \frac{ft^3}{day}$$

$$c_t = (1 \times 10^{-6}) \begin{bmatrix} 5.3909 \\ & 6.8247 \\ & & 8.5796 \\ & & & 7.1843 \\ & & & & 8.397 \\ & & & & & 9.1181 \\ & & & & & & 8.9183 \\ & & & & & & & 9.137 \\ & & & & & & & & 9.1969 \end{bmatrix} psi^{-1}$$

Example 9.5. IMPES solution to two-phase flow in 2D without capillary pressure.—cont'd

$$\vec{P} = \begin{bmatrix} 4335.1 \\ 3347.6 \\ 2844 \\ 2966.2 \\ 2720 \\ 2522.9 \\ 2161.4 \\ 2032.1 \\ 2018.3 \end{bmatrix} \text{psia} \quad \vec{S}_w = \begin{bmatrix} 0.63282 \\ 0.48 \\ 0.28183 \\ 0.44075 \\ 0.30575 \\ 0.21243 \\ 0.23856 \\ 0.2096 \\ 0.20055 \end{bmatrix}$$

At t = 1200 days,

$$\vec{P} = \begin{bmatrix} 4293.9 \\ 3334.1 \\ 2848.9 \\ 2961.1 \\ 2724.4 \\ 2524.8 \\ 2165.2 \\ 2032.4 \\ 2018.3 \end{bmatrix} \text{psia} \quad \vec{S}_w = \begin{bmatrix} 0.641 \\ 0.49405 \\ 0.29488 \\ 0.45571 \\ 0.32222 \\ 0.21655 \\ 0.24643 \\ 0.21284 \\ 0.20083 \end{bmatrix}$$

At t = 1300 days,

$$\vec{P} = \begin{bmatrix} 4255.6 \\ 3319.8 \\ 2852.7 \\ 2954.4 \\ 2727.2 \\ 2526.8 \\ 2169.1 \\ 2032.7 \\ 2018.3 \end{bmatrix} \text{psia} \quad \vec{S}_w = \begin{bmatrix} 0.64822 \\ 0.50647 \\ 0.30783 \\ 0.46909 \\ 0.33825 \\ 0.2215 \\ 0.25478 \\ 0.21676 \\ 0.20122 \end{bmatrix}$$

and at steady state,

Continued

Example 9.5. IMPES solution to two-phase flow in 2D without capillary pressure.—cont'd

$$
\vec{P} = \begin{bmatrix} 3111.2 \\ 2509.2 \\ 2310.2 \\ 2339.9 \\ 2257.4 \\ 2196.1 \\ 2058.7 \\ 2012.5 \\ 2007 \end{bmatrix} \text{psia} \quad \vec{S}_w = \begin{bmatrix} 0.8 \\ 0.8 \\ 0.8 \\ 0.8 \\ 0.8 \\ 0.8 \\ 0.8 \\ 0.8 \\ 0.8 \end{bmatrix}
$$

9.8 Pseudocode for multiphase flow

Here, we present pseudocode for a multiphase, multidimensional reservoir simulator using the IMPES method. The code is easily adaptable to the SS or fully implicit methods. As usual, it is recommended to create several simple subroutines/functions/modules that complete a specific task. Many of the subroutines, or a simpler version of them, should have already been created for single-phase flow and pseudocode was provided in previous chapters. It is recommended that those files are used as starting points and adapted for multiphase flow. It is assumed that subroutines for calculating relative permeability and capillary pressure have already been developed (Chapter 1) and can be used.

9.8.1 Preprocessing

A preprocessing subroutine was created in Chapter 1 and can be adapted as needed here. Input reservoir, fluid, petrophysical, well, and numerical property data are read in from data files or spreadsheets, grids and grid locations are assigned, and well-containing grids are identified. For multiphase flow, PVT data might be approximated as pressure-independent constants (common for undersaturated reservoirs) or a table of PVT properties at various pressures is provided. PVT properties at a given grid pressure can later be computed via interpolation or using curve fit functions developed in this preprocessing step. Initial grid pressures and saturations of all phases are determined using the grid depths and capillary pressure curves as described in Chapter 1. Desired plots, such as the initial pressure/saturation field, well locations, PVT properties versus pressure, petrophysical properties versus saturation, etc., can be made during preprocessing.

9.8.2 Block properties

Block properties are computed for an input block #, i, along with PVT, reservoir, and petrophysical data provided from an input file. If pressure dependent, block PVT properties and viscosities are computed using the known block pressure and interpolation of the PVT table or direct calculation from curve fit equations developed in the preprocessing subroutine. Capillary pressure of the block can be computed using the block saturation and a subroutine for capillary pressure (created in Chapter 1). The accumulation term, A, capillary-corrected formation volume factors, β_α, and total compressibility, c_t, can be computed for each block. Finally, the accumulation terms, C, of the block is computed using Eq. (9.7).

9.8.3 Interblock properties

Any two integers (i_1, i_2) corresponding to two adjacent grid blocks are used as inputs along with reservoir, PVT, petrophysical, and numerical properties of all blocks. The direction of connection (x-, y-, or z-) should also be provided as inputs to the subroutine. First, interblock geometric properties are computed using Eq. (9.17a); the coder should be careful to use the correct directional values of permeability, grid size, etc., for anisotropic systems. Second, the upwinded block (i_1, i_2) is determined by comparing block potentials of the phase ($\Phi_{i1,\alpha}$, $\Phi_{i2,\alpha}$). The phase relative permeabilities of the upwinded block are computed using that block's saturation and a subroutine for relative permeability. The interblock transmissibility of a phase can be computed using Eq. (9.17c).

9.8.4 Well productivity index

The well productivity index in a perforated block is computed using Eq. (9.20). Reservoir, petrophysical, fluid, numerical, and solution properties of the block are required as inputs as well as the well # and perforated block #. In the subroutine, the direction of the well (x-, y-, z-) in the block must be provided or determined. The phase relative permeabilities, using the block saturations, must be calculated by calling the relative permeability function.

9.8.5 Well arrays

Vectors, Q_α, and matrices, $\mathbf{J}_\alpha$, are needed in the matrix equations to describe injection and production from wells and are created by looping through all wells. For each block, i, that is perforated by a well, an entry in row i of the phase vectors, Q_α, is updated for rate wells or an entry in the main diagonal (row i, column i) of the phase matrices, $\mathbf{J}_\alpha$, is made for constant BHP wells (called *Well Productivity Index*) using the equations and procedures described in Section 9.8. For a reservoir model with many more grids than wells, most entries of Q_α and the main diagonal of $\mathbf{J}_\alpha$ will be empty (zero); therefore, sparse storage should be used to optimize computational speed and memory. Q and $\mathbf{J}$ can be computed from the weighted sum of the phase arrays. Note that well productivity indices will also be needed for postprocessing of constant rate wells but these should be stored separately from the $\mathbf{J}$ matrices.

9.8.6 Grid arrays

The matrices and vectors used for solving saturation (e.g., $\mathbf{T}_\alpha$, etc.) and pressure (e.g., $\mathbf{A}$, $\mathbf{c}_t$, $\mathbf{T}$, $\mathbf{G}$, etc.) are created by looping through all grid blocks. For a grid, i, the *Block Properties* subroutine is called to compute the accumulation term, $A_{i,i}$, $c_{t,i,i}$. Interblock phase ($\mathbf{T}_\alpha$) and total ($\mathbf{T}$) transmissibilities are also computed between the block and each of its neighbors by calling the *Interblock Properties* subroutine. As was done in single-phase flow, *if* statements are needed to determine if the block has a neighbor or is adjacent to a boundary.

9.8.7 Main code

In the main code, a loop through time is created and at each timestep the *Well Arrays* and *Grid Arrays* subroutine is called to compute all the well, transmissibility, accumulation, and gravity vectors. Pressure is computed implicitly by solving a system of equations using Eq. (9.12a) and then phase saturations can be computed using Eq. (9.12b). The time is then updated. Pressures and saturations can be replaced in each time step or saved for making plots in the postprocessing code. Saving results at every timestep can be memory intensive; an alternative is to save a subset of times (e.g., every tenth or hundredth of timesteps) predetermined by the user.

9.8.8 Postprocessing

After the code is successfully run and free of errors, the user will want to analyze and visualize the results. For example, time series plots of well rates/bottomhole pressures, water cut/water-oil ratio/gas-oil ratio may be computed using Eqs. (9.19) and (9.23), respectively. Spatially dependent property, such as reservoir pressure and saturation, fields can be plotted in 2D or 3D at specified times.

SUBROUTINE PREPROCESS
INPUTS: data file(s) and/or spreadsheet(s) of all input data
OUTPUTS: reservoir, fluid, petrophysical, numerical, well properties

READ datafile
INIT x,y,z vectors of grid center locations
INIT array of perforated grid blocks for each well
DEF curve fit for pressure-dependent PVT variables (if desired over
interpolation)

INIT reservoir pressure and saturations
FOR i = 1 to N (# of grid blocks)
    CALL INITIALIZE(i)
ENDFOR

PLOT PVT/Petrophysical data and initial pressure/saturation fields
with well locations

SUBROUTINE BLOCK PROPERTIES
INPUTS: Block index (i), reservoir, fluid, petrophysical, numerical,
solution properties
OUTPUTS: fluid properties, A, c_t, β, P_c, C_1, C_2, C_3 (scalars)

CALCULATE block fluid properties (e.g. viscosity, formation volume
factor) using curve fit eqn/interpolation and block pressure
CALL CAPILLARY PRESSURE to obtain $P_c(i)$ and $P_c(i)'$
CALCULATE $\beta_\alpha(i)$ using eqn 9.8
CALCULATE $A(i)=\Delta x(i)\Delta y(i)\Delta z(i)\phi(i)/\Delta t$
CALCULATE $c_t(i)$ using eqn 9.10
CALCULATE C_1, C_2, C_3 (accumulation) coefficients using eqn 9.7

SUBROUTINE INTERBLOCK PROPERTIES
INPUTS: i_1, i_2, direction, reservoir, fluid, petrophysical, numerical,
solution properties
OUTPUTS: Tw, To, Tg (scalars)

CALCULATE harmonic mean geometric properties using eqn 9.17a
DETERMINE upwinded block (i_1 or i_2) for each phase using eqn 9.17b
CALL RELPERM of upwinded block for each phase
CALCULATE interblock phase transmissibilities using eqn 9.17c

SUBROUTINE PRODUCTIVITY INDEX
INPUTS: well #, perforated block #, well, reservoir, fluid, petrophysical, numerical, solution properties
OUTPUTS: Jw, Jo, Jg

DETERMINE direction (x,y,z) of well and determine direction-dependent variables
CALL RELPERM of block i
CALCULATE phase productivity indexes using eqn 9.20

SUBROUTINE WELL ARRAYS
INPUTS: reservoir, fluid, petrophysical, well, numerical, output
OUTPUTS: Jw, Jo, Jg, Qw, Qo, Qg, Q, $Jvec_w$, $Jvec_o$, $Jvec_g$,

FOR k = 1 to # wells
    FOR j = 1 to # perforated blocks
        SET i = jth perforated block # of well k
        CALL PRODUCTIVITY INDEX to obtain $Jvec_w(i)$, $Jvec_o(i)$, $Jvec_g(i)$
        IF constant rate well
            CALCULATE Qw,sc(i), Qo,sc(i), Qg,sc(i), eqn 9.21, 9.22
            SET Q(i) = Q(i) + $\beta w(i)$*Qw,sc(i) + $\beta o(i)$*Qo,sc(i) + βg,sc(i)*Qg(i)
        ELSEIF constant BHP well
            FOR all α phases
                SET $J_\alpha(i,i)$=$Jvec_\alpha(i)$
                SET Q(i)=Q(i)+ $Jvec_\alpha(i)$*$P_{wf}(i)$
            ENDFOR
            SET J(i,i)= Jw(i) + Jo(i) + Jg(i)
        ENDIF
    ENDFOR
ENDFOR

```
SUBROUTINE GRID ARRAYS
INPUTS: numerical, petrophysical, fluid properties, boundary
conditions
OUTPUTS: Tw, To, Tg, T, Jw, Jo, Jg, J, Qw, Qo, Qg, Q, A, ct, G

FOR i = 1 to # grids (N = Nx*Ny*Nz)
    SET A(i,i), ct(i) = BLOCK PROPERTIES(i)
    FOR α = 1 to # phases (e.g., oil, water, gas)

        #Left Boundary
        IF NOT on left (x=0) boundary
            SET Tα(i,i-1) = -INTERBLOCK(i,i-1,'x')[α]
            SET Tα(i,i) = Tα(i,i) - Tα(i,i-1)
        ELSEIF Dirichlet boundary condition
            SET Tx = INTERBLOCK(i,i,'x')[α]
            SET Jα(i,i) = Jα(i,i) + 2*Tx
            SET Qα(i) = Qα(i) + 2*Tx*P_LB
        ELSEIF Neumann boundary condition
            CALCULATE Qα(i) using eqn 4.13
        ENDIF

        #Right Boundary
        IF NOT on right (x=L) boundary
            SET Tα(i,i+1) = -INTERBLOCK(i,i+1,'x')[α]
            SET Tα(i,i) = Tα(i,i) - Tα(i,i+1)
        ELSEIF Dirichlet boundary condition
            SET Tx = INTERBLOCK(i,i,'x')[α]
            SET Jα(i,i) = Jα(i,i) + 2*Tx
            SET Qα(i) = Qα(i) + 2*Tx*P_RB
        ELSEIF Neumann boundary condition
            CALCULATE Qα(i) using eqn 4.13
        ENDIF

        #Bottom Boundary
        IF NOT on bottom (y=0) boundary
            SET Tα(i,i-NX) = -INTERBLOCK(i,i-NX,'y')[α]
            SET Tα(i,i) = Tα(i,i) - Tα(i,i-NX)
        ELSEIF Dirichlet boundary condition
            SET Ty = INTERBLOCK(i,i,'y')[α]
```

```
                SET Jα(i,i) = Jα(i,i) + 2*Ty
                SET Qα(i) = Qα(i) + 2*Ty*PBB
        ELSEIF Neumann boundary condition
                CALCULATE Qα(i) using eqn 4.13
        ENDIF

        #Top Boundary
        IF NOT on top (y=W) boundary
                SET Tα(i,i+NX) = -INTERBLOCK(i,i+NX,'y')[α]
                SET Tα(i,i) = Tα(i,i) - Tα(i,i+NX)
        ELSEIF Dirichlet boundary condition
                SET Ty = INTERBLOCK(i,i,'y')[α]
                SET Jα(i,i) = Jα(i,i) + 2*Ty
                SET Qα(i) = Qα(i) + 2*Ty*PTB
        ELSEIF Neumann boundary condition
                CALCULATE Qα(i) using eqn 4.13
        ENDIF

        #Front Boundary
        IF NOT on front (z=0) boundary
                SET Tα(i,i-NX*NY) = -INTERBLOCK(i,i-NX*NY,'z')[α]
                SET Tα(i,i) = Tα(i,i) - Tα(i,i-NX*NY)
        ELSEIF Dirichlet boundary condition
                SET Tz = INTERBLOCK(i,i,'z')[α]
                SET Jα(i,i) = Jα(i,i) + 2*Tz
                SET Qα(i) = Qα(i) + 2*Tz*PFB
        ELSEIF Neumann boundary condition
                CALCULATE Qα(i) using eqn 4.13
        ENDIF

        #Back Boundary
        IF NOT on back (z=H) boundary
                SET Tα(i,i+NX*NY) = -INTERBLOCK(i,i+NX*NY,'z')[α]
                SET Tα(i,i) = Tα(i,i) - Tα(i,i+NX*NY)
        ELSEIF Dirichlet boundary condition
                SET Tz = INTERBLOCK(i,i,'z')[α]
                SET Jα(i,i) = Jα(i,i) + 2*Tz
                SET Qα(i) = Qα(i) + 2*Tz*PBaB
        ELSEIF Neumann boundary condition
                CALCULATE Qα(i) using eqn 4.13
        ENDIF
    ENDFOR
ENDFOR
SET T = βwTw + βoTo + βgTg
CALCULATE G = (βwTwPcow + βgTgPcog) + [(ρwgβwTw + ρogβoTo + ρggβgTg)]D
SET Q = Q + Qw + Qo + Qg
```

```
MAIN CODE
CALL PREPROCESS
SET t = 0
WHILE t < t_final
      P_old = Po; Sw_old = Sw
      CALL WELL ARRAYS
      CALL GRID ARRAYS
      CALCULATE Po (and oil potential) using eqn 9.12a
      SET Pw = Po-Pcow; Pg = Pg-Pcog
      CALCULATE Sw, So, Sg using eqn 9.12b
      INCREMENT t += dt
ENDWHILE
CALL POSTPROCESS
```

```
SUBROUTINE POSTPROCESS
INPUTS: reservoir, fluid, petrophysical, well, numerical, solution
properties
OUTPUTS: well data, plots

CALCULATE rates and BHP for all wells/times using eqn 9.16
CALCULATE water cut, WOR, GOR for all production wells/times using eqn
9.20

PLOT well rates and BHP vs time
PLOT water cut, WOR, and GOR vs time
PLOT (contour, surface, etc.) block pressure, saturation, etc. at
desired times
VALIDATE against analytical solution (e.g. Buckley Leverett) if
appropriate
```

9.9 Exercises

Exercise 9.1 Derivation of pressure equation. Derive the pressure Eq. (9.2b) by recombining the component balance Eq. (9.1a). In the recombination, multiply the oil balance by a coefficient such that summation with the water and gas balances result in a pressure equation that is free of time derivatives of saturation. Assume that phase pressures are not equal ($p_w \neq p_g \neq p_o$), but that time derivatives of phase pressures are equal,

$$\frac{\partial p_w}{\partial t} = \frac{\partial p_g}{\partial t} = \frac{\partial p_o}{\partial t}$$

Exercise 9.2 Finite difference of multiphase PDEs. Derive Eqs. (9.6a)−(9.6d) by applying finite differences to Eqs. (9.1) and (9.2).

Exercise 9.3 Control volume approach for multiphase flow. Derive Eqs. (9.6a)−(9.6d) by applying a control volume approach to mass conservation on grid blocks.

Exercise 9.4 Balance equations with capillary pressure and gravity. Repeat Example 9.1 but for three-phase flow and include capillary pressure and gravity in the balance equations.

Exercise 9.5 1D IMPES method with gravity. Repeat Example 9.2 but include a dip angle of 30 degrees on the reservoir. The initial oleic phase pressure is $P_o = 3000$ psia at $x = 0$ (the largest depth) and is in static equilibrium. The oleic and aqueous phase densities are 50 and 63 lb_m/ft^3, respectively. Determine the initial pressure for the oleic phase using its density and then neglect capillary pressure. Assume the initial water saturation is uniform, $S_{wi} = 0.2$. Compute the block pressures and saturations at the next three timesteps. Also compute the bottomhole pressure for both wells if they have a well radius of 0.25 ft and no skin.

Exercise 9.6 2D IMPES with gravity. Repeat the 2D Example 9.5, but now the depths are not uniform,

$$D = [7000 \quad 6800 \quad 6600 \quad 6800 \quad 6600 \quad 6300 \quad 6500 \quad 6300 \quad 6000]$$

where the vector, D, represents the depth of blocks in feet at block centers. The water-oil contact line is at 7100 ft, where the initial oleic phase pressure is $P_o = 3000$ psia. The oleic and aqueous phase densities are 50 and 63 lb_m/ft^3, respectively. Use a drainage capillary pressure curve ($p_e = 10$ psi, $\lambda = 2.0$) to initialize the saturations. Use the Brooks–Corey models introduced in Chapter 1 and a capillary pressure scanning curve (epspc $= 0.1$) to transition from drainage to imbibition. Compute the pressure and saturations at the next three timesteps. Compute the bottomhole pressures for the rate well and oil and water rates for the constant BHP well. Both have a well radius of 0.25 ft and no skin.

Exercise 9.7 Three-phase flow. Repeat Example 9.2, but use a bubble point that is equal to the initial pressure (3000 psia), so that a gaseous phase forms near the producer well. Three-phase relative permeabilities can be computed using Stone's I model and the fitting parameters provided in Example 1.2. Capillary pressure (both oil-water and oil-gas) can be neglected. The oil and gas viscosity can be assumed constants (5 and 0.01 cp, respectively). The z-factor ($= 0.8$) can also be assumed constant. Additional PVT properties, R_s and B_o, can be approximated by the following simple functions.

$$R_s \left(\frac{scf}{STB} \right) = \begin{cases} 600 & \text{if} \quad p \geq p_B \\ 0.2p & \text{if} \quad p < p_B \end{cases} \qquad B_o \left(\frac{RB}{STB} \right) = \begin{cases} 1.2 & \text{if} \quad p \geq p_B \\ 4.0E - 4p & \text{if} \quad p < p_B \end{cases}$$

Exercise 9.8 Two-phase IMPES reservoir simulator. Develop a 2D, two-phase reservoir simulator that allows for capillary pressure and gravity and uses the IMPES method. Wells can be injectors or producers, constant rate or

BHP, vertical or horizontal, and operate on a schedule and/or have constraints. Validate your code against the 1D Buckley—Leverett solution (make plots of saturation vs. dimensionless distance and oil recovery versus time) and any of the two-phase examples and exercises. In particular, compare your code against the 3×3 Exercise 9.6, which includes initialization, capillary pressure, gravity, and various well types.

Exercise 9.9 SS and fully implicit reservoir simulator. Generalize your reservoir simulator to use the SS and Fully Implicit methods and 3D flow.

Exercise 9.10 Black oil reservoir simulator. Generalize your reservoir simulator to model three-phase flow when provided PVT data (in a table or as equations). Validate your code against the 4-block Exercise 9.7. Then discretize the 1D reservoir to 100 blocks and use a stable timestep to solve for pressures and saturations. Make plots of the pressure and saturation (S_w and S_o) fields at various times, pressures of both wells versus time, and water cut (%) and gas-oil ratio (GOR) versus time of the producer well.

Exercise 9.11 Multiphase project. Use the reservoir simulator developed in Exercise 9.8 and 9.9 to solve two-phase (aqueous and oleic) flow in 2D for the Thomas oilfield field using any method (IMPES, SS, Fully Implicit). The input files (Thomas.yml, etc.) are provided at https://github.com/mbalhof/Reservoir-Simulation.

The well schedule is provided in the data file. If the bottomhole pressure reaches the bubble point, convert the well to a BHP well, with a constraint equal to the bubble point. If a producer well reaches a water cut $\geq 95\%$, convert the well to a constant rate, water injector with rate specified in the input file. Continue the simulation until the last producer well reaches a water cut of 95%. Make plots of bottomhole pressure and total injection/production rate for all wells and the water cut for wells while they are producers. Make 2D contour or surface plots of the pressure and water saturation field at various times.

Exercise 9.12 3D, three phase multiphase project. Use your multiphase reservoir simulator to solve the project in Exercise 9.11 in three dimensions (use $N_z = 5$ uniform blocks) and allow constant rate production wells to drop below the bubble point but if they reach 100 psia, convert the well to a BHP well of 100 psia. Since the pressure drops below the bubble point, three phases will be present so use a three-phase IMPES simulator. PVT data below the bubble point for the Thomas oilfield can be found in the data file, Thomas.yml. Assume that all grids that are intersected by a well are perforated.

References

Aziz, K., Settari, A., 1979. Petroleum Reservoir Simulation. Applied Science Publishers.

Brandt, A., 1977. Multi-level adaptive solutions to boundary-value problems. Mathematics of Computation 31, 333−390.

Chen, Z., 2007. Reservoir Simulation: Mathematical Techniques in Oil Recovery. Society for Industrial and Applied Mathematics.

Coats, K.H., 1968. Elements of Reservoir Simulation, Lecture Notes. Univ. of Tex., Houston.

Coats, K.H., 2003. IMPES stability: the CFL limit. Spe Journal 8 (03), 291−297.

Douglas, J., Peaceman, D.W., Rachford, H.H., 1959. A method for calculating multi-dimensional immiscible displacement. Transactions of the AIME 216 (01), 297−308.

Franc, J., et al., 2016. Benchmark of different CFL conditions for IMPES. Comptes Rendus Mécanique 344 (10), 715−724.

Liu, J., Pope, G.A., Sepehrnoori, K., 1995. A high-resolution finite-difference scheme for nonuniform grids. Applied Mathematical Modelling 19, 162−172.

Peszyńska, M., Wheeler, M.F., Yotov, I., 2002. Mortar upscaling for multiphase flow in porous media. Computational Geosciences 6 (1), 73−100.

Saad, N., Pope, G.A., Sephrnoori, K., 1990. Application of Higher-Order Methods in Compositional Simulation. SPE Reservoir Engineering, pp. 623−630.

Schmidt, G.H., Jacobs, F.J., 1988. Adaptive local grid refinement and multi-grid in numerical reservoir simulation. Journal of Computational Physics 77 (1), 140−165.

Todd, M.R., O'dell, P.M., Hirasaki, G.J., 1972. Methods for increased accuracy in numerical reservoir simulators. Society of Petroleum Engineers Journal 12 (06), 515−530.

Chapter 10

Numerical solution to multiphase, multicomponent transport

10.1 Introduction

Partial differential equations describing the transport of individual components, or pseudocomponents, were derived in Chapter 7 by performing mass or mole balances. The number of (pseudo)components modeled can be as few as one to as many as tens. In Chapter 8, the finite difference solutions for one or more components in a single phase were presented. The black oil model (Chapter 9) is a unique and simplified multiphase, compositional model that has only three pseudocomponents (gas, oil, and water) and up to three phases (gaseous, oleic, and aqueous), but is not generally classified as a compositional simulator. The black oil model assumes the gas component is partly in the gaseous phase and partly dissolved in the oleic phase, oil is only in the oleic phase, and water only in the aqueous phase. These assumptions and simplifications allow for use of a PVT table and important phase properties (e.g., solution gas ratio, formation volume factor, viscosity) are only functions of pressure.

In this chapter, we introduce a fully compositional model with multiple components that partition between phases. The model accounts for the effect of phase composition, in addition to pressure and temperature, on phase properties such as molar volume and viscosity. Mole fractions are used to describe compositions and molar volumes for inverse density. Correlations are used for phase viscosities. Table 10.1 compares black oil and compositional models.

Compositional models are often used when the composition in the reservoir changes significantly with time, e.g. due to injection of a component or components, and fluid properties are strong functions of the composition. Miscible gas injection, volatile oil and condensate reservoirs, chemical flooding using surfactants and/or polymers, carbon storage, and tracer tests are common applications in which compositional models are preferred. Multiphase, multicomponent transport in subsurface media may refer to (1) hydrocarbon components that partition in the oleic and gaseous, but not aqueous, phase, (2) a component (e.g., CO_2 in carbon capture and storage) or

An Introduction to Multiphase, Multicomponent Reservoir Simulation.
https://doi.org/10.1016/B978-0-323-99235-0.00015-4

TABLE 10.1 Comparison of hydrocarbon (oleic and gaseous) phase properties for black oil and compositional simulators.

	Black oil	Compositional
Composition	R_s (PVT table)	$x_{\kappa,o}, x_{\kappa,g}$ (flash calculation)
Density	B_o, B_g (PVT table)	V_{mg}, V_{mo} (EOS)
Viscosity	μ_o, μ_g (PVT table)	μ_o, μ_g (correlations)
Compressibility	c_o, c_g (PVT table)	c_{og} (EOS)

components that partition between a gaseous (or supercritical) and aqueous phase, (3) a component or components that partition between the oleic and aqueous phase such as a partitioning tracer, (4) components present in all three phases. Here, we focus on the first application (oleic–gaseous) although it has been shown that the second application (aqueous–gaseous) can be modeled in a similar way (Kumar, 2004; Li et al., 2013; Sun et al., 2021). The third application (oleic–aqueous) is discussed briefly at the end of the chapter; it is somewhat easier to implement because flash and EOS calculations are generally not needed. Applications in which components partition into three or more phases are beyond the scope of this book but are described elsewhere (Perschke, 1988; Okuno et al., 2010).

We describe one IMPEC type approach similar to Acs et al. (1985), although fully implicit compositional models are also common (Coats, 1980). The approach used here is intended to provide a general overview of compositional modeling and is not necessarily the most computationally efficient or stable method. In addition to the aforementioned references, the reader is referred to (Young and Stephenson, 1983; Watts, 1986; Chang, 1990; Chang et al., 1990; Coats et al., 1995; Wang et al., 1997; Hustad and Browning, 2010; Farshidi et al., 2013) for advances in compositional modeling. Young (2022) provides a review of compositional simulation.

10.2 Compositional equations for multiphase flow

The partial differential equation that describes mass transport of a component, κ, in a porous medium was derived in Chapter 7, assuming local equilibrium, no chemical reactions, and no adsorption onto the solid surface.

$$\frac{\partial}{\partial t}\left(\phi \sum_{\alpha=1}^{N_p} S_\alpha \rho_\alpha \omega_{\kappa,\alpha}\right) = \frac{\partial(\phi C_\kappa)}{\partial t} = -\nabla \cdot \sum_{\alpha=1}^{N_p}\left[\rho_\kappa \omega_{\kappa,\alpha}\overrightarrow{u}_\alpha - S_\alpha \phi \mathbf{D}_{\kappa,\alpha} \cdot \nabla\left(\rho_\alpha \omega_{\kappa,\alpha}\right)\right]$$

$$+ \sum_{\alpha=1}^{N_p} \rho_\alpha \omega_{\kappa,\alpha}\widetilde{q}_\alpha. \tag{10.1a}$$

Balance equations for compositional models are often written as mole balances. Eq. (10.1b) is the molar form of Eq. (10.1a),

$$\frac{\partial}{\partial t}\left(\phi \sum_{\alpha=1}^{N_p} \frac{S_\alpha x_{\kappa,\alpha}}{V_{m,\alpha}}\right) = \frac{\partial}{\partial t}\left(\frac{\phi x_\kappa(1-S_w)}{V_{m,og}}\right) = -\nabla \cdot \sum_{\alpha=1}^{N_p}\left[\frac{x_{\kappa,\alpha}}{V_{m,\alpha}}\overrightarrow{u}_\alpha - \frac{S_\alpha \phi \mathbf{D}_{\kappa,\alpha}}{V_{m,\alpha}}\cdot \nabla x_{\kappa,\alpha}\right]$$

$$+ \sum_{\alpha=1}^{N_p}\frac{1}{V_{m,\alpha}}x_{\kappa,\alpha}^* \widetilde{q}_\alpha, \tag{10.1b}$$

where S_α and $V_{m,\alpha}$ are the saturation and molar volume, respectively, of phase α and $V_{m,og}$ is the total molar volume of the oleic and gaseous phases. $x_{\kappa,\alpha}$ is the mole fraction of component κ in phase α, and x_κ is the overall mole fraction of component κ in the two hydrocarbon phases. $\mathbf{D}_{\kappa,\alpha}$ is the dispersion coefficient (tensor) of component κ in phase α defined by Eq. 7.7b, and u_α is the velocity of phase α. The source term, $\widetilde{q}_\alpha$, is the rate (volume/time)/volume of phase α generated or consumed. We assume here that hydrocarbon components, κ, are not present in the aqueous phase. A separate balance equation (Chapter 9) is used for the water component which is only present in the aqueous phase.

An overall conservation, or pressure, equation is found through a recombination of all component equations and is given by Eq. (10.2),

$$\phi c_t \frac{\partial p}{\partial t} = -V_{m,og}\sum_{\alpha=1}^{N_p}\nabla \cdot \left[\frac{1}{V_{m,\alpha}}\overrightarrow{u}_\alpha\right] + V_{m,og}\sum_{\alpha=1}^{N_p}\frac{1}{V_{M\alpha}}\widetilde{q}_\alpha \tag{10.2}$$

The accumulation term has been expanded to give a time derivative in terms of pressure.

The pressure Eq. (10.2) can be coupled with N_c-1 component Eq. (10.1) plus the water balance. Solution requires several auxiliary equations. The sum of mole fractions of component κ in each phase as well as overall is 1.0,

$$\sum_{\kappa=1}^{N_c} x_{\kappa,\alpha} = 1; \quad \sum_{\kappa=1}^{N_c} x_\kappa = 1. \tag{10.3}$$

The ratio of the mole fractions between two phases (here, gaseous and oleic) is referred to as the equilibrium coefficient or *K-value* and is generally a function of pressure, temperature, and composition,

$$K_\kappa = \frac{x_{\kappa,g}}{x_{\kappa,o}} = f(P,T,x_\kappa). \tag{10.4}$$

The phases must be in thermodynamic equilibrium, meaning that the component *fugacities*, f_κ, of all components in the gaseous and oleic phases must be equal,

$$f_{\kappa,g} = f_{\kappa,o}. \tag{10.5}$$

The phase molar volumes, the inverse of molar densities, and phase viscosities are functions of pressure, temperature, and composition,

$$V_{m,\alpha} = \frac{1}{\rho_{M,\alpha}} = f\left(P, T, \vec{x}_\alpha\right), \tag{10.6}$$

$$\mu_\alpha = f\left(P, T, \vec{x}_\alpha\right). \tag{10.7}$$

The overall molar volume of the hydrocarbon phases is given by a weighted sum of phase molar volumes,

$$V_{m,og} = V_{m,o}(1-v) + V_{m,g}v = \left[\frac{1}{1-S_w}\left(\frac{S_o}{V_{m,o}} + \frac{S_g}{V_{m,g}}\right)\right]^{-1}, \tag{10.8}$$

where v is the mole fraction of the gaseous phase for the hydrocarbon mixture. Compressibility of the hydrocarbon mixture is given by,

$$c_{og} = -\frac{1}{V_{m,og}} \frac{\partial V_{m,og}}{\partial p}. \tag{10.9}$$

A change in pressure results in a change in composition and phase mole fraction which makes evaluation of Eq. (10.9) cumbersome.

The phase velocities are described by the multiphase form of Darcy's law (Eq. 1.21) with phase relative permeabilities (Chapter 1). Finally, the sum of all phase saturations sum to 1.0,

$$\sum_{\alpha=1}^{N_p} S_\alpha = 1. \tag{10.10}$$

We ignore capillary pressure in the development of the compositional simulator here but can be included in a manner similar to Chapter 9.

10.3 Finite difference equations

Although fully implicit formulations are available in many simulators, one type of IMPEC approach is presented here. The overall balance Eq. (10.2) is discretized to obtain an implicit, algebraic pressure equation. The matrix equations for pressure are similar to that for single-phase flow and the black oil model,

$$\left(\mathbf{T} + \mathbf{J} + \mathbf{Ac}_t\right)\vec{P}^{n+1} = \mathbf{Ac}_t \vec{P}^n + \vec{Q} + \mathbf{J}\vec{P}_{wf}. \tag{10.11a}$$

The component balance Eq. (10.1) can be discretized to explicit algebraic equations,

$$\vec{x}_\kappa^{n+1} = \vec{x}_\kappa^n + (\mathbf{I} - diag\{S_w\})^{-1}\mathbf{A}^{-1}V_{m,og}\left[\mathbf{T}_\kappa \vec{P}^{n+1} + \mathbf{U}_\kappa \vec{x}^n + \vec{Q}_\kappa + \mathbf{J}_\kappa\left(\vec{P}_{wf} - \vec{P}^{n+1}\right)\right], \tag{10.11b}$$

where $\mathbf{U}_\kappa$ and $\mathbf{T}_\kappa$ are component dispersion and transmissibility matrices, respectively, as introduced in Chapter 8 and described in more detail later in this chapter. Compressibility terms are neglected in the component equation. The IMPEC approach involves solving the pressure Eq. (10.11a) implicitly, and then the component balance Eq. (10.11b) explicitly. Aqueous phase saturation can also be updated explicitly as was done in Chapter 9. The matrices and vectors in Eqs. (10.11a) and (10.11b) must be computed based on the auxillary equations described by Eqs. (10.3)−(10.10). The overall numerical solution method is described next.

10.4 Solution method

Given initial conditions of aqueous phase saturation (S_w), pressure (P), temperature (T), and overall composition of components (x_κ) in every grid block, the steps involved in IMPEC simulation, for the application of hydrocarbon components partitioned in the oleic and gaseous phases, are as follows: (1) use a *flash calculation* to determine molar compositions of each phase ($x_{\kappa,g}$ and $x_{\kappa,o}$), and mole fraction of hydrocarbon that is gaseous phase (v). This step requires an estimate of K-values; (2) use an *Equation of State* (EOS) to compute molar volumes of the gaseous (V_{mg}) and oleic (V_{mo}) phase. Also, use the EOS to compute new K-values and repeat steps 1 and 2 until convergence; (3) using known gaseous phase mole fraction (v) and molar volumes of each phase, calculate hydrocarbon phase saturations (S_o, S_g); (4) compute total compressibility, c_t, in each grid block; (5) calculate viscosities of each hydrocarbon phase (μ_o, μ_g) using correlations and the known block pressure, temperature, and phase compositions; (6) calculate relative permeabilities (k_{rw}, k_{ro}, k_{rg}) using phase saturations, and then interblock phase (and total) transmissibilities (T_w, T_o, T_g, T); (7) Compute the source vector and productivity index arrays for the wells; (8) solve system of equations (pressure equation; 10.11a) implicitly to obtain new block pressures (P), update the overall compositions, x_κ, explicitly, using Eq. (10.11b), and update water saturation explicitly; (9) proceed to the next timestep and repeat steps 1−8.

The method is described for multiple components partitioned between the oleic and gaseous phase with an inert aqueous phase. However, the method can be easily adapted to applications such as CO_2 in an aqueous and gaseous/supercritical phase or components (such as tracers) partitioned between an oleic and aqueous phase. The following sections provide details on these steps.

10.4.1 Flash calculations

10.4.1.1 K-values

Flash calculations (Fig. 10.1) begin with selecting K-values (equilibrium ratios). The simplest estimate for selecting K-values is to assume an ideal liquid

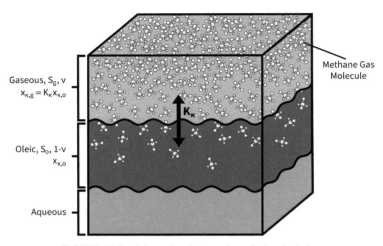

Gaseous, S_g, v
$x_{\kappa,g} = K_\kappa x_{\kappa,o}$

Methane Gas Molecule

K_κ

Oleic, S_o, 1-v
$x_{\kappa,o}$

Aqueous

FIGURE 10.1 Schematic of a two-phase flash calculation.

and gas, in which case Raoult's law and Dalton's law of partial pressures apply, which leads to,

$$K_\kappa = \frac{P_\kappa^{sat}}{p}, \tag{10.12}$$

where P_κ^{sat} is the vapor pressure of component κ. The vapor pressure is a function of temperature and can be estimated from various methods. However, mixtures are far from ideal at reservoir conditions, so Eq. (10.12) is not recommended for reservoir simulation, even as an initial guess. A better estimate can be found from Wilson's (1969) equation,

$$K_\kappa = \frac{p_{c,\kappa}}{p} \exp\left[5.37(1 + \omega_\kappa)\left(1 - \frac{T_{c,\kappa}}{T}\right)\right], \tag{10.13}$$

where p and T are the absolute temperature and pressure of the grid block, $p_{c,\kappa}$ and $T_{c,\kappa}$ are the critical pressure and temperature, respectively, of the component, and ω_κ is the acentric factor of the component. Eq. (10.13) may be used as an initial guess for the first timestep. However, in a time-dependent reservoir simulation, the K-value from the previous timestep may give a very good estimate. An even better estimate may come from extrapolation from the previous timesteps,

$$K_\kappa^n = \frac{p_i^n}{p_i^{n-1}} K_\kappa^{n-1}. \tag{10.14}$$

Regardless of the initial guess, an iterative approach will be required that uses an EOS as described in Section 10.4.2.

Once K-values are determined (or estimated), the number of phases present can be determined. The criteria for the existence of two phases are,

$$\sum_{\kappa=1}^{N_c} K_\kappa x_k > 1 \text{ and } \sum_{\kappa=1}^{N_c} \frac{x_k}{K_\kappa} > 1. \tag{10.15}$$

If the first criterion is not met, then only an oleic phase is present, and if the second criterion is not met, then only a gaseous phase is present. If there are two phases, a flash calculation must be performed to determine the compositions of each phase and the mole fraction of each phase.

10.4.1.2 Rachford–Rice flash

If two hydrocarbon phases are present, then the component and overall balances are utilized to obtain,

$$\sum_{\kappa=1}^{N_c} x_{\kappa,\alpha} = \sum_{\kappa=1}^{N_c} \frac{x_\kappa}{1 + \upsilon(K_\kappa - 1)} = 1. \tag{10.16}$$

Eq (10.16) is the Rachford–Rice equation (Rachford and Rice, 1952), a nonlinear equation in terms of υ (Fig. 10.2) that can be solved using root-finding methods such as bisection (bound between 0 and 1) or Newton–Raphson. Newton's method is generally the faster of the two but may converge to a nonphysical answer ($v = 0$) or diverge if a poor guess is provided (value in previous timestep is recommended). More advanced solution methods have been developed that are fast and guaranteed to converge (Michelsen, 1998).

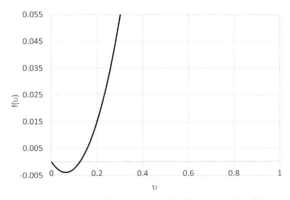

FIGURE 10.2 Nonlinear Rachford–Rice equation for Example 10.1. The root is $\upsilon = 0.1252$.

Upon solution, the oleic phase compositions can be computed explicitly using Eq. (10.15) and then gaseous phase compositions using K-values. Pseudocode is shown below of a subroutine for a flash calculation. Example 10.1 demonstrates a flash calculation in a single grid block for a three-component mixture.

Example 10.1 Flash calculation.

Three hydrocarbon components, methane (C1), n-butane (C4), and n-decane (C10) with overall molar composition of 0.5, 0.2, and 0.3, respectively, are partitioned between an oleic and gaseous phase in a reservoir grid block.

(a) Estimate K values for each component using Wilson's (1969) equation at a pressure and temperature 2000 psia and 150 F, respectively. The critical properties are $T_c = [343; 765; 1111.7]$ °R and $p_c = [666.4; 550.6; 305.2]$ psia. The acentric factors are $\omega = [0.008; 0.193; 0.4868]$.

(b) Using improved K-values of C1 $= 2.1251$, C4 $= 0.2916$, and C10 $= 0.0137$, determine the mole fraction of hydrocarbons that are gaseous, v, and the compositions of each phase ($x_{\kappa,o}, x_{\kappa,g}$).

Solution

(a) Using Wilson's equation,

$$K_{C1}^{n+1} = \frac{666.4}{2000} \exp\left[5.37(1 + 0.008)\left(1 - \frac{343}{(150 + 460)}\right)\right] = 3.5618$$

$$K_{C4}^{n+1} = \frac{550.6}{2000} \exp\left[5.37(1 + 0.193)\left(1 - \frac{765}{(150 + 460)}\right)\right] = 0.0541$$

$$K_{C10}^{n+1} = \frac{305.2}{2000} \exp\left[5.37(1 + 0.4868)\left(1 - \frac{1111.7}{(150 + 460)}\right)\right] = 0.0002146$$

The computed K- values are of similar in magnitude to the actual values provided in part (b), but do have significant error. Advanced methods for computing K-values are discussed in Section 10.4.2.

(b) A flash calculation is used to determine, v.

$$\sum_{\kappa=1}^{N_c} \frac{x_\kappa}{1 + v(K_\kappa - 1)} = 1$$

Plugging in the overall molar compositions and improved K-values,

$$f(v) = \frac{0.5}{1 + v(2.1251 - 1)} + \frac{0.2}{1 + v(0.2916 - 1)} + \frac{0.3}{1 + v(0.0137 - 1)} - 1 = 0$$

Solution using Newton–Raphson or Bisection gives (see Example 6.2),

$$v = 0.1252 \frac{\text{moles gaseous}}{\text{moles gaseous} + \text{oleic}}$$

The compositions of the oleic and gaseous phases can then be computed,

$$x_{\kappa,o} = \frac{x_\kappa}{1 + v(K_\kappa - 1)}; \quad x_{\kappa,g} = K_\kappa \cdot x_{\kappa,o}$$

$$x_{C1,o} = \frac{0.5}{1 + 0.1252(2.1251 - 1)} = 0.4382; \quad x_{C1,g} = 2.1251 \cdot 0.4382 = 0.9313$$

$$x_{C4,o} = \frac{0.2}{1 + 0.1252(0.2916 - 1)} = 0.2195; \quad x_{C4,g} = 0.2916 \cdot 0.2195 = 0.0640$$

$$x_{C10,o} = \frac{0.3}{1 + 0.1252(0.0137 - 1)} = 0.3423; \quad x_{C10,g} = 0.0137 \cdot 0.3423 = 0.0047$$

The gaseous phase is mostly the light component, methane, and the oleic phase is mostly the heavy component, decane.

10.4.2 Equations of state

An equation of state (EOS) relates the pressure, volume, and temperature of a phase. As discussed in Chapter 1, the simplest EOS is the ideal gas law, $pV_{mg} = RT$, which is only valid at low pressures (e.g., 1 atm; 14.7 psia) and not valid at typical reservoir conditions of high pressure and temperature. A compressibility factor for each phase, $z_\alpha = pV_{m,\alpha}/RT$, is often employed to account for nonideality, which is a function of pressure, temperature, and composition.

10.4.2.1 Cubic EOS

Many researchers have attempted to correct the ideal gas law for nonideal conditions. One important class of EOS, that is commonly used in reservoir simulation, is cubic; it has the general cubic polynomial form,

$$z_\alpha^3 + \alpha z_\alpha^2 + \beta z_\alpha + \gamma = 0, \tag{10.17}$$

where α, β, and γ are coefficients of the cubic equation. Although many cubic EOSs exist in the literature, three common ones are presented here: Van der Waals (1873), Soave-Redlich-Kwong (SRK) (1972), and Peng-Robinson (PR) (1976). The Van der Waals EOS is rarely used in practice but is useful for theoretical understanding of vapor–liquid equilibria. Many reservoir simulators use the SRK or PR EOS and the two models generally give similar results, although the SRK EOS often overpredicts liquid–phase molar volume unless a correction is used. Table 10.2 compares the coefficients Eq. (10.16) for the three models. Péneloux et al. (1982) added a volume shift correction, applicable to both the SRK and PR equations of state, to improve the liquid—phase molar volume. Although the Péneloux correction is used in most compositional reservoir simulators, it is not discussed further here.

In Table 10.2 the parameters A and B are defined as,

$$A = \frac{ap}{(RT)^2}; \quad B = \frac{bp}{RT},$$

(10.18)

and T_c and p_c are the critical temperature and pressure, respectively, and ω_κ is the acentric factor of the component, all of which can be measured experimentally and are tabulated for many common fluids. T_r and p_r are the pseudoreduced temperature (T/T_c) and pressure (p/p_c), respectively. Table 10.2 and Eq. (10.18) can be used to compute the constants in the cubic Eq. (10.17) but require mixing rules to be applied to multicomponent mixtures.

10.4.2.2 Mixing rules

Multicomponent fluids, such as reservoir fluids, require mixing rules to compute the coefficients of the cubic EOS. The mixing rules are a weighted average of the component constants,

$$b = \sum_\kappa x_{\kappa,\alpha} b_\kappa$$

$$a = \sum_\kappa \sum_\lambda x_{\kappa,\alpha} x_{\lambda,\alpha} \left(1 - \delta_{\kappa,\lambda}\right) \left(a_\kappa a_\lambda\right)^{1/2}$$

(10.19)

where $\delta_{\kappa,\lambda}$ is a binary interaction parameter between components κ and λ and is less than 1.0. For components very similar in molecular structure the parameter can be close or equal to 0.0. Binary interaction parameters between many common components are tabulated in the literature (Sandler, 2006).

TABLE 10.2 Coefficients for Van der Waals (VDW), Soave—Redlich—Kwong (SRK), and Peng—Robinson EOS.

	VDW	SRK	Peng—Robinson
α	$-1-B$	-1	$-1+B$
β	A	$A-B-B^2$	$A-3B^2-2B$
γ	$-AB$	$-AB$	$-AB+B^2+B^3$
a_κ	$\frac{27R^2 T_c^2}{64 p_c}\sigma_\kappa$	$0.42748\frac{R^2 T_c^2}{p_c}\sigma_\kappa$	$0.45724\frac{R^2 T_c^2}{p_c}\sigma_\kappa$
σ_κ	1	$\left[1 + m\left(1 - T_r^{1/2}\right)\right]^2$	$\left[1 + m\left(1 - T_r^{1/2}\right)\right]^2$
m_κ		$0.480 + 1.574\omega - 0.176\omega^2$	$0.37464 + 1.54226\omega - 0.26992\omega^2$
b_κ	$\frac{RT_c}{8p_c}$	$0.08664\frac{RT_c}{p_c}$	$0.077796\frac{RT_c}{p_c}$

10.4.2.3 Solution to the cubic EOS

Table 10.2 and Eqs. (10.18), (10.19) provide the coefficients of the cubic EOS. For multicomponent, two-phase mixtures, two cubic equations are formed, one for the gaseous phase and one for the oleic phase. Solution to the cubic equations results in the z-factors, z_g and z_o, for the gaseous and oleic phases, respectively. The molar volumes of each phase, V_{mg} and V_{mo}, can then be computed from $V_{m,\alpha} = z_\alpha RT/p$.

Solution for each cubic equation can produce up to three roots which can be solved numerically or analytically; analytical solution is preferred because of computational speed. The analytical solution is given by,

$$z_j = -\frac{1}{3}\left(\alpha + v_j C + \frac{\Delta_0}{v_j C}\right); \quad j = 1, 2, 3, \tag{10.20}$$

where,

$$v_1 = 1; \quad v_2 = \frac{-1 + i\sqrt{3}}{2}; \quad v_3 = \frac{-1 - i\sqrt{3}}{2},$$

$$C = \sqrt[3]{\frac{\Delta_1 + \sqrt{-27\Delta}}{2}},$$

$$\Delta_0 = \alpha^2 - 3\beta$$

$$\Delta_1 = 2\alpha^3 - 9\alpha\beta + 27\gamma.$$

The discriminant, Δ, is defined as,

$$\Delta = 18\alpha\beta\gamma - 4\alpha^3\gamma + \alpha^2\beta^2 - 4\beta^3 - 27\gamma^2. \tag{10.21}$$

Up to three solutions per equation can be found depending on the value of the discriminant:

1. $\Delta < 0$. There is one real root and two complex roots. The real root is used.
2. $\Delta > 0$. There are three real roots. If the equation is for the oleic phase, the smallest of the three roots is correct (lowest molar volume) and if the equation is for the gaseous phase, the largest of the three roots is used (largest molar volume).
3. $\Delta = 0$. There is one real, multiple root. This occurs where there is only one hydrocarbon phase, i.e., supercritical fluid.

Fig. 10.3 shows the cubic polynomial for both oleic and gaseous phases described by Examples 10.1 and 10.2. In each phase, only one real root exists and is therefore used for the respective phase.

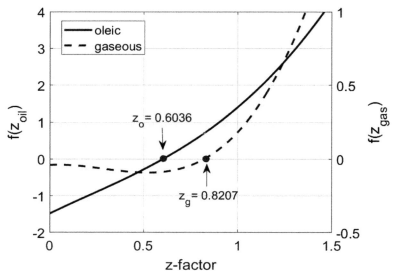

FIGURE 10.3 P-R Cubic EOS for both gaseous and oleic phases in Example 10.2. Roots occur at $f(z) = 0$. Only one real root exists for both phases in this example.

10.4.2.4 K-value calculation

The cubic EOS can also be used to compute accurate K-values, an improvement over the initial guesses described in Section 10.4.1.1. The fugacities of each component must be equal in the oleic and gaseous phases ($f_{\kappa,g} = f_{\kappa,o}$). Using the definitions of fugacities,

$$\phi_{\kappa,g} x_{\kappa,g} p = \phi_{\kappa,o} x_{\kappa,o} p, \qquad (10.22)$$

where $\phi_{o,\kappa}$ and $\phi_{g,\kappa}$ are the fugacity coefficients of component κ in the oleic and gaseous phases, respectively. From Eq. (10.22), the K-values are the ratios of fugacity coefficients,

$$K_\kappa = \frac{x_{\kappa,g}}{x_{\kappa,o}} = \frac{\phi_{\kappa,o}}{\phi_{\kappa,g}}. \qquad (10.23)$$

The fugacity coefficients can be calculated from the EOS parameters (McCain, 1990; Sandler, 2006),

$$\ln \phi_{\kappa,\alpha} = -\ln(z_\alpha - B) + (z_\alpha - 1)B'_\kappa - \frac{A}{2^{1.5}B}\left(A'_\kappa - B'_\kappa\right)\ln\left[\frac{z_\alpha + \left(2^{1/2} + 1\right)B}{z_\alpha - \left(2^{1/2} - 1\right)B}\right].$$

$$(10.24)$$

where A and B are given in Eq. (10.17) and,

$$A'_\kappa = \frac{1}{a}\left[2a_\kappa^{1/2}\sum_\lambda y_\lambda a_\lambda^{1/2}\left(1 - \delta_{\kappa,\lambda}\right)\right]$$

$$B'_\kappa = \frac{b_\kappa}{b}$$

If the estimated K-values used to perform the flash calculation match the values using Eq. (10.23), the phase behavior calculation is complete in the grid block for the current timestep and the phase compositions ($x_{\kappa,o}$ and $x_{\kappa,g}$), mole fraction of gaseous phase (v), and phase molar volumes (V_{mg} and V_{mo}) are defined. If the new K-values are different, iterations are required and flash/EOS calculations must be continued until convergence is obtained, e.g.,

$$\varepsilon_\kappa = \frac{\left(K_\kappa^{new} - K_\kappa^{old}\right)^2}{K_\kappa^{new}K_\kappa^{old}} < tol, \tag{10.25}$$

where *tol* is a predetermined tolerance. Example 10.2 demonstrates the use of an EOS to determine phase molar volumes and K values.

Example 10.2 EOS calculation.
Using the phase compositions computed in Example 10.1, determine the (a) molar volumes (V_{mg}, V_{mo}) of each hydrocarbon phase and (b) K-values of each component. The pressure and temperature are 2000 psia and 150 °F, respectively. Use the Peng–Robinson Equation of State and binary interaction parameters, $\delta_{C1\text{-}C10} = 0.04$; $\delta_{C4\text{-}C10} = 0.008$. The component critical properties and acentric factors were given in Example 10.1.

Solution
(a) First, calculate the component properties using equations in Table 10.2,

$$m_{C1} = 0.37464 + 1.54226\cdot(0.008) - 0.26992(0.008)^2 = 0.3870$$
$$m_{C4} = 0.37464 + 1.54226\cdot(0.193) - 0.26992(0.193)^2 = 0.6622$$
$$m_{C10} = 0.37464 + 1.54226\cdot(0.4868) - 0.26992(0.4868)^2 = 1.0614$$

$$a_{C1} = 0.45724\frac{\left(10.732\frac{\text{psia} - \text{ft}^3}{\text{lbmole} - °R}\right)^2 (343 °R)^2}{666.4 \text{ psia}}\left[1 + 0.3870\left(1 - \sqrt{\frac{610R}{343}}\right)\right]^2$$

$$= 7052\frac{\text{psia} - \text{ft}^6}{\text{lbmole}^2}$$

Continued

Example 10.2 EOS calculation.—cont'd

$$a_{C4} = 0.45724 \frac{\left(10.732 \frac{\text{psia} - \text{ft}^3}{\text{lbmole} - {}^\circ\text{R}}\right)^2 (765 \ {}^\circ\text{R})^2}{550.6 \ \text{psia}} \left[1 + 0.6622\left(1 - \sqrt{\frac{610R}{765}}\right)\right]^2$$

$$= 64191 \frac{\text{psia} - \text{ft}^6}{\text{lbmole}^2}$$

$$a_{C10} = 0.45724 \frac{\left(10.732 \frac{\text{psia} - \text{ft}^3}{\text{lbmole} - {}^\circ\text{R}}\right)^2 (1111.7 \ {}^\circ\text{R})^2}{305.2 \ \text{psia}} \left[1 + 1.0614\left(1 - \sqrt{\frac{610R}{1111.7}}\right)\right]^2$$

$$= 346769 \frac{\text{psia} - \text{ft}^6}{\text{lbmole}^2}$$

$$b_{C1} = 0.077796 \frac{10.73 \cdot 343}{664.4} = 0.430 \frac{\text{ft}^3}{\text{lbmole}}$$

$$b_{C4} = 0.077796 \frac{10.73 \cdot 765}{550.6} = 1.160 \frac{\text{ft}^3}{\text{lbmole}}$$

$$b_{C10} = 0.077796 \frac{10.73 \cdot 1111.7}{305.2} = 3.041 \frac{\text{ft}^3}{\text{lbmole}}$$

Next use the mixing rules for both the oleic and gaseous phases. Use compositions computed in Example 10.1

$$b_o = \sum_\kappa x_{\kappa,o} b_\kappa = 0.4382 \cdot 0.430 + 0.2195 \cdot 1.160 + 0.3423 \cdot 3.041 = 1.4839 \frac{\text{ft}^3}{\text{lbmole}}$$

$$a_o = \sum_\kappa \sum_\lambda x_{\kappa,\alpha} x_{\lambda,\alpha} \left(1 - \delta_{\kappa,\lambda}\right) (a_\kappa a_\lambda)^{\frac{1}{2}} = 0.4382 \cdot 0.4382 \cdot (1 - 0)(7052 \cdot 7052)^{\frac{1}{2}}$$

$$+ \ 0.4382 \cdot 0.2195 \cdot (1 - 0)(7052 \cdot 64191)^{\frac{1}{2}} + 0.4382 \cdot 0.3423 \cdot (1 - 0.04)(7052 \cdot 346769)^{\frac{1}{2}}$$

$$+ \ 0.2195 \cdot 0.4382 \cdot (1 - 0)(64191 \cdot 7052)^{\frac{1}{2}} + 0.2195 \cdot 0.2195 \cdot (1 - 0)(64191 \cdot 64191)^{\frac{1}{2}}$$

$$+ \ 0.2195 \cdot 0.3423 \cdot (1 - 0.008)(64191 \cdot 346769)^{\frac{1}{2}} +$$

$$0.3423 \cdot 0.4382 \cdot (1 - 0.04)(346769 \cdot 7052)^{\frac{1}{2}}$$

$$+ \ 0.3423 \cdot 0.2195 \cdot (1 - 0.008)(346769 \cdot 64191)^{\frac{1}{2}}$$

$$+ \ 0.3423 \cdot 0.3423 \cdot (1 - 0)(346769 \cdot 346769)^{\frac{1}{2}} = 85575 \frac{\text{psia} - \text{ft}^6}{\text{lbmole}^2}$$

$$b_g = \sum_\kappa x_{\kappa,g} b_\kappa = 0.9313 \cdot 0.430 + 0.0640 \cdot 1.160 + 0.0047 \cdot 3.041 = 0.4887 \frac{\text{ft}^3}{\text{lbmole}}$$

$$a_g = \sum_\kappa \sum_\lambda x_{\kappa,g} x_{\lambda,g} \left(1 - \delta_{\kappa,\lambda}\right) (a_\kappa a_\lambda)^{\frac{1}{2}} = 0.9313 \cdot 0.9313 \cdot (1 - 0)(7052 \cdot 7052)^{\frac{1}{2}}$$

$$+ \ 0.9313 \cdot 0.064 \cdot (1 - 0)(7052 \cdot 64191)^{\frac{1}{2}}$$

$$+ \; 0.9313 \cdot 0.0047 \cdot (1 - 0.04)(7052 \cdot 346769)^{\frac{1}{2}}$$

$$+ \; 0.064 \cdot 0.9313 \cdot (1 - 0)(64191 \cdot 7052)^{\frac{1}{2}}$$

$$+ \; 0.064 \cdot 0.064 \cdot (1 - 0)(64191 \cdot 64191)^{\frac{1}{2}}$$

$$+ \; 0.064 \cdot 0.0047 \cdot (1 - 0.008)(64191 \cdot 346769)^{\frac{1}{2}} +$$

$$0.0047 \cdot 0.9313 \cdot (1 - 0.04)(346769 \cdot 7052)^{\frac{1}{2}}$$

$$+ 0.0047 \cdot 0.064 \cdot (1 - 0.008)(346769 \cdot 64191)^{\frac{1}{2}}$$

$$+ 0.0047 \cdot 0.0047 \cdot (1 - 0)(346769 \cdot 346769)^{\frac{1}{2}} = 9400.8 \frac{\text{psia} - \text{ft}^6}{\text{lbmole}^2}$$

Compute the coefficients of the cubic polynomial,

$$A_o = \frac{a_o P}{(RT)^2} = \frac{85,575 \cdot 2000}{(10.73 \cdot 610)^2} = 3.9935; \quad B_o = \frac{b_o P}{RT} = \frac{1.4839 \cdot 2000}{10.73 \cdot 610} = 0.4533$$

$$A_g = \frac{a_g P}{(RT)^2} = \frac{9400.8411 \cdot 2000}{(10.73 \cdot 610)^2} = 0.4387; \quad B_g = \frac{b_g P}{RT} = \frac{0.4887 \cdot 2000}{10.73 \cdot 610} = 0.1493$$

$$\alpha_o = -1 + B_o = -1 + 0.4533 = -0.5467;$$

$$\beta_o = A_o - 3B_o^2 - 2B_o = 3.9935 - 3 \cdot 0.4533^2 - 2 \cdot 0.4533 = 2.4704;$$

$$\gamma_o = -A_o B_o + B_o^2 + B_o^3 = -3.9935 \cdot 0.4533 + 0.4533^2 + 0.4533^3 = -1.5117$$

$$\alpha_g = -1 + B_g = -1 + 0.1493 = -0.8507;$$

$$\beta_g = A_g - 3B_g^2 - 2B_g = 0.4387 - 3 \cdot 0.1493^2 - 2 \cdot 0.1493 = 0.0732;$$

$$\gamma_g = -A_g B_g + B_g^2 + B_g^3 = -0.4387 \cdot 0.1493 + 0.1493^2 + 0.1493^3 = -0.0399$$

which gives cubic polynomials for the oleic and gaseous phases,

$$z_o^3 - 0.5467 z_o^2 + 2.4704 z_o - 1.5117 = 0$$

$$z_g^3 - 0.8507 z_g^2 + 0.0732 z_g - 0.0399 = 0$$

The roots can be solved analytically. The discriminant of both phases:

$$\Delta_o = 18 \alpha_o \beta_o \gamma_o - 4\alpha_o^3 \gamma_o + \alpha_o^2 \beta_o^2 - 4\beta_o^3 - 27\gamma_o^2 = -84.422$$

$$\Delta_g = 18 \alpha_g \beta_g \gamma_g - 4\alpha_g^3 \gamma_g + \alpha_g^2 \beta_g^2 - 4\beta_g^3 - 27\gamma_g^2 = -0.09421$$

Continued

Example 10.2 EOS calculation.—cont'd

The oleic and gaseous phases each have one real root and two imaginary roots because the discriminant is negative. Using Eq. (10.20), the roots of both equations can be found and the real roots used,

$$z_o = 0.6036; \quad z_g = 0.8207$$

$$V_{mo} = \frac{z_o RT}{p} = \frac{0.6036 \cdot 10.73 \cdot 610}{2000} = 1.9756 \frac{ft^3}{lbmole}$$

$$V_{mg} = \frac{z_g RT}{p} = \frac{0.8207 \cdot 10.73 \cdot 610}{2000} = 2.6863 \frac{ft^3}{lbmole}$$

(b) The new K values for the three components can be computed using Eqs (10.23) and (10.24)

$$A'_{\kappa,o} = \frac{1}{a_o} \left[2a_\kappa^{1/2} \sum_\lambda x_{\lambda,o} a_\lambda^{1/2} (1 - \delta_{\kappa,\lambda}) \right] = [0.5581 \quad 1.7314 \quad 4.0184]$$

$$A'_{\kappa,g} = \frac{1}{a_g} \left[2a_\kappa^{1/2} \sum_\lambda x_{\lambda,g} a_\lambda^{1/2} (1 - \delta_{\kappa,\lambda}) \right] = [1.7284 \quad 5.2373 \quad 11.7462]$$

$$B'_{\kappa,o} = \frac{b_\kappa}{b_o} = [0.2896 \quad 0.7818 \quad 2.0495]$$

$$B'_{\kappa,g} = \frac{b_\kappa}{b_g} = [0.8793 \quad 2.3736 \quad 6.2227]$$

Plugging into the equations gives:

$$K_{C1} = \frac{\phi_{C1,O}}{\phi_{C1,G}} = 2.1251$$

$$K_{C4} = \frac{\phi_{C4,O}}{\phi_{C4,G}} = 0.2916$$

$$K_{C10} = \frac{\phi_{C10,O}}{\phi_{C10,G}} = 0.0137$$

K-values did not change significantly from the values provided in Example 10.1, part (b). If they did change, we would iterate on the flash calculation and EOS until convergence.

10.4.3 Phase saturation

Using the definition of phase saturation as the ratio of the volume of the phase to the total pore volume gives the hydrocarbon phase saturations,

$$S_g = \frac{V_{mg} v}{V_{mg} v + V_{mo} (1 - v)} (1 - S_w)$$

$$S_o = \frac{V_{mo} (1 - v)}{V_{mg} v + V_{mo} (1 - v)} (1 - S_w), \tag{10.26}$$

where the initial water saturation in each grid block is defined and can be updated explicitly at the end of the timestep in a manner similar to described in Chapter 9. Example 10.3 demonstrates the calculation of phase saturations.

Example 10.3

Compute the saturations of the gaseous (S_g) and oleic (S_o) phase for the compositions described in Example 10.1 and pressure/temperature in Example 10.2. Assume a water saturation, S_w, of 10%.

Solution

$$S_o = \frac{V_{mo}(1-v)}{V_{mg}v + V_{mo}(1-v)}(1-S_w) = \frac{1.9756(1-0.1252)}{2.6863 \cdot 0.1252 + 1.9756(1-0.1252)}(1-0.1) = 0.7533$$

$$S_g = \frac{V_{mg}v}{V_{mg}v + V_{mo}(1-v)}(1-S_w) = \frac{2.6863 \cdot 0.1252}{2.6863 \cdot 0.1252 + 1.9756(1-0.1252)}(1-0.1) = 0.1467$$

10.4.4 Two-phase compressibility

Compressibility was defined in Chapter 1 as a relative change in volume with respect to a change in pressure. Computation of multicomponent, two-phase compressibility is challenging because a change in pressure also results in a change of composition. The equation for compressibility of the entire hydrocarbon mixture (both the oleic and gaseous) was given in Eq. (10.9), where the molar volume is a weighted average of phase molar volumes, Eq. (10.8). The partial derivative in Eq. (10.9) can be computed analytically but is cumbersome (Firoozabadi et al., 1988). Therefore, we show a numerical approach that uses a finite difference approximation,

$$c_{og} = -\frac{1}{V_{m,og}}\frac{\partial V_{m,og}}{\partial p} \approx -\frac{1}{V_{m,og}}\frac{V_{m,og}(p+\delta p) - V_{m,og}(p)}{\delta p}, \tag{10.27}$$

where $V_{m,og}$ can be computed using Eq. (10.8) and the known phase molar volumes and mole fractions. Then the flash and EOS calculations can be performed again at a small, perturbed pressure, $p + \delta p$. Flash and EOS calculations are computationally expensive; a faster approximation uses the values in the grid block in the previous timestep but can lead to errors if the pressure changes significantly.

The total compressibility is a weighted sum of phase compressibilities and the formation compressibility,

$$c_t = c_{og}(1 - S_w) + c_w S_w + c_f \tag{10.28}$$

Example 10.4 demonstrates the calculation of compressibility in a grid block.

Example 10.4

Compute the total compressibility of the system described in Examples 10.1–10.3. Use finite differences with $\delta p = 20$ psi to determine the hydrocarbon compressibility. The aqueous phase and formation compressibility are both 1E-06 psi^{-1}.

Solution

Use Eq. (10.26). At the new pressure, estimate the K-values of all three components at 2020 psia,

$$K_i(2000 + 20) = K_i \frac{2000}{2000 + 20} = [2.0819 \quad 0.28872 \quad 0.013642]$$

After iteration using flash calculations and the EOS, the following K-values are obtained,

$$K_{C1} = 2.1083; \quad K_{C4} = 0.2934; \quad K_{C10} = 0.0141$$

Perform a flash calculation at 2020 psia using the new K-values,

$$v = 0.1191 \frac{\text{moles gaseous}}{\text{moles gaseous + oleic}}$$

$$x_{\kappa,o} = [0.4382 \quad 0.2195 \quad 0.3423]; \quad x_{\kappa,g} = [0.9313 \quad 0.0640 \quad 0.0047]$$

Re-solve Peng-Robinson EOS to get new molar volumes,

$$V_{mo}(2000 + 20) = 1.9689 \frac{\text{ft}^3}{\text{lbmole oleic}}; \; V_{mg}(2000 + 20) = 2.6575 \frac{\text{ft}^3}{\text{lbmole gaseous}}$$

Calculate the new molar volume of the oleic and gaseous phase at 2020 psia using Eq. (10.8),

$$V_{m,og}(2000 + 20) = \left[\left(1.9689 \frac{\text{ft}^3}{\text{lbmole oleic}} \right)(1 - 0.1191) \right.$$
$$\left. + \left(2.6575 \frac{\text{ft}^3}{\text{lbmole gaseous}} \right)(0.1191) \right] = 2.0510 \frac{\text{ft}^3}{\text{lbmole HC}}$$

Calculate the compressibility of the two-phase HC mixture using finite difference approximation

$$c_{og} = -\frac{1}{V_{m,og}} \frac{\partial V_{m,og}}{\partial p} \approx -\frac{1}{V_{m,og}} \frac{V_{m,og}(p + \varepsilon) - V_{m,og}(p)}{\varepsilon}$$

$$= -\frac{1}{2.0646} \frac{2.0510 - 2.0646}{20} = 3.30E - 04 \text{psi}^{-1}$$

Finally, calculate the total compressibility,

$$c_t = S_w c_w + (1 - S_w)c_{og} + c_f = 0.1 \cdot 1E - 6 + 0.9 \cdot 3.3048E - 4 + 1E - 6$$
$$= 2.975E - 04 \text{ psi}^{-1}$$

10.4.5 Phase viscosity

The viscosity of a phase is dependent upon the pressure, temperature, and composition of that phase. Correlations are generally used to estimate multi-component viscosities; the Lorenz-Bray-Clark (LBC) (1964) correlation is presented here and can be used for both the oleic and gaseous phases,

$$\left[(\mu_\alpha - \mu_\alpha^*)\xi_\alpha + 10^{-4} \right]^{1/4} = a_1 + a_2\rho_{r,\alpha} + a_3\rho_{r,\alpha}^2 + a_4\rho_{r,\alpha}^3 + a_5\rho_{r,\alpha}^4, \quad (10.29)$$

where $a_1 = 0.10230$, $a_2 = 0.023364$, $a_3 = 0.058533$, $a_4 = -0.040758$, $a_5 = 0.0093324$. The reduced phase density, $\rho_{r,\alpha}$, is a ratio of the molar volume of the phase and critical molar volume of the mixture:

$$\rho_{r,\alpha} = \frac{V_{mc}}{V_{m,\alpha}} = \frac{\sum_{\kappa=1}^{N_c} x_{\kappa,\alpha} V_{mc,\kappa}}{V_{m,\alpha}}. \quad (10.30)$$

The viscosity-reducing parameter, ξ_a, is defined as,

$$\xi_\alpha = \left(\sum_{\kappa=1}^{N_c} x_{\kappa,\alpha} T_{c,\kappa} \right)^{1/6} \left(\sum_{\kappa=1}^{N_c} x_{\kappa,\alpha} M_\kappa \right)^{-1/2} \left(\sum_{i=1}^{N_c} x_{\kappa,\alpha} P_{c,\kappa} \right)^{-2/3}, \quad (10.31)$$

where $T_{c\kappa}$ is the critical temperature of the component in Kelvin, $p_{c\kappa}$ is the critical pressure of the component in atm, and M_κ is the molecular weight of the component.

Finally, μ_α^* is the phase viscosity of the mixture at low pressures, 1 atm (14.7 psia). Lee and Eakin (1964) proposed a simple correlation,

$$\mu_\alpha^* = \frac{10^{-4}(17.94 + 0.0321M)T^{3/2}}{1.8T + 75.4 + 13.9M}, \quad (10.32)$$

where the temperature is in Kelvin, molecular weight in lb_m/lbmole, and viscosity in cp. The average molecular weight is a mole-fraction weighted value as was done in Eq. (10.30). Herning and Zipperer (1936) proposed a different formula for the phase mixture at low pressures,

$$\mu_\alpha^* = \frac{\sum_{\kappa=1}^{N_c} x_{\kappa,\alpha} \mu_\kappa^* M_\kappa^{1/2}}{\sum_{i=1}^{N_c} x_{\kappa,\alpha} M_\kappa^{1/2}}. \quad (10.33)$$

The low-pressure viscosity of the component can be calculated as (Stiel and Thodos, 1961),

$$\mu_\kappa^* = \begin{cases} 34 \times 10^{-5} \dfrac{T_{r,\kappa}^{0.94}}{\xi_\kappa} & \text{if} \quad T_{r,\kappa} < 1.5 \\[3mm] 17.78 \times 10^{-5} \dfrac{(4.58T_{r,\kappa} - 1.67)^{5/8}}{\xi_\kappa} & \text{if} \quad T_{r,\kappa} > 1.5 \end{cases}, \quad (10.34)$$

where $T_{r,\kappa}$ is the reduced temperature $(T/T_{c,\kappa})$ and the component reduced viscosity parameter is defined by Eq. (10.31) but for the single component. The component critical temperature and pressure in Eq. (10.34) are in Kelvin and atm, respectively.

Brine (aqueous phase) viscosity is usually measured in a viscometer or estimated with a correlation (Numbere et al., 1977) as a function of temperature, pressure, and salinity. Example 10.5 demonstrates the calculation of oleic and gaseous phase viscosity for a three-component mixture. Figs. 10.4A and 10.4B shows the predicted oleic phase and gaseous phase viscosity,

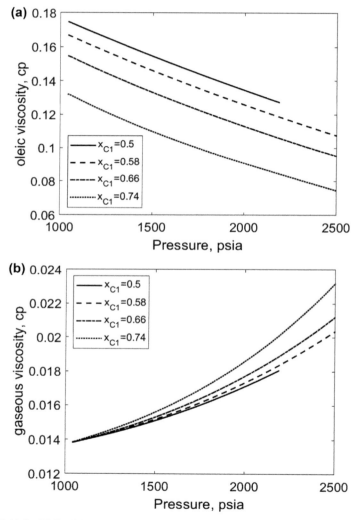

FIGURE 10.4 LBC (1964) prediction of viscosity versus pressure for (a) oleic and (b) gaseous phase at various methane compositions and T=150 F. Each curve represents a constant overall composition of C1 = 0.5, 0.58, 0.66, and 0.74, C4=0.2, and the balance C10. The composition of each phase varies with pressure even at the same overall composition.

respectively, versus pressure of a three component mixture at various overall compositions.

Example 10.5

Compute viscosities of the oleic and gaseous phase at $p = 2000$ psia and $T = 150\ °F$ and the phase compositions determined in Example 10.1. Use the Lorenz–Bray–Clark (LBC) correlation.

Solution

$$\left[(\mu_\alpha - \mu_\alpha^*)\xi_\alpha + 10^{-4}\right]^{1/4} = a_1 + a_2\rho_{r,\alpha} + a_3\rho_{r,\alpha}^2 + a_4\rho_{r,\alpha}^3 + a_5\rho_{r,\alpha}^4$$

For the oleic phase,

$$\rho_{r,o} = \frac{\sum\limits_{\kappa=1}^{N_c} x_{\kappa,\alpha} V_{mc,\kappa}}{V_{m,o}} = \frac{0.4382 \cdot 1.59 + 0.2195 \cdot 4.0774 + 0.3423 \cdot 9.6418}{1.9756} = 2.4762$$

$$\xi_o = \frac{\left(0.4382 \cdot \frac{343°}{1.8}\text{K} + 0.2195 \cdot \frac{765°}{1.8}\text{K} + 0.3423 \cdot \frac{1111.7°}{1.8}\text{K}\right)^{1/6}}{(0.4382 \cdot 16 + 0.2195 \cdot 58 + 0.3423 \cdot 142)^{1/2} \left(0.4382 \cdot \frac{666.4}{14.7}\text{atm} + 0.2195 \cdot \frac{550.6}{14.7}\text{atm} + 0.3423 \cdot \frac{305.2}{14.7}\text{atm}\right)^{2/3}} = 0.0304$$

$$M_o = \sum\limits_{\kappa=1}^{N_c} x_{\kappa,o} M_\kappa = 0.4382 \cdot 16 + 0.2195 \cdot 58 + 0.3423 \cdot 142 = 68.35\ \frac{\text{lb}_m}{\text{lb}_{mole}}$$

$$\mu_o^* = \frac{10^{-4}(17.94 + 0.0321 \cdot 68.35)\left(\frac{610°}{1.8}\text{K}\right)^{1/2}}{1.8\frac{610}{1.8} + 75.4 + 13.9 \cdot 68.35} = 0.0077\text{cp}$$

which can be substituted into Eq. (10.29) obtain the viscosity of the oleic phase,

$$\mu_o = \frac{\left(\left[a_1 + a_2\rho_{r,o} + a_3\rho_{r,o}^2 + a_4\rho_{r,o}^3 + a_5\rho_{r,o}^4\right]^4 - 10^{-4}\right)}{\xi_o} + \mu_o^* = 0.1350\ \text{cp}$$

For the gaseous phase.

$$\rho_{r,g} = \frac{\sum\limits_{\kappa=1}^{N_c} x_{\kappa,\alpha} V_{mc,\kappa}}{V_{m,g}} = \frac{0.9313 \cdot 1.59 + 0.0640 \cdot 4.0774 + 0.0047 \cdot 9.6418}{2.6863} = 0.6652$$

$$\xi_g = \frac{\left(0.9313 \cdot \frac{343°}{1.8}\text{K} + 0.0640 \cdot \frac{765°}{1.8}\text{K} + 0.0047 \cdot \frac{1111.7°}{1.8}\text{K}\right)^{1/6}}{(0.9313 \cdot 16 + 0.0640 \cdot 58 + 0.0047 \cdot 142)^{1/2} \left(0.9313 \cdot \frac{666.4}{14.7}\text{atm} + 0.0640 \cdot \frac{550.6}{14.7}\text{atm} + 0.0047 \cdot \frac{305.2}{14.7}\text{atm}\right)^{2/3}} = 0.044$$

Continued

Example 10.5—cont'd

$$M_g = \sum_{\kappa=1}^{N_c} x_{\kappa,g} M_\kappa = 0.9313 \cdot 16 + 0.0640 \cdot 58 + 0.0047 \cdot 142 = 19.2793 \frac{\text{lb}_m}{\text{lb}_{\text{mole}}}$$

$$\mu_g^* = \frac{10^{-4}(17.94 + 0.0321 \cdot 19.2793)\left(\frac{610^\circ}{1.8}K\right)^{3/2}}{1.8\frac{610}{1.8} + 75.4 + 13.9 \cdot 19.2793} = 0.0121 \text{ cp}$$

which can be substituted into Eq. (10.29) obtain the viscosity of the gaseous phase,

$$\mu_g = \frac{\left(\left[a_1 + a_2\rho_{r,g} + a_3\rho_{r,g}^2 + a_4\rho_{r,g}^3 + a_5\rho_{r,g}^4\right]^4 - 10^{-4}\right)}{\xi_g} + \mu_g^* = 0.0171 \text{ cp}$$

10.4.6 Relative permeability and transmissibility

The relative permeability values can be computed as a function of saturation using correlations and methods (e.g., Stone I's model) described in Chapter 1. Interblock transmissibilities can be computed as was done in Chapter 9 (Eq. 9.12) with the geometric term computed using a harmonic mean and fluid (relative permeability and viscosity) by upwinding from the block of higher potential. The total transmissibility is a weighted sum of phase transmissibilities. Component, phase, and total transmissibilities of each component in each phase can be computed,

$$T_{\kappa,\alpha,i\pm\frac{1}{2}} = x_{\kappa,\alpha,i\pm\frac{1}{2}} T_{\alpha,i\pm\frac{1}{2}}; \quad T_{\kappa,i\pm\frac{1}{2}} = \sum_{\alpha=1}^{N_p} T_{\kappa,\alpha,i\pm\frac{1}{2}}; \quad T_{i\pm\frac{1}{2}} = V_{m,og,i}\left(T_{o,i\pm\frac{1}{2}} + T_{g,i\pm\frac{1}{2}}\right) + B_{w,i}T_{w,i\pm\frac{1}{2}}$$

(10.35)

and the hydrocarbon phase transmissibility is defined with a molar volume[1] instead of formation volume factor as previous chapters. For 1D,

---

1. The phase transmissibilities in Eq. (10.36) could be further multiplied by a molar volume at standard conditions and then the total transmissibility divided by molar volume at standard conditions in Eq. (10.35), which would yield the same units as previous chapters. V_{mg} at standard conditions is known (379.4 ft^3/lbmole), but V_{mo} at standard conditions would need additional calculation or measurement. However, the values would cancel in calculation of total transmissibility, so they value used is of no consequence. For the aqueous phase we use the traditional definition using formation volume factor.

$$T_{\alpha,i+\frac{1}{2}} = \left(\frac{k_{r,\alpha}}{\mu_\alpha V_{m,\alpha}}\right)_{i+\frac{1}{2}} \left(\frac{ka}{\Delta x}\right)_{i+\frac{1}{2}} \tag{10.36}$$

Importantly, the units of phase transmissibilities in Eq. (10.36) are different than in previous chapters because of the inclusion of molar volume instead of formation volume factor. The mole fraction of each component and the molar volume of each phase are upwinded from the block of higher phase potential in Eqs. (10.35) and (10.36).

Dispersive transmissibilities can also be computed,

$$U_{\kappa,\alpha,i\pm\frac{1}{2}} = \left(\frac{S_\alpha \phi D_{\kappa,\alpha} a}{V_{m,\alpha} \Delta x}\right)_{i\pm\frac{1}{2}} ; U_{\kappa,i\pm\frac{1}{2}} = \sum_{\alpha=1}^{N_p} U_{\kappa,\alpha,i\pm\frac{1}{2}}, \tag{10.37}$$

where the phase saturation, dispersion coefficient, and molar volume are upwinded from the block of higher potential. The dispersive transmissibility matrix is tridiagonal in 1D. Example 10.6 demonstrates a calculation of interblock transmissibilities.

Example 10.6
Compute relative permeabilities for all phases with saturations computed in Example 10.3 using Stone I three-phase model with parameters $S_{wr} = 0.1$, $S_{orw} = 0.4$, $S_{org} = 0.2$, $S_{gr} = 0.05$, $k_{rw}^0 = 0.3$, $k_{row}^0 = 0.8$, $k_{rog}^0 = 0.8$, $k_{rg}^0 = 0.3$, $N_w = 2.0$, $N_g = 2.0$, $N_{ow} = 2.0$, $N_{og} = 2.0$.

Then compute the transmissibility of all phases for a grid block with permeability 100 mD, cross-sectional area of 1000 ft², and grid block length of 250 ft. The water viscosity is 0.5 cp.

Solution
The relative permeabilities are computed using Stone I's model (Eqs. 1.24 and 1.26) as was done in Example 1.2,

$$k_{rw} = k_{rw}^0 S_{wD}^{N_w} = 0.0; \quad k_{rg} = k_{rg}^0 S_{gD}^{N_g} = 0.0066$$

$$k_{ro}(S_w, S_o, S_g) = \frac{S_o^* k_{row} k_{rog}}{k_{row}^0 (1 - S_w^*)(1 - S_g^*)} = 0.5797$$

The geometric portion of transmissibility is,

$$\left(\frac{ka}{\Delta x}\right)_{i\pm\frac{1}{2}} = \frac{100 \text{ mD} \cdot \left(1000 \text{ ft}^2\right)}{250 \text{ ft}} = 400 \text{ mD} - \text{ft}$$

Then the phase transmissibilities are,

$$T_{g,i\pm\frac{1}{2}} = \left(\frac{k_{rg}}{V_{mg}\mu_g}\right)_{i\pm\frac{1}{2}} \left(\frac{ka}{\Delta x}\right)_{i\pm\frac{1}{2}} = \frac{0.0066}{2.6863\frac{\text{ft}^3}{\text{lbmole}}0.0171\text{cp}} \cdot 400\text{mD} - \text{ft} = 57.75\frac{\text{mD} - \text{ft}}{\text{cp}}\frac{\text{lbmole}}{\text{ft}^3}$$

$$T_{o,i\pm\frac{1}{2}} = \left(\frac{k_{ro}}{V_{mo}\mu_o}\right)_{i\pm\frac{1}{2}} \left(\frac{ka}{\Delta x}\right)_{i\pm\frac{1}{2}} = \frac{0.5797}{1.9756\frac{\text{ft}^3}{\text{lbmole}}0.135\text{cp}} \cdot 400\text{mD} - \text{ft} = 869.21\frac{\text{mD} - \text{ft}}{\text{cp}}\frac{\text{lbmole}}{\text{ft}^3}$$

$$T_{w,i\pm\frac{1}{2}} = \left(\frac{k_{rw}}{B_w\mu_w}\right)_{i\pm\frac{1}{2}} \left(\frac{ka}{\Delta x}\right)_{i\pm\frac{1}{2}} = \frac{0.000}{1 \cdot 0.5\text{cp}} \cdot 400\text{mD} - \text{ft} = 0.0\frac{\text{mD} - \text{ft}}{\text{cp}}\frac{\text{scf}}{\text{ft}^3}$$

Continued

Example 10.6—cont'd

The component transmissibilities,

$$T_{C1,g,i\pm\frac{1}{2}} = x_{C1,g,i\pm\frac{1}{2}} T_{g,i\pm\frac{1}{2}} = 0.9313 \cdot 57.75 = 53.7856 \frac{\text{lbmole C1}}{\text{ft}^3} \frac{\text{mD} - \text{ft}}{\text{cp}}$$

$$T_{C4,g,i\pm\frac{1}{2}} = x_{C4,g,i\pm\frac{1}{2}} T_{g,i\pm\frac{1}{2}} = 0.064 \cdot 57.75 = 3.6959 \frac{\text{lbmole C1}}{\text{ft}^3} \frac{\text{mD} - \text{ft}}{\text{cp}}$$

$$T_{C10,g,i\pm\frac{1}{2}} = x_{C10,g,i\pm\frac{1}{2}} T_{g,i\pm\frac{1}{2}} = 0.0047 \cdot 57.75 = 0.2711 \frac{\text{lbmole C1}}{\text{ft}^3} \frac{\text{mD} - \text{ft}}{\text{cp}}$$

$$T_{C1,o,i\pm\frac{1}{2}} = x_{C1,o,i\pm\frac{1}{2}} T_{o,i\pm\frac{1}{2}} = 0.4382 \cdot 869.21 = 380.9305 \frac{\text{lbmole C1}}{\text{ft}^3} \frac{\text{mD} - \text{ft}}{\text{cp}}$$

$$T_{C4,o,i\pm\frac{1}{2}} = x_{C4,o,i\pm\frac{1}{2}} T_{o,i\pm\frac{1}{2}} = 0.2195 \cdot 869.21 = 190.7699 \frac{\text{lbmole C1}}{\text{ft}^3} \frac{\text{mD} - \text{ft}}{\text{cp}}$$

$$T_{C10,o,i\pm\frac{1}{2}} = x_{C10,o,i\pm\frac{1}{2}} T_{o,i\pm\frac{1}{2}} = 0.3423 \cdot 869.21 = 297.5173 \frac{\text{lbmole C1}}{\text{ft}^3} \frac{\text{mD} - \text{ft}}{\text{cp}}$$

$$T_{C1} = T_{C1,g} + T_{C1,o} = 434.72 \frac{\text{lbmole C1}}{\text{ft}^3} \frac{\text{mD} - \text{ft}}{\text{cp}}$$

$$T_{C4} = T_{C4,g} + T_{C4,o} = 194.47 \frac{\text{lbmole C1}}{\text{ft}^3} \frac{\text{mD} - \text{ft}}{\text{cp}}$$

$$T_{C10} = T_{C10,g} + T_{C1,o} = 297.79 \frac{\text{lbmole C1}}{\text{ft}^3} \frac{\text{mD} - \text{ft}}{\text{cp}}$$

10.4.7 Wells and source terms

Two types of wells are considered here: (1) injector wells of constant surface rate and specified composition and (2) constant bottomhole pressure producer wells. A constant rate injector with surface rate Q^{sc} is assumed to be either water or gas. Water injectors are treated similarly to Chapter 9; the injection rate is used along with formation volume factor in the pressure equation and also to update water saturation explicitly.

A gas injector will have a specified injected composition, $x_{K,inj}$. The molar injection rate (lbmoles/day) of each component is,

$$Q_{inj,\kappa} = \frac{x_{K,inj}}{V_{m,g}^{sc}} Q_{inj}^{sc}, \tag{10.38}$$

where $V^{sc}_{m,g} = 379.4$ ft^3/lbmole for any gas at standard conditions (14.7 psia, 60 °F). A source vector, $Q_{inj,\kappa}$, can be computed using Eq. (10.38) for all injector wells and used in the component balance Eq. (10.11b). The total injection volumetric rate (ft^3/day), used in the pressure equation, at reservoir conditions is then,

$$Q_{inj,i}^{rc} = Q_{inj,i}^{sc} \frac{V_{mog}}{379.4} \tag{10.39}$$

where the molar volumes and phase compositions are those of the well block, i. A source vector, Q, can be computed using Eq. (10.34) for all injector wells.

The phase and total productivity indices of a well are computed the usual way,

$$J_\alpha = \frac{2\pi k h}{\ln\left(\frac{r_{eq}}{r_w}\right)}\frac{k_{r,\alpha}}{\mu_\alpha}; \quad J = \sum_{\alpha=1}^{N_p} J_\alpha, \tag{10.40}$$

where the relative permeability is computed using the well block saturation for a producer with equivalent radius, r_{eq}, computed using the methods described in Chapter 5. A productivity index matrix, $\mathbf{J}$, for producer wells can be created for use in the pressure equation. Component productivity indices of each phase can also be computed,

$$J_{\kappa,\alpha} = \frac{x_{\kappa,\alpha}}{V_{m,\alpha}}J_\alpha; \quad J_\kappa = \sum_{\alpha=1}^{N_p} J_{\kappa,\alpha} \tag{10.41}$$

The units of the component productivity index are different than the phase or total productivity index. Component productivity matrices can be included in the component balances. Example 10.7 demonstrates calculations for constant rate and constant BHP wells.

Example 10.7

For the grid block described by Example 10.6, compute (a) the component, phase, and overall source vectors for an injector well of 0.1 MMscf/day methane and (b) component, phase, and overall productivity indices for a producer well.

Solution
Computing the source vectors,

$$Q_{C1} = x_{inj,C1}\frac{Q^{sc}}{379.4} = 1.0\,\frac{\text{lbmole C1}}{\text{lbmole gas}}\cdot\frac{1E05\,\text{scf}/\text{day}}{379.4\,\text{scf}/\text{lbmole gas}} = 263.6\,\frac{\text{lbmoles C1}}{\text{day}}$$

$$Q_{C4} = Q_{C10} = 0$$

$$Q^{rc} = Q^{sc}\left(\frac{V_{mog}}{379.4}\right) = 1.0E05\,\frac{\text{scf}}{\text{day}}\left(\frac{2.0646\,\frac{\text{ft}^3\text{HC}}{\text{lbmoles HC}}}{379.4\,\frac{\text{scf}}{\text{lbmole HC}}}\right) = 543.92\,\frac{\text{ft}^3}{\text{day}}$$

Continued

Example 10.7—cont'd

Computing the productivity indices,

$$J_\alpha = \frac{2\pi k h}{\ln\left(\frac{0.2\Delta x}{r_w}\right)} \frac{k_{r\alpha}}{\mu_\alpha} = \frac{2\pi 100 \cdot 10}{\ln\left(\frac{0.2\cdot 250}{0.25}\right)} \frac{k_{r\alpha}}{\mu_\alpha} = 1186 \frac{k_{r\alpha}}{\mu_\alpha}$$

$$J_w = 0.0; \quad J_o = 1186 \frac{0.5797}{0.135} = 5091 \frac{mD - ft}{cp}; \quad J_g = 11856 \frac{0.0066}{0.0171} = 460.0 \frac{mD - ft}{cp}$$

$$J = J_w + J_o + J_g = 5551 \frac{mD - ft}{cp}$$

$$J_{\kappa,\alpha} = \frac{x_{\kappa,\alpha} J_\alpha}{V_{m,\alpha}}$$

$$J_{C1,o} = \frac{0.4382 \cdot 5091}{1.9756} = 1129; \quad J_{C4,o} = \frac{0.2195 \cdot 509}{1.9756} = 565.6;$$

$$J_{C10,o} = \frac{0.3423 \cdot 5091}{1.9756} = 882.1 \frac{lbmole\ \kappa - mD}{cp \cdot ft^2}$$

$$J_{C1,g} = \frac{0.9313 \cdot 460.0}{2.6863} = 159.0; \quad J_{C4,g} = \frac{0.064 \cdot 460.0}{2.6863} = 10.96;$$

$$J_{C10,g} = \frac{0.0047 \cdot 460.0}{2.6863} = 0.8037 \frac{lbmole\ \kappa - mD}{cp \cdot ft^2}$$

10.4.8 Pressure and composition solution

In the IMPEC approach, grid pressures are computed implicitly using Eq. (10.11a) and then component mole fractions are computed explicitly using Eqs. (10.11b). The aqueous phase is assumed to not contain any hydrocarbon components. However, the aqueous phase saturation can change and is updated explicitly using Eq. (9.12b). Once new grid pressures and compositions are determined, the simulation can move to the next time level and the steps outlined above can be repeated. Example 10.8 demonstrates calculation of pressures and new compositions of a three-component, three-phase (oleic, gaseous, plus aqueous), four-grid block system in 1D.

Example 10.8

A reservoir $1000 \times 100 \times 10$ ft has a homogenous permeability and porosity of 100 mD and 20%, respectively. The initial pressure is 2000 psia and the temperature 150 °F. The composition is uniform, C1 = 0.5, C4 = 0.2, and C10 = 0.3. The initial water saturation is 10%, aqueous phase viscosity is 0.5 cp, and water compressibility is 1E-06 psi^{-1}. The formation compressibility is also 1.0E-06 psi^{-1}.

An injector well is placed at $x = 0$ in which 0.1 MMscf/day of methane is injected and a producer well, operated at 800 psia, is located at $x = L$. Solve the pressure equation to obtain block pressures, P, and explicitly solve for the new block compositions for one timestep. Also, update the water saturation explicitly. Use $N = 4$ grids in 1D and a timestep of 0.1 days. Assume flow is advection dominated, so physical dispersion can be neglected.

Solution

All reservoir, fluid, and component properties are homogeneous and uniform; therefore all grid blocks have the same transmissibilities (and no upwinding required) and accumulation matrices at $t = 0$. Many of the necessary calculations were performed in Examples 10.1−10.7.

The phase transmissibility matrices use the values from Example 10.6,

$$\mathbf{T}_g = \begin{pmatrix} 57.75 & -57.75 & & \\ -57.75 & 115.5 & -57.75 & \\ & -57.75 & 115.5 & -57.75 \\ & & -57.75 & 57.75 \end{pmatrix} \frac{md-ft}{cp} \frac{lbmole}{ft^3};$$

$$\mathbf{T}_o = \begin{pmatrix} 869.2 & -869.2 & & \\ -869.2 & 1738.4 & -869.2 & \\ & -869.2 & 1738.4 & -869.2 \\ & & -869.2 & 869.2 \end{pmatrix} \frac{md-ft}{cp} \frac{lbmole}{ft^3}; \quad \mathbf{T}_w = 0.0 \frac{md-ft}{cp} \frac{scf}{ft^3}$$

where the water transmisibility is defined with formation volume factor

$$\mathbf{T} = \mathbf{V}_{m,og} \left(\mathbf{T}_g + \mathbf{T}_o \right) + \mathbf{B}_w \mathbf{T}_w = \begin{pmatrix} 1912.9 & -1912.9 & & \\ -1912.9 & 3825.7 & -1912.9 & \\ & -1912.9 & 3825.7 & -1912.9 \\ & & & 1912.9 \end{pmatrix} \frac{mD-ft}{cp}$$

Continued

Example 10.8—cont'd

The elements of the accumulation matrix, $A_{ii} = hw\Delta x\phi/\Delta t = 500,000 \frac{ft^3}{day}$. The total compressibility of each block was computed in Example 10.4.

$$\mathbf{Ac}_t = \begin{pmatrix} 500000 & & & \\ & 500000 & & \\ & & 50000 & \\ & & & 5000 \end{pmatrix} \begin{pmatrix} 2.975 & & & \\ & 2.975 & & \\ & & 2.975 & \\ & & & 2.975 \end{pmatrix} \times 10^{-4}$$

$$= \begin{pmatrix} 148.8 & & & \\ & 148.8 & & \\ & & 148.8 & \\ & & & 148.8 \end{pmatrix} \frac{ft^3}{psi-day}$$

There is a constant rate injector well in block #1 and a constant BHP producer well in block #4,

$$\vec{Q} = \begin{pmatrix} 543.9 \\ 0 \\ 0 \\ 0 \end{pmatrix} \frac{ft^3}{day}$$

$$\mathbf{J} = \mathbf{J}_w + \mathbf{J}_o + \mathbf{J}_g = diag\begin{pmatrix} 0 \\ 0 \\ 0 \\ 0 \end{pmatrix} + diag\begin{pmatrix} 0 \\ 0 \\ 0 \\ 5091 \end{pmatrix} + diag\begin{pmatrix} 0 \\ 0 \\ 0 \\ 460.0 \end{pmatrix}$$

$$= diag\begin{pmatrix} 0 \\ 0 \\ 0 \\ 5551 \end{pmatrix} \frac{mD-ft}{cp}; \quad \vec{P}_{wf} = \begin{pmatrix} 0 \\ 0 \\ 0 \\ 800 \end{pmatrix} psia$$

Solving the system of equations implicitly,

$$(\mathbf{T} + \mathbf{J} + \mathbf{Ac}_t)\vec{P}^{n+1} = \mathbf{Ac}_t\vec{P}^n + \vec{Q} + \mathbf{J}\vec{P}_{wf}$$

$$\mathbf{P}^n = \begin{pmatrix} 2000 \\ 2000 \\ 2000 \\ 2000 \end{pmatrix} psia \Rightarrow \mathbf{P}^{n+1} = \begin{pmatrix} 2003.3 \\ 1999.2 \\ 1984.8 \\ 1783.9 \end{pmatrix} psia$$

As expected, the pressures increase near the injector well and decrease near the producer. The compositions can then be updated explicitly,

$$\vec{x}_{\kappa}^{n+1} = \vec{x}_{\kappa}^{n} + (\mathbf{I} - diag\{\mathbf{S}_w\})^{-1}\mathbf{A}^{-1}\mathbf{V}_{mog}\left[\mathbf{T}_{\kappa}\vec{P}^{n+1} + \dot{\mathbf{U}}_{\kappa}\vec{x}^{n} + \vec{Q}_{\kappa} + \mathbf{J}_{\kappa}\left(\vec{P}_{wf} - \vec{P}^{n+1}\right)\right]$$

$$\vec{x}_{C1} = \begin{pmatrix} 0.5005 \\ 0.5 \\ 0.5001 \\ 0.5019 \end{pmatrix}; \vec{x}_{C4} = \begin{pmatrix} 0.1998 \\ 0.2 \\ 0.2 \\ 0.1994 \end{pmatrix}; \vec{x}_{C10} = \begin{pmatrix} 0.2997 \\ 0.3 \\ 0.2999 \\ 0.2987 \end{pmatrix}$$

As expected, there is an increase in the concentration of C1 in all blocks because C1 is injected, but a net decrease in C4 and C10 concentration.

$$\vec{S}_{w}^{n+1} = \vec{S}_{w}^{n} + \mathbf{A}^{-1}\mathbf{c}_t^{-1}\left[-\mathbf{T}_w^n\left(\vec{P}^{n+1}\right) + B_w\vec{Q}_w^{sc} + \mathbf{J}_w\left(\vec{P}_{wf} - \vec{P}^{n+1}\right)\right]$$
$$- \vec{S}_{w}^{n}(\mathbf{c}_w + \mathbf{c}_f)\left(\vec{P}^{n+1} - \vec{P}^{n}\right)$$

$$\vec{S}_w = \begin{pmatrix} 0.1 \\ 0.1 \\ 0.1 \\ 0.1 \end{pmatrix}$$

There is negligible change in water saturation because no water is injected or produced and water transmissibility is zero.

10.5 Oleic–aqueous bipartitioning components

Most hydrocarbons have negligible solubility in the aqueous phase[2] and water has negligible solubility in the oleic phase. However, components such as alcohols and perflourocarbons are soluble in both the oleic and aqueous phases. Although alcohols and perflorocarbons do not usually appear naturally in appreciable quantities in subsurface reservoirs, a common application is the injection of these tracers for the purpose of reservoir characterization (e.g., determining sweep efficiency and residual oil saturation (Dean et al., 2016)). Tracers are usually injected at small concentrations and have minimal effect on phase viscosity or density. They also do not affect the phase saturations significantly, although they may partition between the oleic and aqueous

---

2. The solubility of oil in an aqueous phase is generally low (< 100 ppm) which is neglible for the purposes of producing hydrocarbons. However, the solubility may be important for aquifer remediation strategies.

phases. The tracer K-values are measured experimentally or determined from empirical models for use in the reservoir simulator. They are usually assumed constants for the life of the test. Since phase saturations, densities, viscosities, and compressibilities do not depend on the tracer concentration, many of the steps in Section 10.4 can be avoided. Overall composition is updated explicitly in each timestep using Eq. (10.12b).

Conservative tracer tests are common and involve the injection of one or more tracers that remain in a single phase, e.g. chloride or bromide ion in the aqueous phase. A *partioning tracer test* involves at least two tracers, one that is soluble in both oleic and aqueous phases, such as an alcohol or per-flurocarbon (Dugstad et al., 1999), and a conservative one that is soluble only in the aqueous phase. The injected tracer concentrations and masses are specified at each injector well. At the producer well, the concentrations are monitored and the breakthrough times (first appearance of each tracer at producer well) and residence time distributions are recorded. Fig. 10.5 illustrates a partitioning tracer test for the injection of a pulse.

One can analytically derive expressions of the breakthrough times of each tracer for 1D, advection-dominated two-phase flow (dispersion is neglected), and homogenous reservoir properties at residual oil saturation,

$$t_{Dbt,1} = \frac{qt_{bt,1}}{V_b\phi} = (1 - S_{or}) + KS_{or}$$

$$t_{Dbt,2} = \frac{qt_{bt,2}}{V_b\phi} = 1 - S_{or}$$

$$(10.42)$$

where $t_{Dbt,1}$ and $t_{Dbt,2}$ are dimensionless breakthrough times. A reservoir simulator can be used to model more complicated tracer transport when the assumptions of 1D, homogeneity, and/or no dispersion are not valid.

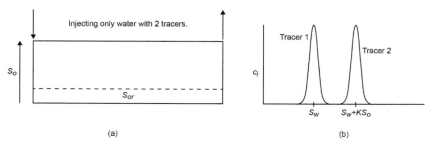

(a)   (b)

FIGURE 10.5 (A) Schematic of a tracer test with two tracers: one that is miscible in both oleic and aqueous phases, and the other miscible only in the aqueous phase; and (B) concentration profiles of both tracers, at the producer well with time.

Tracer tests are *inverse* problems, meaning that reservoir properties, e.g. S_{or} and ϕ, are inferred from measured data, i.e. breakthrough times and residence time distributions (Shook et al., 2009). Eq. (10.42) can be solved simultaneously to obtain porosity and residual oil saturation given the measured breakthrough times. The reservoir simulator solves a *forward* problem; residence time distributions are predicted based on reservoir inputs. However, the simulator can be used to solve inverse problems, such as this one, by iterating on unknown reservoir properties until predicted breakthrough times match the measured ones. This is an example of *history matching*, a broader class of problems in which models/simulators are used to determine reservoir inputs by minimizing the difference in measured data and simulator prediction.

10.6 Pseudocode for multiphase, multicomponent transport

The main pseudocode for multiphase, multicomponent transport of components partitioned between an oleic and gaseous phase is shown below. Pseudocode for for three-phase relative permeability was presented in Chapter 1. Subroutines for preprocessing, postprocessing, and creation of well and block arrays can be adapted from those in Chapter 9. Finally, the main code presented here can be easily adapted for other multiphase applications such as gaseous—aqueous phases in carbon storage or bipartitioning tracers as described in Section 10.5.

```
SUBROUTINE FLASH
INPUTS: P, T, xₖ, K-values
OUTPUTS: υ, xₒ,xg

DEF f(υ) using eqn 10-15
COMPUTE root of f(υ) using bisection or Newton Raphson

FOR κ = 1 to Nc (number components)
     COMPUTE xₖ,ₒ using eqn 10-15
     COMPUTE xₖ,g = Kₖxₖ,ₒ
ENDFOR
```

```
SUBROUTINE EOS
INPUTS: T, P, x, component properties, phase (oleic/gaseous), EOS (PR
or SRK)
OUTPUTS: Vₘα (molar volumes of oleic/gaseous phase), K-values

FOR i = 1 to Nc (number components)
     COMPUTE mₖ,σₖ,aₖ using Table 10.2
ENDFOR
```

```
COMPUTE a,b using mixing rules in eqn 10-19
COMPUTE A,B using eqn 10-18
COMPUTE α,β,γ using Table 10.2

COMPUTE roots, z, of cubic polynomial, eqn 10-17
IF oleic phase
     z = smallest real root
ELSEIF gaseous phase
     z = largest real root
ENDIF

COMPUTE molar volume, Vₘₐ= zₐRT/p

FOR κ = 1 to Nc
     COMPUTE A'ₖ, B'ₖ
     COMPUTE φₖ of oleic phase using eqn 10-24
     COMPUTE φₖ of gaseous phase using eqn 10-24
     COMPUTE Kₖ using eqn 10-23
ENDFOR
```

SUBROUTINE TWO-PHASE COMPRESSIBILITY
INPUTS: P, T, x, component properties, $V_{m,og}$, δP
OUTPUTS: c_{og}

```
CALCULATE K-values at P + δP
CALL FLASH at P + δP
CALL EOS(oleic phase) at P + δP
CALL EOS(gaseous phase) at P + δP

COMPUTE Vₘₒg at P and P+δP using eqn 10-8
COMPUTE cₒg using eqn 10-27
```

SUBROUTINE VISCOSITY
INPUTS: P, T, x_α, phase(α), component properties
OUTPUTS: phase viscosity

```
CALCULATE reduced phase density using eqn 10-30
CALCULATE viscosity reducing parameter using eqn 10-31
CALCULATE low-pressure viscosity using eqn 10-32 or 10-33
CALCULATE phase viscosity using eqn 10-29
```

```
MAIN CODE
CALL PREPROCESS
WHILE t < t_final
     FOR i = 1 to N (number grids)
          INIT tolerance, error
          WHILE error > tolerance
               CALL FLASH
               CALL EOS (oleic phase)
               CALL EOS (gaseous phase)
               CALCULATE error in K-values using eqn 10-25
          ENDWHILE

          CALCULATE phase saturations using eqn 10-26
          CALL TWO-PHASE COMPRESSIBILITY
          CALL VISCOSITY (oleic)
          CALL VISCOSITY (gaseous)
          CALL 3-PHASE RELATIVE PERMEABILITY
     ENDFOR
     CALL GRID ARRAYS
     CALL WELL ARRAYS
     CALCULATE pressure using eqn 10.11a
     CALCULATE composition using eqn 10.11b
     CALCULATE water saturation using eqn 9.7b
     INCREMENT t = t + Δt
ENDWHILE
CALL POSTPROCESS
```

10.7 Exercises

Exercise 10.1. Derivation of finite difference equations. Derive the explicit, finite difference equations (Eq. 10.11b) for component mole fractions, x_κ.

Exercise 10.2. Compositional calculation. Repeat Examples 10.1 through 10.8 but with $P = 1500$ psia, $T = 250\,°F$, $S_{w,init} = 0.2$, and compositions C1 $= 0.4$, C4 $= 0.35$, and C10 $= 0.25$. Use the SRK EOS. To make the problem easier, use PVT software to perform the flash and EOS calculations.

Exercise 10.3. Flash calculation computer code. Write a code (subroutine) to compute phase compositions and mole fractions of each phase given a P, T, and composition using a two-phase flash calculation. Validate the code against Example 10.1 and/or Exercise 10.2.

Exercise 10.4. EOS computer code. Write a code (subroutine) to compute molar volumes of each phase given pressure, temperature, and phase composition using an EOS. For additional flexibility, allow for the user to choose the Van der Waals, Peng–Robinson or SRK EOS. Validate the code against Example 10.2 and/or Exercise 10.2.

Exercise 10.5. Two-phase compressibility computer code. Write a code to numerically estimate the two-phase hydrocarbon compressibility and total compressibility. Validate the code against Example 10.4 and/or Exercise 10.2.

Exercise 10.6. Viscosity of multicompnent mixtures. Write a code to compute the phase viscosity of a mixture given temperature, pressure, composition, and component properties (molecular weight, critical properties). Validate the code against Example 10.5 and/or Exercise 10.2 for both the oleic and gaseous phases.

Exercise 10.7. Multicomponent, multiphase reservoir simulator. Develop a reservoir simulator to perform an IMPEC calculation for a 1D multicomponent, multiphase system. Your code should call the subroutines/functions developed in Exercises 10.3−10.6 as well as new subroutines to compute grid and well arrays adapted from previous chapters. For an additional challenge, make your code more flexible to work for 2D or 3D, include gravity, and dispersion. Validate your code against Example 10.8 and/or Exercise 10.2. Then scale up to $N = 100$ uniform grids in 1D and make plots of producer total rate (reservoir conditions) versus time and component mole fractions of the producer versus time.

Exercise 10.8. Multicomponent, multiphase project. Consider the 2D, two-phase (oleic/aqueous) heterogeneous Thomas oilfield reservoir with the same petrophysical and fluid properties introduced in Exercise 9.11. A constant rate injector well of 1000 STB/day (5615 scf/day) of water is placed at $(x = 0, y = 0)$ and a constant BHP producer of 1000 psia is placed at $(x = L, y = w)$. You may assume dispersion is negligible.

(a) Run your multiphase, oil/water simulator until residual saturation (average $S_w > 0.98(1\text{-}S_{or})$) and steady-state pressure is nearly reached and save the pressure/saturation field.

(b) Adapt your simulator developed in Exercise 10.7 for the case of two-phase (oleic/aqueous) bipartitioning components. Using part (a) as an initial condition, inject two tracers: one miscible in only in the aqueous phase and the other in both phases with $K = x_o/x_w = 1.5$. The tracers should be injected as a pulse. i.e. over a short time period, $t_D = 0.0001$ pore volumes.

(c) Run your simulator and determine the breakthrough times of both tracers. Make plots of the concentration field of tracer #1 and #2 in the aqueous phase at $t_D = 0.5$ PVI. Also make plots of tracer concentrations at the producer well versus dimensionless time.

References

Acs, G., Doleschall, S., Farkas, E., 1985. General purpose compositional model. Society of Petroleum Engineers Journal 25 (04), 543–553.

Chang, Y., 1990. Development and application of an equation of state compositional simulator. PhD Dissertation. The University of Texas at Austin.

Chang, Y.-B., Pope, G.A., Sepehrnoori, K., 1990. A higher-order finite-difference compositional simulator. Journal of Petroleum Science and Engineering 5 (1), 35–50.

Coats, K.H., 1980. An equation of state compositional model. SPE Journal v20 (5), 363–376.

Coats, K.H., Thomas, L.K., Pierson, R.G., 1995. Compositional and black oil reservoir simulation. SPE Reservoir Simulation Symposium. OnePetro.

Dean, R.M., et al., 2016. Use of partitioning tracers to estimate oil saturation distribution in heterogeneous reservoirs. SPE Improved Oil Recovery Conference. OnePetro.

Dugstad, Ø., et al., 1999. Application of tracers to monitor fluid flow in the Snorre field: a field study. SPE Annual Technical Conference and Exhibition. OnePetro.

Farshidi, S.F., et al., 2013. Chemical reaction modeling in a compositional reservoir-simulation framework. SPE Reservoir Simulation Symposium. OnePetro.

Firoozabadi, A., Nutakki, R., Wong, T., Aziz, K., 1988. EOS predictions of compressibility and phase behavior in systems containing water, hydrocarbons, and CO_2. SPE Reservoir Engineering, pp. 673–684.

Herning, F., Zipperer, L., 1936. Calculation of the viscosity of technical gas mixtures from the viscosity of the individual gases. *Gas u.* Wasserfach 79, 69.

Hustad, O.S., Browning, D.J., 2010. A fully coupled three-phase model for capillary pressure and relative permeability for implicit compositional reservoir simulation. Spe Journal 15 (04), 1003–1019.

Kumar, A., 2004. A simulation study of carbon sequestration in deep saline aquifers. MS Thesis. University of Texas at Austin.

Lee, A., Eakin, B., 1964. Gas viscosity of hydrcarbon mixtures. SPE Journal 247–249.

Li, B., Tchelepi, H.A., Benson, S.M., 2013. Influence of capillary-pressure models on CO_2 solubility trapping. Advances in Water Resources 62, 488–498.

Lorenz, J., Bray, B., Clark, C., 1964. Calculating viscosities of reservoir fluids from their compositions. Journal of Petroleum Technology 1171–1176.

McCain Jr., W., 1990. The Properties of Petroleum Fluids, secondedition. PennWell Publishing Company, Tulsa, Oklahoma.

Michelsen, M.L., 1998. Speeding up two-phase PT-flash, with applications for calculation of miscible displacement. Fluid Phase Equilibria 143, 1–12.

Numbere, D., Brigham, W.E., Standing, M.B., 1977. Correlations for physical properties of petroleum reservoir brines. Petroleum Research Institute, Stanford University.

Okuno, R., Johns, R.T., Sepehrnoori, 2010. Three-phase flash in compositional simulation using a reduced method". SPE Journal 689–703.

Péneloux, A., Rauzy, E., Fréze, R., 1982. In: A consistent correction for Redlich-Kwong-Soave volumes, vol. 8. Fluid Phase Equilibria, p. 7e23.

Peng, D., Robinson, D., 1976. A new two-constant equation of state. Industrial and Engineering Chemistry 15 (1), 59–64.

Perschke, D.R., 1988. *Equation of state phase behavior modeling for compositional simulation.* The University of Texas at Austin.

Rachford Jr., H.H., Rice, J.D., 1952. Procedure for use of electronic digital computers in calculating flash vaporization hydrocarbon equilibrium. Journal of Petroleum Technology 4 (10), 19.

Sandler, S., 2006. Chemical, biochemical, and engineering thermodynamics. John Wiley&Sons.

Shook, G.M., Pope, G.A., Asakawa, K. 2009. Determining reservoir properties and flood performance from tracer test analysis. SPE Annual Technical Conference and Exhibition. OnePetro.

Soave, G., 1972. Equilibrium constants from a modified Reflich-Kwong equation of state. Chemical Engineering Science v27 (6), 1197−1203.

Stiel, L.I., George, T., 1961. The viscosity of nonpolar gases at normal pressures. AIChE Journal 7 (4), 611−615.

Sun, Z., Xu, J., Espinoza, D.N., Balhoff, M.T., 2021. Optimization of subsurface CO_2 injection based on neural network surrogate modeling. Computational Geosciences (published online).

Van der Waals, J.H., 1873. On the continuity of the gases and liquid state. Doctoral Dissertation, Leiden University.

Wang, P., et al., 1997. A new generation EOS compositional reservoir simulator: Part I-formulation and discretization. SPE Reservoir Simulation Symposium. OnePetro.

Watts, J.W., 1986. A compositional formulation of the pressure and saturation equations. SPE Reservoir Engineering 1 (03), 243−252.

Wilson, G.M., 1969. A modified Redlich-Kwong equation of state, application to general physical data calculations, vol 15. 65th National AIChE Meeting, Cleveland, OH.

Young, L.C., 2022. Compositional reservoir simulation: A review. SPE Journal 1−47.

Young, L.C., Stephenson, R.E., 1983. A generalized compositional approach for reservoir simulation. Society of Petroleum Engineers Journal 23 (05), 727−742.

Index

'*Note:* Page numbers followed by "f" indicate figures and "t" indicate tables.'

A

Accumulation matrix, 113
Adaptive meshing techniques, 222—224
Advective flux, 175
Advective transport, 175
Aqueous phase
  compressibility, 12
  viscosity, 12—14
Aquifer remediation strategies, 3

B

Black oil model, 3
  capillary pressure, 233
  capillary pressure and relative
      permeability, 232
  hydrocarbon phase properties, 284t
  oleic phase pressure, 233
  PDEs, 231—232
  phase balance equations, 232
  relative phase mobilities, 231—232
Block mass balances, 206—207
Brooks—Corey model, 17
Bubble point, 4—5
Buckley—Leverett solution, 189,
    191—192
Bulk compressibility, 6
Bulk concentration, 3—4

C

Capillary pressure, 30
  definition, 20—21
  drainage process, 20
  imbibition process, 20
  models, 21
  vs. saturation curve, mixed-wet medium,
      20, 20f
  scanning curves, 22—23
Cell-centered approach, 204
Centered finite difference formulas, 66t
Complex gridding algorithms, 95
Component (κ), 2—3

Component transport, porous media
  advection, 175
  1D cartesian ADE
    constant concentration, 186—187
    constant velocity, 186
    diffusion-dominated flow without
        advection, 187
    diffusion-dominated flow without
        diffusion/dispersion, 187
    dimensionless ADE, 186
    dimensionless distance, 186
    Laplace transforms, 187
    mixed boundary condition, 187—188
    Peclet number, 186
    semi-infinite porous medium, 186f
  hydrodynamic dispersion, 176—182
  mass balance equations, 183—186
  reactive transport, 182
  semianalytical solution
    breakthrough time and oil recovery,
        194—195
    Buckley—Leverett solution, 189, 191—192
    capillary pressure, 196
    fractional flow, 189—190
    pseudocode, 196—198
    shock fronts, 192—193
  transport mechanisms, 175—182
Compositional simulators, 2—3, 284t
Compressible gases, diffusivity equation,
    45—47
Constant rate wells, 102, 258—261
Continuity equation
  accumulation terms, 40—41
  mass balance, 40
Control-volume approach, 37, 38f, 98
Convective transport, 175
Crank—Nicholson (C—N) methods, 77—83
Critical point (CP), 4—5
Cubic equation of state
  K-value calculation, 294—298
  mixing rules, 292
  Peng—Robinson, 292t

Cubic equation of state (*Continued*)
 Soave—Redlich—Kwong (SRK), 292t
 solution, 293
Cylindrical coordinates, 51—54

D

Darcy's law, 14—16, 37, 155—156
 one dimentional, 14—15
 permeability, 15
 porous media, 14—15
Density, 3, 7
Dew point, 4—5
1D heat equation
 finite medium, 48—50
 semi-infinite medium, 50—51
Diagonal matrix, 142—143
Diffusive transport, 176—177
Diffusivity equation, 37
 analytical solutions, 48—54
 compressible gases, 45—47
 general multiphase flow, 41—42
 single-phase flow, 42—47
Dirichlet boundary condition, 71, 103—104,
  211—212, 211f
Drawdown pressure, 1—2

E

Effective diffusion coefficient, 177
Effective Peclet number, 221
Empirical models, 16—19
Endpoint permeability, 16—17, 109
Enhanced oil recovery (EOR), 1—2

F

Finite difference solutions
 boundary and initial conditions
  Dirichlet boundary condition, 71
  linear systems of equations, 83—84
  Neumann boundary condition, 71
  Robin boundary conditions, 72
 higher-order approximations, 85—88
 multiphase and multicomponent transport,
  286—287
 multiphase flow
  capillary pressure, 234—235
  mass balance equations, 235—237
  pressure equation, 233—234
  variables, 234
 parabolic diffusivity (heat) equation
  1D diffusivity, 70
  discretization, 68

1D porous medium, 69f
 spatial derivatives, 69
 time interval, 69
 pseudocode, 88—89
 solution methods
  Crank—Nicholson (C—N) methods,
   77—83
  explicit method, 72—76
  implicit method, 76—77
  mixed methods, 77—83
 stability and convergence, 84, 85f
 Taylor series expansion
  first-order backward difference
   approximation, 60
  first-order forward difference
   approximation, 59—60
  higher-order approximations, 64—68
  higher-order derivative approximations,
   61—64
  second derivatives, 61
Finite element method (FEM), 96
Finite volume method (FVM), 96
Flash calculation
 K-values, 287—289
 Rachford—Rice flash, 289—291
 two-phase flash calculation, 288f
Forchheimer flow, 155—156
Formation compressibility, 6, 6f
Formation volume factor, 7—8
Forward/backward finite difference formulas,
 66t
Fully compositional model, 283
Fully implicit method
 black oil model
  interblock transmissibilities and
   upwinding, 248—255
  mass balance equations, 247
  Newton—Raphson approach, 248
  nonlinearities, 247
  saturation and pressure—dependent
   variables, 247
  SS formulation, 247—248
  unconditionally stable, 248
 single-phase component transport,
  218—219

G

Gas compressibility, 8
Gaseous phase
 properties, 7—9
 single—phase flow
  compressibility, 152

Klinkenberg effect, 153
  permeability, 153
  viscosity, 9, 152–153

H

Horizontal wells and anisotropy
  constant rate conditions, 140
  constrained bottomhole pressure, 140
  horizontal well traversing multiple grids,
    139f
  Peaceman's model, 139–140
  productivity index, 140–142
Hydrodynamic dispersion
  diffusive transport, 176–177
  dimensionless dispersion, 179f
  dispersion tensor, 180
  empirical equations, 178–179
  fluxes, 181–182
  mechanical dispersion, 178
  molecular diffusion and mechanical
    dispersion, 176

I

Implicit pressure, Explicit concentration
    ((IMPEC), 212–214
Implicit Pressure, Explicit Saturation
    (IMPES), 237–243
Implicit Pressure, Implicit Concentration
    (IMPIC), 214–218
Implicit Pressure, Explicit Concentration
    (IMPEC), 212–213, 215–218,
    225
Inactive grids, 96–97
Infinite-acting flow, 52–54
Interblock transmissibilities
  block pressures, 255t
  countercurrent flow, 249
  1D displacement process, 249–251
  relative permeabilities and viscosities, 249
  single–phase flow, 248–249
  upwinding, 249
Interfacial tension, 1–2
Irreducible water saturation, 16–17

J

Jacobian matrix, 165

K

Klinkenberg permeability, 153
Krylov subspace methods, 83

L

Linear algebra software libraries, 83–84
Linear systems of equations, 83–84
Logarithmic grids
  2D reservoir, 95f
  radial flow problems, 130f

M

Mass balance equations
  black oil model, 247
  component transport
    compositional equations, 184–186
    control volume, 183
    coordinate system and dimensions, 184
    mass accumulation, 183
    one–dimensional component, 183
    single–phase flow, 184
Mechanical dispersion, 176
Microemulsions, 2–3
Molecular diffusion, 176
Multidimensional Newton's method
  Jacobian matrix, 165
  nonlinear gas flow, 166–169
  numerical finite difference approximations,
    166
  Taylor series, 164
  vector of pressure values, 164–165
Multidimensional problems
  constant rate wells, 102
  corner blocks, 104–107
  Dirichlet conditions, 103–104
  gridding and block numbering
    2D and 3D domains, 94–95
    1D simulation models, 93
    grid dimensions, 95
    irregular geometry and inactive grids,
      96–97
  initial conditions, 107
  matrix arrays
    accumulation and compressibility, 113
    gravity, 115–120
    source terms, 114–115
    transmissibility, 114
  Neumann boundary conditions, 102–103
  pseudocode
    grid arrays, 121
    interblock transmissibility, 120–121
    main code, 121–122
    postprocessing, 122–124
    preprocessing, 120
    well arrays, 121
  reservoir heterogeneities, 107–113

Multidimensional problems (*Continued*)
single-phase flow
accumulation, 99—100
control—volume approach, 98
endpoint relative permeability, 98
flux terms, 100—101
general diffusivity equation, 98
mass balance and flux, 98—99
sources and sinks, 101
wells and boundary conditions, 102—107
Multiphase and multicomponent transport
compositional equations
balance equations, 285
component equations, 285
fugacities, 285
K—factor, 285
mole fractions, 285
partial differential equation, 284
phase molar volumes and viscosities, 286
finite difference equations, 286—287
oleic—aqueous bipartitioning components, 311—313
psuedocode, 313—315
solution method
Equation of State (EOS), 287, 291—298
flash calculation, 287—291
phase saturation, 298—299
phase viscosity, 301—304
pressure and composition solution, 308—311
relative permeability and transmissibility, 304—306
two-phase compressibility, 299—300
vapor pressure, 288
wells and source terms, 306—308

N

Neumann boundary conditions, 71, 102—103
Newton's method
1D, 161—164, 162f
multidimensions, 164—169
Newton—Raphson equation, 163, 219, 248
pseudocode, 169—170
Taylor series, 162
Nonlinearities
numerical methods
Newton's method, 161—169
Picard iteration, 157—161
reservoir/fluid properties, 157

unknown block pressure functions, 156—157
single-phase flow problems, 151—156
compressibility, 151
Forchheimer flow, 155—156
gas flow, 152—153
geochemical reactions and geomechanics, 151
non-Newtonian flow, 153—155
permeability, 151
pressure/time-dependent variables, 151
Non-Newtonian flow
apparent/effective shear rate, 154
interblock transmissibilities, 155
power-law model, 153—154
viscosity *versus* shear rate, 154f
Normalized saturation, 17—18
Numerical dispersion
adaptive meshing techniques, 222—224
advection-dominated flow, 221
1D, homogeneous advection-dispersion equation, 223t
dimensionless concentration *versus* distance, 222f
effective Peclet number, 221
local Peclet and grid number, 221—222
numerical dispersion coefficient, 221
spatial differencing scheme, 221
Numerical solution
black oil model
capillary pressure, 233
capillary pressure and relative permeability, 232
finite difference equations, 233—237
oleic phase pressure, 233
PDEs, 231—232
phase balance equations, 232
pseudocode, 272—279
relative phase mobilities, 231—232
solution methods, 237—248
stability, 256
wells and well models, 256—272
single-phase component transport, 225—226
block mass balances, 206—207
cell-centered approach, 204
channeling, 224
concentration change, 206—207
constant concentration, 211—212
diffusivity/pressure equation, 201
discretized mass balance, 204—206
finite difference solution, 201—203

implicit pressure, 212—214
matrices, 205—206
matrices and vectors, 211
multicomponents and multidimensions,
    225—226
no flux boundary condition, 207—211
numerical dispersion, 221—224
pseudocode, 226—228
stability, 219—220
tridiagonal (in 1D) dispersivity matrix,
    225—226
upwinding, 205
viscous fingering, 224—225
wells and boundary conditions,
    206—212

O

Oil compressibility, 11
Oil formation volume factor, 9—10
Oil pseudocomponent, 3
Oil-water and oil-gas relative permeabilities,
    18—19
Oil-water capillary pressure, 20—21
Oleic—aqueous bipartitioning components,
    311—313
Oleic phase properties, 9—11
  density, 32
  molecular weight, 32
Original gas-oil contact line (OGOC), 23
Original water-oil contact line (OWOC), 23
Orthogonal grids, 96

P

Partial differential equations (PDEs), 37,
    127—128
Peaceman's model, 139—140
Petrophysical properties
  capillary pressure, 20—21
  Darcy's law, 14—16
  relative permeability, 16—19
Phase behavior
  bubble point, 4—5
  critical point (CP), 4—5
  dew point, 4—5
  formation properties, 6—7
  gaseous phase properties, 7—9
  multicomponent hydrocarbon mixture, 5f
  oleic phase properties, 9—11
  reservoir pressure, 5
Phase mass balances
  cartesian coordinates, 38—40

control volume, 37, 38f
Phases, 2, 2f
Phase saturation, 298—299
Phase viscosity, 301—304
Picard iteration, 157—161
Pore compressibility, 6
Pore concentration, 3—4
Porosity, 3
Pressure and composition solution, 308—311
Pressure-dependent reservoir properties
  gaseous phase, 9f
  oleic phase, 10f
Pressure equation, 42
Primary fluid production, 1—2
Productivity index, 143—146
Pseudocomponents, 2—3
Pseudo-steady-state (PSS) flow, 52

R

Radial flow equations
  depletion flow, 129
  diffusivity equation
    discretization, 130—135
    gridding, 129—130
  drainage boundary, 129
  partial differential equations, 127—128
  practical considerations, 147
  production wells, 128, 128f
  pseudocode, single-phase flow, 147
  pseudo-stead state, 129
  stabilized flow, 129
  transitional flow, 129
  wellbores, 127, 128f
  wells and well models
    cartesian grids, 135—142
    constraints, 135—136
    diagonal matrix, 142—143
    1D reservoir, 143—146
    2D reservoir, 143—146
    horizontal wells and anisotropy,
        139—142
    inclusion, 142—146
    mass balance, 137—139
    matrix equations, 142—146
    productivity index, 143—146
    steady-state radial flow, 136—137
Reactive transport, 182
Relative permeability, 16—19, 29—30,
    304—306
Reservoir engineering principles, 1—2
Reservoir geometry, cubic control volume,
    38f

Reservoir heterogeneities
  1D reservoir, 107–108
  fluid properties, 109
  geometric properties, 109–113
    accumulation terms, 113
    harmonic mean, 109–113
    interblock transmissibilities, 107
    interblock phase transmissibility, 108
Reservoir initialization
  aqueous phase pressure, 23
  initial phase pressures and saturations, 23
  original gas-oil contact (OGOC) line, 24
  original water-oil contact line
    (OWOC), 24
Reservoir pressure, 5
Reservoir simulation, 1
Residual saturation, 16–17
Restricted diffusion coefficient, 177
Retrograde condensation, 5
Robin boundary conditions, 72

S

Secondary fluid production, 1–2
Shock fronts, 192–193
Simultaneous solution (SS) method
  block equations, 243
  relative permeability, 244–245
  saturations, 243
  two-phase flow, 245–247
  unknown variables, 243–244
Single-phase flow
  component mass balance equations, 184
  diffusivity equation, 42–47
    near-wellbore flow, 44f
    slightly compressible liquids, 43–44
Single-point upwinding approach, 205
Solution gas-oil ratio, 10
Specific gravity
  gases phase, 7
  oil, 9
Stability
  black oil model, 256
  numerical solution, 219–220

Standard conditions (SC), 3
Stationary methods, 83
Steady-state radial flow, 136–137
Stokes–Einstein equation, 176
Subsurface porous media
  component (κ), 2–3
  phases, 2
Supercritical fluids, 2–3

T

Time-dependent well constraints, 261–272
Tortuosity, 177
Total compressibility, 14
Tracer tests, 313
Transport arrays, 226
Transport mechanisms
  advection, 175
  hydrodynamic dispersion, 176–182
  reactive transport, 182
Two-phase compressibility, 299–300
Two-phase flash calculation, 288f

U

Upwinding approach, 205

V

Viscosity, 1–2, 11
Viscous fingering, 224–225

W

Waterflooding, 1–2
Well productivity index, 273
Wells and well models
  black oil model
    constant BHP injector wells, 261
    constant BHP producer wells, 261
    constant rate injector wells, 258–261
    time-dependent well constraints,
      261–272
    well rate, 257–258
  single-phase, 206–212